AF719133

WITHDRAWN

INTERNATIONAL

REVIEW OF CYTOLOGY

VOLUME 42

ADVISORY EDITORS

H. W. BEAMS
HOWARD A. BERN
W. BERNHARD
GARY G. BORISY
ROBERT W. BRIGGS
R. COUTEAUX
B. DAVIS
N. B. EVERETT
DON FAWCETT
MICHAEL FELDMAN
CHARLES J. FLICKINGER
WINFRID KRONE
K. KUROSUMI
MARIANO LA VIA
GIUSEPPE MILLONIG
MONTROSE J. MOSES
ANDREAS OKSCHE
VLADIMIR R. PANTIĆ
D. C. REANNEY
LIONEL I. REBHUN
JEAN PAUL REVEL
WILFRED STEIN
ELTON STUBBLEFIELD
H. SWIFT
DENNIS L. TAYLOR
J. B. THOMAS
TADASHI UTAKOJI
ROY WIDDUS
A. L. YUDIN

In connection with Dr. James F. Danielli's editorial responsibilities, please note that, effective May 1, 1975, the postal address for

INTERNATIONAL REVIEW OF CYTOLOGY

will be:

Worcester Polytechnic Institute, Worcester, Massachusetts 01609

for *all new* manuscripts and correspondence pertaining thereto.

INTERNATIONAL
Review of Cytology

EDITED BY

G. H. BOURNE
Yerkes Regional Primate Research Center
Emory University
Atlanta, Georgia

J. F. DANIELLI
Worcester Polytechnic Institute
Worcester, Massachusetts

ASSISTANT EDITOR
K. W. JEON
Department of Zoology
University of Tennessee
Knoxville, Tennessee

VOLUME 42

ACADEMIC PRESS New York San Francisco London 1975
A Subsidiary of Harcourt Brace Jovanovich, Publishers

Copyright © 1975, by Academic Press, Inc.
ALL RIGHTS RESERVED.
NO PART OF THIS PUBLICATION MAY BE REPRODUCED OR TRANSMITTED IN ANY FORM OR BY ANY MEANS, ELECTRONIC OR MECHANICAL, INCLUDING PHOTOCOPY, RECORDING, OR ANY INFORMATION STORAGE AND RETRIEVAL SYSTEM, WITHOUT PERMISSION IN WRITING FROM THE PUBLISHER.

ACADEMIC PRESS, INC.
111 Fifth Avenue, New York, New York 10003

United Kingdom Edition published by
ACADEMIC PRESS, INC. (LONDON) LTD.
24/28 Oval Road, London NW1

Library of Congress Catalog Card Number: 52-5203

ISBN 0–12–364342–2

PRINTED IN THE UNITED STATES OF AMERICA

Contents

Regulators of Cell Division: Endogenous Mitotic Inhibitors of Mammalian Cells

BISMARCK B. LOZZIO, CARMEN B. LOZZIO, ELENA G. BAMBERGER, AND STEPHEN V. LAIR

Ultrastructure of Mammalian Chromosome Aberrations

B. R. BRINKLEY AND WALTER N. HITTELMAN

Computer Processing of Electron Micrographs: A Nonmathematical Account

P. W. HAWKES

Cyclic Changes in the Fine Structure of the Epithelial Cells of Human Endometrium

Mildred Gordon

The Ultrastructure of the Organ of Corti

Robert S. Kimura

Endocrine Cells of the Gastric Mucosa

Enrico Solcia, Carlo Capella, Gabriele Vassallo, and Roberto Buffa

Membrane Transport of Purine and Pyrimidine Bases and Nucleosides in Animal Cells

Richard D. Berlin and Janet M. Oliver

List of Contributors

Numbers in parentheses indicate the pages on which the authors' contributions begin.

ELENA G. BAMBERGER (1), *The University of Tennessee Memorial Research Center and Hospital, Knoxville, Tennessee*

RICHARD D. BERLIN (287), *Department of Physiology, School of Medicine, The University of Connecticut Health Center, Farmington, Connecticut*

B. R. BRINKLEY (49), *Division of Cell Biology, Department of Human Biological Chemistry and Genetics, The University of Texas Medical Branch, Galveston, Texas*

ROBERTO BUFFA (223), *Institute of Pathological Anatomy, The University of Pavia, Pavia, Italy, and Histopathology, Histochemistry and Ultrastructure Center, The University of Pavia-Varese, Varese, Italy*

CARLO CAPELLA (223), *Institute of Pathological Anatomy, The University of Pavia, Pavia, Italy, and Histopathology, Histochemistry and Ultrastructure Center, The University of Pavia-Varese, Varese, Italy*

MILDRED GORDON (127), *Department of Obstetrics and Gynecology, Yale University School of Medicine, New Haven, Connecticut*

P. W. HAWKES (103), *The Cavendish Laboratory, University of Cambridge, Cambridge, England*

WALTER N. HITTELMAN (49), *Department of Developmental Therapeutics, The University of Texas, M. D. Anderson Hospital and Tumor Institute, Houston, Texas*

ROBERT S. KIMURA (173), *Department of Otolaryngology, Massachusetts Eye and Ear Infirmary, Boston, Massachusetts, and Department of Otolaryngology, Harvard Medical School, Boston, Massachusetts*

STEPHEN V. LAIR (1), *The University of Tennessee Memorial Research Center and Hospital, Knoxville, Tennessee*

BISMARCK B. LOZZIO (1), *The University of Tennessee Memorial Research Center and Hospital, Knoxville, Tennessee*

CARMEN B. LOZZIO (1), *The University of Tennessee Memorial Research Center and Hospital, Knoxville, Tennessee*

JANET M. OLIVER (287), *Department of Physiology, School of Medicine, The University of Connecticut Health Center, Farmington, Connecticut*

ENRICO SOLCIA (223), *Institute of Pathological Anatomy, The University of Pavia, Pavia, Italy, and Histopathology, Histochemistry and Ultrastructure Center, The University of Pavia-Varese, Varese, Italy*

GABRIELE VASSALLO (223), *Institute of Pathological Anatomy, The University of Pavia, Pavia, Italy*

INTERNATIONAL

REVIEW OF CYTOLOGY

VOLUME 42

Regulators of Cell Division: Endogenous Mitotic Inhibitors of Mammalian Cells

BISMARCK B. LOZZIO, CARMEN B. LOZZIO, ELENA G. BAMBERGER, AND STEPHEN V. LAIR

The University of Tennessee Memorial Research Center and Hospital, Knoxville, Tennessee

I. Introduction

Normal tissues and organs maintain a uniform mass throughout the adult life of a mammal. This implies that the rate of cell renewal in mammalian tissues is in balance with that of cell death, and the number of functional cells is directly proportional to the functional demand. It has long been suspected that cell division in mammalian tissues must be under some kind of hormonelike control. For these reasons the attention of several cell biologists has been directed toward the isolation of one or more diffusible substances regulating the processes of mitosis and cell differentiation. These substances may have several practical applications, such as the control of malignant cell proliferation.

Although the search for tissue-specific (chalones) and nonspecific endogenous mitotic inhibitors has been intensified in the last few years, a large body of literature is frequently forgotten or not known. Indeed, natural regulators of cell division have been detected in and/or isolated from a variety of tissues of vertebrates and invertebrates. It is difficult, however, for a worker in this field to become acquainted with the results obtained by other investigators, because the information gathered during nearly 50 years is dispersed in more than 300 articles published in a dozen languages. For these reasons a comprehensive review appears to be warranted.

This article deals with the antimitotic substances found in a variety of tissues and sera of humans and animals, as well as with inhibitors produced by cultured cells. A previous review dealt with inhibitors of hematopoietic cell proliferation (Lozzio, 1973).

As a tribute to their work, we mention briefly some accomplishments of the pioneers in this field of research. It was probably Walton (1914), in England, who first reported that a liver extract had an inhibitory effect on cultured cells. Carrel and Ebeling (1921), in the United States, investigated the effect of serum on fibroblasts grown in culture. Woglom (1929), in the United States, reported the results of nearly 30 years of studies on the immunological function of the spleen in tumor rejection and metastatic infiltration. Unfortunately, he could not draw a firm conclusion on the role of the spleen in tumor growth. Watson (1960), in Canada, reported results of unpublished studies made in 1928. He found that the injection of a spleen extract into a patient with lymphatic leukemia produced atrophy of the infiltrated lymph nodes and regression of the disease. Carrel (1930) also presented evidence on an epidermal chalonelike substance. Roffo (1937a,b,c), in Argentina, achieved the first successful *in vivo* tumor regression by giving striated and heart muscle extracts to rats bearing transplantable carcinomas and sarcomas.

The results of these early studies and other important findings from several other investigators are discussed herein.

II. Antimitotic Substances in Cells and Tissues

A. Cultured Cells and Conditioned Media

Considerable attention has lately been given to the limitation of growth of many mammalian cell cultures at high cell densities having a concomitant reduction in nucleic acid synthesis. Some of the factors involved have been well discussed by Stoker (1969). The growth of normal cultured fibroblasts may be controlled by physical contact (contact inhibition), protein factors present in the serum added to the medium, and attachment to rigid surfaces. Stoker (1969) showed that fibroblasts are unable to grow in semisolid agar or methylcellulose gel, where they become spherical and do not divide. Their growth in a liquid medium depends on being anchored to glass or plastic surfaces, where cells spread with probable changes in membrane configuration. However, the growth of a monolayer on rigid surfaces is limited by other factors, such as contact inhibition when the cells are present at higher densities.

The results of studies on reduced growth rate of Chinese hamster cells under crowded conditions suggest that a component of the medium from confluent cultures limits their exponential growth (Froese, 1971). Adrenaline was found to act synergistically with this inhibitor, as indicated by a marked increase in growth inhibitory action at a concentration of 0.5 μg/ml of the medium. The inhibitor reported by Froese (1971) appeared to have a molecular weight of over 13,000, because it was retained in a standard dialysis tube. Since only Chinese hamster cells were used in these experiments, the cell specificity of the compound is not known.

It has also been reported, by Bürk (1966, 1967), that an inhibitor of cell division is produced by cell lines obtained from Syrian hamster kidney cells. The inhibitor's effect was neutralized by the addition of serum to the medium. It is interesting to note that normal hamster kidney cells (BHK21) release an inhibitor (anomin) into the medium, whereas their polyoma-transformed derivatives (Bürk, 1967) do not produce anomin. Further, the growth of polyoma-infected cells was inhibited by all the extracts obtained from normal cells of human and animal origin, but not by extracts from a variety of tumor cells of the same origin. Anomin is therefore a nonspecific inhibitor of cancer cell proliferation produced by normal cells. It has a molecular weight of 1000 to 2000, and is neither ethanolamine nor heparin. Bellanger and Harel (1969) found an inhibitor of protein synthesis in the supernatant of cultured BHK21 polyoma-transformed cells in the stationary phase. In these experiments the incorporation of alanine-^{14}C into protein was inversely proportional to the cell concentration. The inhibitor appears to be a thermostable low-molecular-weight compound. It remains to be demonstrated whether or not this inhibitor is similar or identical to that reported by Bürk (1967).

Experiments made by García-Giralt and Macieira-Coelho (1969) and García-Giralt *et al.* (1970) suggest that exhaustion of the division potential of WI-38 human fibroblasts is due to the progressive accumulation of an inhibitor. This inhibitor has not yet been characterized either in terms of specificity or chemical properties. The same investigators demonstrated that the synthesis of DNA of WI-38 fibroblasts was inhibited by "conditioned" culture media from the same cells. Evidence for a specific inhibitor (chalone) of WI human fibroblast growth has been presented by Houck *et al.* (1972). A fibroblast extract or dialyzed conditioned medium of WI-38 cells markedly inhibited thymidine-^{3}H uptake by fibroblasts, but did not inhibit that of lymphocytes or HeLa cells. The molecular weight of this material was between 30,000 and 50,000. The activity was destroyed by trypsin and by heating at 58°C for 30 minutes. The crude fibroblast chalone described by Houck and co-workers (1972) is most likely to be very similar or identical to the inhibitor found previously by García-Giralt *et al.* (1970), using the same line of human fibroblasts.

The results of a recent study by Engelhardt (1971) demonstrate that an inhibitor of protein synthesis is produced by monkey kidney cells (Vero M3) that have been grown to high cell density and have entered the stationary phase of growth. The inhibitor was not found in low-density cultures of Vero M3 and HeLa S3 cells. The substance appears to remain intracellular, because its inhibitory action was detectable only after cell disruption by freezing and thawing.

Some characteristics and properties of an inhibitor of DNA synthesis produced by human lymphoblast cell lines have been reported by Smith *et al.* (1970). The inhibitor found in the culture medium of lymphoblasts suppressed DNA synthesis and strongly inhibited the incorporation of uridine-^{3}H by normal human leukocytes stimulated by phytohemagglutinin (PHA). It appeared to affect the phases of cell differentiation, resulting in blastogenesis and usually in mitosis under the influence of PHA.

B. Embryo, Placenta, and Ovaries

On the assumption that tissues with the highest rate of growth require a greater concentration of a hypothetical regulator of cell division to ensure orderly growth, Murphy and Sturm (1933, 1934a,b) explored the effect of crude extracts from murine embryo skin and placenta on a variety of homologous tumors. The administration of these extracts to mice produced a definite retarding action on the growth of two transplantable carcinomas, but they were without effect on sarcomas. The intraperitoneal injection of extracts of dessicated embryo skin and placenta also decreased markedly the rate of postoperative local recurrence after surgical removal of spontaneous mammary carcinomas of mice. Tumor autografts either failed to grow, or their subsequent growth was significantly retarded after a short period of contact with extracts *in vitro*. Approximately two-thirds of spontaneous and transplantable tumors ceased to grow after treatment with these extracts. Since in previous experiments an extract of chicken tumors exerted an inhibitory action only on sarcomas, and because embryo and placenta extracts mainly affected carcinomas, Murphy and Sturm postulated that the action of these different extracts "was not species limited but appears to be tissue limited." This was probably the second time an inhibitor similar to a chalone was mentioned in the literature. Although the nature of the inhibitor was not known, it did not seem to be a simple proteolytic enzyme. Recently, a concentrated ultrafiltrate of chick embryo was found to inhibit the growth of homologous explants of normal liver cells (fibroblasts) by Coogan *et al.* (1968, 1969). The inhibitors in the concentrated ultrafiltrate were low-molecular-weight compounds stable to heat, cold, lyophilization, ultraviolet irradiation, and electrolytic desalting, as well as to mild acid and alkaline hydrolysis. Purification of these inhibitors by Sephadex gel and ion-exchange column chromatography yielded two inhibitory fractions, one soluble in acetone and the other soluble in absolute methanol. Both were soluble in distilled water. The two fractions were further purified by electrophoresis and chromatography. The inhibitors, all of which were localized in a single ultraviolet absorbing peak (260 nm) from each fraction had identical electrophoretic mobilities and R_f values. The acetone fraction was relatively pure. Complete inhibition occurred at less than $1 \times 10^{-6}\,M$ concentration. The inhibitors have not yet been identified, but evidence suggests that they belong to a group of "minor" nucleic acid derivatives (Coogan *et al.*, 1968, 1969). Other low-molecular-weight substances have been isolated from chick embryos by Kagen and Linder (1972). The three partially purified compounds were of very low molecular weight (from 500 to 700), heat-stable, and resistant to treatment with RNase, DNase, trypsin, and pronase. Each of the three inhibitors, obtained by molecular seiving through Diaflo membranes (Amicon) and gel filtration, blocked the uptake of isoleucine, as well as lysine, by cells obtained by homogenization of whole chick embryos. None of them inhibited DNA synthesis, but one suppressed uridine incorporation. Further characterization of these inhibitors is not available as yet.

In a recent paper, Baden (1973) reported the presence of a protein in crude extracts of human amniotic and chorionic membranes which are rich in connective tissue (see also Section II,F). It had a relative molecular weight of about 70,000 and inhibited DNA synthesis and cell division of human lymphocytes, fibroblasts, and epidermal cells. The DNA synthesis of rat skin cells, SV40-transformed fibroblasts, and lymphocytes stimulated with PHA was also suppressed by the placental extract. Its activity was destroyed by heating, trypsination, or prolonged storage in an aqueous solution. This protein therefore had neither tissue nor species specificity, and its capacity for arresting mitosis could not be enhanced by the addition of epinephrine to the culture medium.

Matsumara and Ishida (1954) extracted a water-soluble substance called chorionin from human placenta. It inhibited the growth of some transplanted solid tumors in mice, but no further characterization was made.

A search for antimitotic substances has also been made in molluscs and amphibians. The presence of an accelerator and a retarding cleavage factor for sea urchin eggs in the ovaries of the same animal has been reported by Menkin (1959). The accelerator was identified as a dinucleotide, whereas the retarding factor was a polynucleotide. Since heparin is known to inhibit cell division, and the jelly of sea urchin eggs is one of the materials richest in heparinlike substances, extracts of the eggs were prepared by Heilbrunn *et al.* (1951). The extract from sea urchin eggs had a mild effect in retarding the cleavage of fertilized homologous eggs. The ovaries of the starfish, however, contain a very powerful inhibitor of cell division of heterologous eggs of marine animals. The active substance acts like heparin in keeping the protoplasmic fluid in the interior of the cells and, if present in sufficient concentration, prevents mitotic activity.

The low protein synthesis of cell-free preparations from frog embryos and larvae was related to the presence of a soluble heat-labile component of high molecular weight, which also inhibited strongly amino acid incorporation by a cell-free system from frog liver (Strittmatter, 1968).

Another specific inhibitor of rRNA was claimed by Shiokawa and Yamada (1967) to be present in the conditioned medium of dissociated amphibian blastulas. Its existence has been denied by Van Snick and Brachet (1971) who, using autoradiographic techniques, were unable to demonstrate the presence of such an inhibitor.

C. Kidneys

The presence of inhibitory substances in the kidneys was demonstrated almost 40 years ago (McJunkin and Hartman, 1933). The inhibitor(s) were extracted with dilute acids and by acid alcohol. The administration of these crude extracts to normal rats reduced the number of mitoses as determined on microscopic sections. In subsequent experiments, McJunkin and Henry (1935) partially purified a lipoid inhibitor from kidneys, liver, and myocardium. These extracts had marked antimitotic activity for kidneys and liver tissues and, when given intraperitoneally to rats in large doses, completely suppressed cell division. The evidence pointed to phospholipids as the inhibiting agent.

Mature amphibian mesonephric kidneys contain a mitotic inhibitor which has several properties similar to those of a chalone (Simnett and Chopra, 1969; Chopra and Simnett, 1969). The Colcemid metaphase-arrest technique indicated that crude extracts of *Xenopus laevis* inhibit mitosis in cultured larval pronephros but not in epidermal cells. Studies on the synergistic activity with adrenal hormones demonstrated that cortisone and adrenaline enhance the inhibitory effect of kidney extracts on the pronephros. The inhibitor therefore appears to be tissue-specific. The results of Chopra and Simnett (1969) suggest that chalonelike substances may also limit cell division during larval development. Evidence for the presence of other inhibitors in the kidneys of humans and animals has been given previously (Lozzio, 1973).

D. Liver

Since shortly after the turn of the century the liver has been known to contain substances that inhibit the growth of cells in tissue culture. It was probably Walton (1914) who first reported that liver extracts had an inhibitory effect on cultured cells. Similar findings were reported in subsequent years with various aqueous or alcoholic liver extracts (Heaton, 1926, 1929; Shibuya *et al.*, 1935).

Brues *et al.* (1936) prepared saline extracts from rat, mouse, chicken, and chick embryo livers. All extracts, except those from chick embryo livers, were found to be inhibitory to freshly explanted fibroblasts. Alcoholic extracts from chicken, rat, lamb, and bovine livers were also prepared, and all were inhibitory. The inhibitor(s) was most effective on fibroblasts, and had an equal effect on various embryonic fibroblasts and malignant fibroblasts of mouse sarcoma 180. The inhibitor(s) was effective on both homologous and heterologous tissues. A further attempt at isolation and identification of the inhibitor(s) was also made by Brues *et al.* (1940). As a result of this work, it was concluded that the same liver substance was extracted by saline or alcohol, but that in the saline extract the inhibitory material was intimately associated with proteins or other colloids. Extraction of liver with ether, benzene, or acetone yielded much smaller quantities of the inhibitor(s) than saline or alcohol extraction. One inhibitor was isolated and identified as ethanolamine. However, its biological properties differed from the bulk of the inhibitory material present in the liver. Its activity was pH-dependent and was much greater on normal tissues than on malignant ones. Several salts, amino acids, and vitamins known to be normal constituents of the liver were tested in tissue cultures, and none of these was inhibitory to cell growth. The inhibitory material markedly inhibited the radial growth of normal or malignant fibroblasts in tissue cultures, and mitoses were only 1–2% as numerous as in the controls. All the extracts were found to have the following common properties, which suggested that the inhibitor(s) might have significance in the regulation of growth. All the extracts were inhibitory to cell growth but did not produce irreversible cell damage as determined by normal pulsation of heart muscle fragments. Normal contractions of heart muscle *in vitro* were observed, even at concentrations twice that necessary to arrest cell division completely. All the extracts were reversible in action. The inhibitor(s) was present in adult liver in concentrations approximately equal to those that inhibited growth *in vitro,* while it was present in much lower concentrations in embryonic livers.

Evidence suggesting the existence of a liver chalone was provided by Saetren (1956) and Stich and Florian (1958). Saetren demonstrated that the intraperitoneal injection of a liver homogenate decreased the mitotic index of regenerating adult rat liver. The mitotic index of the liver of normal young rats was also decreased by injection of the homogenate. This inhibitor was highly specific for hepatocytes. Stich and Florian reported essentially the same findings when they studied the effect of serum and liver homogenates on the mitotic rate of regenerating liver in partially hepatectomized rats. The serum and liver homogenates of normal (intact) adult rats were found to inhibit the onset of mitosis, but the serum and liver homogenates of partially hepatectomized adult rats had no effect on the mitotic rate. On the basis of these results, Stich and Florian (1958) postulated that an organ-specific inhibitor of mitosis (chalone) was present in the serum and in the liver. The inhibitory effect of adult rat serum lasted for only 1 day, suggesting that the compound was rapidly metabolized, hence it must be continually synthesized to prevent hepatocyte division.

Scaife (1970) reported that normal rat serum contained a high-molecular-weight specific inhibitor (probably protein) which retarded the growth of rat embryo liver cells *in vitro*. The activity of the substance decreased after partial hepatectomy and was essentially absent from the serum of newborn rats. Since the inhibitor did not affect the growth of embryo kidney cells, it was speculated that it might be a liver chalone which was released into the serum.

Verly *et al.* (1971) reported the purification of a chalone from rabbit liver. The chalone was purified 450-fold by a combination of ethanol precipitation, Sephadex G-50 chromatography, and ultrafiltration. The inhibitor was characterized as a polypeptide of low molecular weight. The action of the chalone on liver DNA synthesis was found to be tissue-specific. The dose–effect relationship was in agreement with the theory of receptor sites on the hepatocytes with a low affinity for the chalone.

In contrast to the above findings, Wayss *et al.* (1973) prepared liver extracts from adult rats and failed to find a tissue-specific inhibiting factor. Although the liver extracts were found to inhibit DNA synthesis of regenerating adult rat liver, tissue extracts from kidney and lung exhibited the same degree of inhibition. This study therefore failed to demonstrate the presence of a hepatic chalone.

Other inhibitors have been isolated from liver tissue that do not fit the criteria for a chalone. One of the protein inhibitors repeatedly isolated from liver is the enzyme arginase.

In recent years the effect of various inhibitors isolated from liver on DNA, RNA, and protein synthesis has been extensively investigated. Otsuka and Terayama (1966) reported that the cell sap from normal rat liver inhibited the incorporation of orotate-^{14}C into the RNA of ascites hepatoma cells (AH-414). The inhibitory effect was dependent on the concentration of the inhibitor and was removed by washing the cells. The active material was thermolabile and displayed proteinlike features. Activity was found at pH 4.5, suggesting that the substance was not related to DNase. The cell sap from regenerating liver was active, but the cell sap from ascites hepatoma cells was inactive. The inhibitor was subsequently shown to be active only on the DNA synthesis of intact cells (Otsuka, 1967). No inhibition of DNA synthesis was observed in a cell-free system, thus the cell sap did not contain inhibitors of thymidylate kinases or DNA polymerase.

Frank (1968) reported that the incorporation of radioactive uridine, adenosine, orthophosphate, and L-leucine into rat embryo cells was reversibly inhibited by a protein fraction from adult rat liver. Ultracentrifugation of the fraction indicated a molecular weight of about 125,000. The decreased incorporation of RNA and protein precursors was shown not to be due to degradation of the medium (eliminating arginase as the possible inhibitor), inhibition of their uptake into the cells, a deficiency of ATP, or an increased rate of RNA degradation. After incubation of the cultures with the fraction, the cells were shown to contain an RNA fraction in which the incorporation of radioactive uridine was selectively inhibited.

Nilsson (1970) found that human liver supernatant contained an inhibitor that affected DNA and RNA synthesis of HeLa cells. This inhibitor seemed to lack any arginase activity. The inhibitor weakly inhibited penetration of DNA precursors, and strongly inhibited the phosphorylation and incorporation of thymine into DNA. Nilsson suggested that the primary inhibition target might be DNA polymerase, since extracts from control HeLa cells had a higher DNA polymerase activity than extracts from inhibited cells. The mechanism of RNA synthesis was also studied, and impaired phosphorylation of UMP was observed with quantities of the inhibitor sufficient to cause growth inhibition.

Henderson (1970) reported that an extract of normal rat liver inhibited DNA, RNA, and protein synthesis in normal and neoplastic liver cell cultures. Maximum inhibition was observed with low concentrations of the inhibitor, and less inhibition at higher concentrations.

Miyamoto and Terayama (1971) reported that an adult rat liver cell extract inhibited the incorporation of thymidine-^{14}C into the DNA of ascites hepatoma cells. At least two inhibitory components were found, one of which was arginase. Arginase was not present in the liver of newborn rats, while the thermolabile inhibitor component was present in both newborn and adult rat liver. The experimental evidence suggested that the latter inhibitor was thymine hydrolase and/or phosphorylase. Malignant hepatomas are known to lack both arginase and thymine-decomposing enzymes.

Aujard *et al.* (1973) investigated the inhibition of DNA synthesis by liver extracts from normal adult rats. The inhibition was studied in synchronous cultures of rat hepatoma cells, and was shown not to be due to direct action of the liver extracts in the S phase but to be the consequence of their action in the G_1 phase.

Several substances have been isolated from liver tissue, which are inhibitory to a variety of tumor cells. Suzuki (1959) tested a liver extract of normal rats on rat ascites hepatoma cells and on Yoshida sarcoma cells of rats. The inhibitor(s) not only retarded growth of the cells, but also caused a marked decrease in the number of cells within the cultures. Dialysis of the liver extract yielded inhibitory substances in both the high- and low-molecular-weight fractions. Trypsin digestion of the high-molecular-weight fraction transformed part of this fraction into low-molecular-weight substances with full inhibitory activity. The inhibiting substances in the low-molecular-weight part of the liver extract were investigated and were found to be heat-stable, acid- and alkali-stable, and soluble in 95% alcohol and chloroform–methanol (1 : 1).

Herbut and Kraemer (1960) prepared saline extracts from livers of several species and tested them for tumor inhibitory activity against Gardner lymphosarcoma 6C3HED carried subcutaneously by C3H mice. Complete regression of tumor growth was obtained with an extract of guinea pig liver; marked but incomplete regression of the tumor was observed with extracts from sheep, hog, and rabbit liver; slight and inconsistent regression was found using horse and bovine liver; and no retardation of tumor growth was observed with extracts from human livers. However, these investigators cautioned that the discrepancy in the results might have been due to limitations of the dosage or perhaps to the age of the host supplying the liver.

Hori and Ukita (1962) isolated an active principle from saline and salt-free water extracts of bovine liver, which inhibited the growth of rat ascites tumor cells *in vitro*. The principle was concentrated by stepwise salting out with ammonium sulfate and by removal of inactive heat-denatured proteins. The active protein principle was then separated on DEAE-cellulose columns and was found to inhibit the growth of both rat ascites hepatoma cells (AH-130) and Yoshida sarcoma cells. The inhibitory activity was not repressed by the addition of arginine, thus the inhibitor was concluded to be different from arginase.

Oftebro *et al.* (1963) isolated two inhibitors from ox livers by aqueous phenol extraction. These two substances completely arrested the mitotic activity of HeLa cells and one, containing both dialyzable and nondialyzable material, also inhibited the mitotic activity of Chang liver cells. Most of the material contained polypeptides, indicating that the inhibitors were polypeptides or low-molecular-weight components attached to a polypeptide.

Sugihara and Araki (1964b) found that the precipitate of saline extracts of bovine liver was separable into two fractions, a mitosis-promoting and a mitosis-inhibiting factor. The factors were tested on cultures of Ehrlich ascites tumor cells.

Nakahara and Fukuoka (1961) reported the isolation of a carcinostatic factor from mouse liver. When this factor was allowed to act on Ehrlich ascites carcinoma cells *in vitro*, it completely destroyed the transplantability of the cells into susceptible mice. The carcinostatic factor was not affected by digestion by pancreatic RNase, as the antitumor activity remained intact even though the RNA content was reduced to almost 10% (Nakahara *et al.*, 1962). On the basis of these results, it was concluded that the factor was totally unrelated to RNA.

Strong (1968) reported that aqueous suspensions of alcoholic liver extracts inhibited spontaneous tumors of mammary gland origin in mice. One moiety of the liver extracts was water-soluble and was found to decrease in effectiveness with the age of the mouse. The other moiety was alcohol-soluble, and both were found to be inhibitory to tumor growth.

Chany and Frayssinet (1971) prepared liver extracts from rats, mice, hamsters, and cows and found them inhibitory to the growth of hepatoma LF cells, KB cells (only the rat liver extract was tested), MKSV-transformed mouse cells, and hamster transformed cells (TSV clone 2). Arginase was not responsible for the inhibition, as the addition of arginine to the medium did not alter the inhibitory effect.

E. Malignant Tumors

In this section we consider only those substances extracted from tumors and organs of tumor-bearing animals that produce inhibition of the growth of homologous or heterologous normal and neoplastic cells.

An inhibitor of chicken fibroblast growth was found in a sarcoma of rats induced by benzopyrene (Werner, 1945; Doljanski *et al.*, 1944). Apparently, two substances were present in this extract. One was a growth inhibitor precipitable with 96% ethanol and soluble in an acetone–petroleum ether mixture. It was probably lipoid in nature. The second was an unidentified growth-promoting agent whose activity became apparent when the lipoid antagonistic factor was removed from the preparation (Werner, 1945).

Modification of the survival of tumor allografts in mice has been the subject of numerous studies which have been discussed by Kaliss and Snell (1951). Prior treatment of the host with extracts from normal and neoplastic tissues either significantly increased (large doses) or decreased (small doses) the number of "takes" of tumor allografts (Kaliss, 1952).

An inhibitor of mouse tumors, with the characteristics of a lipid, was detected in necrotic mouse sarcoma 180, as well as in incubated normal mouse liver and spleen cells. This substance was heat-stable, insoluble in water, and soluble in organic solvents. It could be stored for a long period of time with little loss in activity. The lipid substance was effective in preventing growth of transplantable mammary gland tumors and, when mixed with viable sarcoma 180, Ehrlich ascites carcinoma and mammary cells, also prevented the development of each of these tumors (Miller and Kimsey, 1967).

Acetone-soluble compounds from human tissues are mainly formed by saturated and unsaturated glycerides and phosphatides. The proportion of each one varies from normal to neoplastic human tissues in which the amount of saturated glycerides increases by a factor of two (Guidetti, 1964). The water-soluble phase of acetonic extracts contains sugar, peptides, and amino acids. The therapeutic utilization of these extracts was proposed by Guidetti (1953, 1956), who also examined the lipid composition of some human tumors (Guidetti and Castoldi, 1956). A 10-year survey of 2500 patients suffering from advanced cancers and treated with acetonic extracts in the terminal phase of illness revealed favorable results on the tumor itself or on the general health in about 20% of the cases. No reaction at all was observed in 55%. The investigators speculated that the beneficial effect observed in some patients was related to the antigenic properties of polypeptides or the content of lipids, in particular lipoid haptenes of the cytolipine-H and -G type (Graf and Rapport, 1961).

Proteins and polypeptides have also been extracted from tumors and found effective in preventing neoplastic growth. The development of a subcutaneous Ehrlich ascites tumor was inhibited by a polypeptidelike substance (see also Section II,H) obtained from a homologous tumor (Sugihara and Araki, 1963). A search for mitotic inhibitors in organs of mice bearing a subline of Ehrlich carcinoma was also made by Sugihara and Araki (1964a). They found growth-promoting and -inhibiting effects in an extract of visceral organs of mice with the tumor and in the tumor itself. The inhibitory activity was present in two of several fractions prepared. The inhibitor from normal tissues of tumor-bearing mice had a lower molecular weight than that from the tumor. Another cytotoxic octapeptide has been isolated and crystallized from various human and animal tumor fluids (Holmberg, 1968). Chemical analyses indicate that the biological activity may be related to a unique amino acid sequence: Tyr-Cys-Tyr.

There are eight reports indicating that specific inhibitors (chalones) of malignant cell multiplication can be extracted from tumors of rats, mice, and hamsters. These include: epidermal carcinoma (Bullough and Laurence, 1968c), keratinizing epidermal carcinoma (Bullough and Deol, 1971), melanoma (Mohr *et al.*, 1968; Bullough and Laurence, 1968a,d), lymphoma (Bullough and Laurence, 1970b), and mylogenous leukemia (Rytömaa and Kiviniemi, 1968, 1970). Extracts of homologous and heterologous normal tissues (except melanomas) from the same or a different species also inhibited the growth of the tumors mentioned above. In most experiments the chalone treatment was equally effective *in vitro* and *in vivo*, and some cases produced a temporary or permanent regression of the disease (e.g., chloroleukemia of the mouse).

F. Muscles and Connective Tissues

Since the incidence of primary tumors and metastases in skeletal muscles is extremely low, Roffo (1926) studied the effect of extracts from the heart and skeletal muscles on transplantable adenocarcinoma of the breast and fusiform cell sarcoma of rats. These tumors developed spontaneously in some rats of his colony, and the first passage to a normal host was made about 1910. Two types of extracts were prepared from bovine, rat, and dog muscles under aseptic conditions. They consisted of either spontaneous tissue lysis for 15–20 days (autolyzates) or acid hydrolysis of the muscles (hydrolyzates) for a similar period of time. In both cases the tissues were maintained in tissue culture media and the supernatant was tested in rats with transplantable tumors. In subsequent publications (Roffo, 1927a,b, 1937a,b,c, 1938; Roffo and García-Velloso, 1927; Roffo and Lopez-Ramirez, 1930), it was demonstrated that crude extracts of bovine heart and striated muscles produced complete inhibition of the growth of the transplantable carcinomas and sarcomas mentioned above (Fig. 1). The treatment, which consisted of an injection (0.5 ml) of the "lyzates" every 48 hours, was begun when the tumor had progressed for at least 19 days. Both extracts produced regression of the tumors which had extensive areas of necrosis. Finally, the tumors were reabsorbed, leaving no trace of injury in the place of inoculation. Control rats receiving no treatment died within 40 days, whereas the tumor-bearing rats treated with muscle hydrolyzates lived for 100 days, when the experiment was terminated without noticeable tumor recurrence. The effect of muscle autolyzates and hydrolyzates on adenocarcinoma and sarcoma cells in culture was also determined (Roffo, 1927a; Roffo and Villanueva, 1927a,b, 1930).

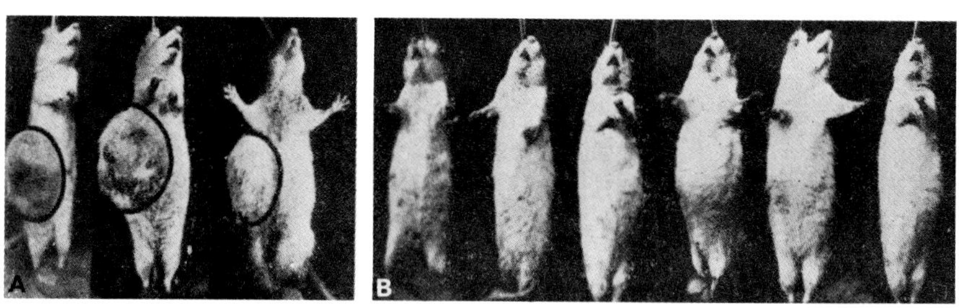

FIG. 1. Effect of the injection of a hydrolyzate of bovine heart muscle on the growth of transplanted carcinomas. (A) Controls uninjected. (B) Injected rats killed 40 days after transplantation of the growth. From Roffo (1938). Reproduced with permission from *The Lancet*.

The extracts were markedly cytotoxic to malignant cells at a concentration of 0.1% of the culture medium, whereas fibroblasts grew at the normal rate when the concentration of either extract was increased to 10%. This finding indicates that the antimitotic substance in muscle extracts was 100 times more active on neoplastic than on normal cells. Some of the components of muscle autolyzates and hydrolyzates were studied by Roffo and Correa (1929). They found proteins, polypeptides, amino acids, and lipids, but the active anticancer substance was not identified.

Cevese and Ferro (1951) confirmed and extended the results obtained by Roffo and his co-workers. Extracts from muscles were found to inhibit the growth of a rat transplantable sarcoma originally developed by treatment of rats with benzopyrene. Extracts from other organs were also as effective as muscle extract in producing complete and permanent tumor regression when given subcutaneously to 20 rats.

The effect of aqueous extracts of skeletal muscle on the growth of Ehrlich ascites tumor cells *in vitro* was also studied by Sato and Grob (1966). Their results indicate that unknown constituents present in saline muscle extract inhibited the growth of tumor cells by approximately 40%. Similar extracts from spleen and liver had no effect on the division of ascitic carcinoma cells.

The inability of an extract from rat heart muscle to stimulate the growth of chicken fibroblast was due to the presence of inhibitor(s) which masked the activity of the growth-promoting substance. The inhibitor(s) was probably of lipoid nature (Werner, 1944).

An inhibitor of the growth of normal chicken fibroblasts was detected in the supernatant of a tryptic digest of autologous tendon by Simms and Stillman (1937a). By using similar proteolytic procedures (Parshley, 1965; Parshley and Mandl, 1963, 1965; Parshley *et al.*, 1965), crude extracts of tendons, muscles, and aorta of chick and cow, and dog organs were obtained by Parshley and co-workers. Partially purified extracts from the same tissue were obtained by methanol and ethanol fractionation followed by calcium chloride precipitation and further purification on a DEAE-cellulose column. The major active component eluted from the column contained 50% protein and 10% RNA. Despite this, about 40% of inhibitor could not be accounted for. It was speculated that this constituent of normal tissues was a protein complex with mucopolysaccharide and/or nucleic acids. The nondialyzable macromolecule (stable at 58°C) was found to be highly active against a variety of cells from human tumors. At doses of 0.05 mg/ml of the culture medium, it mainly inhibited the growth of human cancer cells of mesenchymal origin. The inhibitory effect on normal fibroblasts was much less marked than that on malignant cells.

Cornin, an antimitotic protein, has been isolated from beef cornea and rabbit muscle (Nisida and Murakami, 1965a). On bioassay cornin has an antimitotic action on the early development of sea urchin eggs (Nisida *et al.*, 1964). The substance was isolated by boiling beef cornea or rabbit muscle, followed by alcoholic fractionation; the cornin was present in the fraction from 70–90% alcohol. Cornin isolated from cornea was undialyzable and differed in certain properties from the dialyzable substance isolated from muscle. Cornea cornin was separable into three fractions by DEAE-cellulose, all of which had antimitotic activity, but especially fractions II and III. These fractions were nucleoproteins with adenine as the base. When muscle cornin was fractionated, only fractions II and III were active. Fraction II was a nucleoprotein. Both cornins decreased the incorporation of ^{32}P into nucleic acids and DNA synthesis in sea urchin eggs with hypoxanthine as the base. The cornins also inhibited the polymerization of nucleic acids during the development of sea urchin eggs and inhibited the increase of sulfhydryl groups before cleavage of sea urchin eggs. In addition, both cornins were found to depress the incorporation of ^{32}P into DNA and rRNA of regenerating rat liver (Nisida and Murakami, 1965b).

G. Skin (Epidermal Chalone)

The fact that the epidermis maintains a constant thickness under normal conditions (Bullough, 1972) and that it reconstitutes itself rapidly after injury implies the existence of a particular mechanism(s) controlling the division of epidermal cells.

More than 40 years ago, Carrel (1930) found that the application of dog subcutaneous tissue to autologous or homologous skin wounds delayed cicatrization. This observation was probably the first evidence of a chalonelike mechanism controlling skin cicatrization. The same year, Dvorak and Byram (1930) reported that the healing of skin wounds was neither accelerated nor inhibited when they were treated with extracts from various macerated tissues such as liver, kidney, and spleen. However, McJunkin and Matsui (1931) observed some stimulating effect on the regeneration of epidermis in cutaneous wounds by macerated epidermis, especially that of fetal origin.

The study of the regulatory mechanisms of epidermal cell proliferation received a strong impetus during the past decade, when Bullough and Laurence (1960a) rejected the idea of a stimulatory wound hormone and introduced the concept of a chalone—a tissue-specific internal secretion capable of depressing mitotic activity *in vivo* and *in vitro*. The process of wound healing of the skin therefore would be the result of a local reduction in concentration of a normally present mitotic inhibitor, termed epidermal chalone. This concept agrees with the work of Weiss and Kavanau (1957), who suggested that growth regulation occurs automatically through a negative feedback mechanism involving specific diffusible compounds termed antitemplates or blocking templates, specific key compounds that catalyze the growth of the generative mass of an organ. As the result of a series of studies (Bullough, 1962; Bullough and Laurence, 1960b, 1964a,b, 1966; Bullough *et al.*, 1964, 1967; Califano, 1962; Frankfurt, 1971; Moskalewski, 1971), the existence and properties of the epidermal chalone were established. It was characterized as a diffusible inhibitor, most probably a basic protein, produced within the epidermal cells, which is evidently tissue-specific but not species-specific. This factor was soluble in water, nondialyzable, precipitable by alcohol, and heat-labile. Its activity could not be preserved in aqueous solution at −20°C, was rapidly destroyed at 37°C, and remained fairly stable after lyophilization (Elgjo, 1969). The inhibitory action on epidermal cells was demonstrated in the *in vitro* system used, as well as in the skin of normal mice (Bullough and Laurence, 1964b).

Tissue-specific antimitotic chalones have been found even in tissues that have a common embryonic origin, such as melanocytes (Bullough and Laurence, 1968b), sebaceous glands (Bullough and Laurence, 1970a), eccrine sweat glands (Bullough and Deol, 1972), eye lens (Voaden, 1968), tongue epithelium (Laurence and Randers-Hansen, 1972), and rodent stomach squamous epithelium (Frankfurt, 1971). Recently, Chopra *et al.* (1972) demonstrated that human skin contains tissue-specific mitotic inhibitory factors similar to those found in nonhuman tissues. Chopra and Flaxman (1973) also showed that the mitosis of psoriatic epidermal cells was inhibited by extracts from normal human skin.

Chalone activity was regularly found in the aqueous supernatant of skin homogenates. Preliminary attempts to purify a chalone-containing water extract by alcoholic precipitation (Bullough *et al.*, 1964) indicated that more than 80% of the chalone was recovered in the portion precipitable between 71 and 80% alcohol. The active ethanol fraction of pig rind was purified sevenfold by column electrophoresis followed by lyophilization of the effluent fractions preceding the main absorbancy at 280 nm (Hondius-Boldingh and Laurence, 1968). A large amount of inactive contaminants was removed by dialysis against water, leaving a final product purified nearly 2000 times as compared with the initial crude water extract. This highly purified (probably still heterogenous) epidermis-specific mitotic inhibitor shows the characteristic biological activity of a chalone both *in vitro* and *in vivo*.

The purified chalone appears to be an antigenic macromolecule with an apparent molecular weight of 30,000 to 40,000, as determined by gel-filtration, and has an isoelectric point between 5 and 6. It is either a pure glycoprotein or a mixture of one or more proteins, glycoproteins, and/or polysaccharides, which is stable at low pH values but is inactivated at pH 9. It appears to be resistant to pepsin, but the activity is destroyed by trypsin.

Marrs and Voorhees (1971a) described a reproducible extraction procedure and a reliable bioassay for chalone activity in epidermal extracts. They used heat-isolated, pure, newborn rat epidermis as a source of aqueous homogenates, and their preparation containing an active mitotic inhibitor showed complete destruction of cell and organelles with electron microscopy. Preliminary characterization of this inhibitor (Marrs and Voorhees, 1971b) indicated that it appears to be similar to the epidermal chalone. By polyacrylamide gel electrophoresis the aqueous extract exhibited 16 protein bands which diminished to 8 after precipitation with ethanol. The highest inhibitory activity was found in this supernatant and probably was related to the protein content.

The mechanism(s) of action of the epidermal chalone has also been studied extensively. Thus Bullough and Laurence (1964b) found that the chalone prolonged the duration of the mitotic cycle *in vitro*, while Iversen *et al.* (1965) reported that the mitotic index was lowered but the length of the mitotic cycle remained almost unaffected *in vivo*.

The *in vitro* action of the epidermal chalone required the presence of both adrenaline and hydrocortisone, even though their combined actions were not additive. The addition of both hormones proved to be unnecessary *in vivo*, probably as a result of the amount of catecholamines and glucocorticoids readily available in the animals.

An *in vivo* study (Laurence and Randers-Hensen, 1971) showed different mitotic depression of tongue epithelium and ear epidermis (17 and 53%, respectively) by a combination of epidermal chalone, adrenaline and hydrocortisone in adrenalectomized and in intact mice. This seems to indicate that the chalone may not act directly with adrenaline to produce a reduction in the mitotic rate. These results led to the hypothesis of the existence of an "epidermal chalone antagonist" (Laurence *et al.*, 1972). Furthermore, it was postulated that this chalone antagonist must be a mitotic stimulant which promotes cell replacement and is in balance with the chalone.

The action of catecholamines and glucocorticoids may be explained by assuming that the adrenaline inhibits the action of the chalone antagonist, and that hydrocortisone suppresses its production and accounts for the slower action of the chalone antagonist.

An interesting question was to determine the mechanism(s) of action of the chalone at different stages of the cell cycle. Bullough (1965) suggested that possible sites of action of the chalone–adrenalin suppression are early prophase and anaphase, but his histological method of determining mitotic rates does not allow a choice between the two sites. Baden and Sviokla (1968) studied the effect of the chalone on the incorporation of radioactive thymidine into epidermal cells to determine where the inhibition takes place. In *in vitro* experiments DNA synthesis in rat skin cells proceeded normally in the presence of the chalone-adrenaline complex; thus their conclusion was that action occurs later in the mitotic cycle than DNA synthesis. Iversen (1969) found that addition of the chalone to tissue cultures of HeLa cells immediately led to a reduction in the incorporation of thymidine-^{3}H, whereas the incorporation of leucine-^{3}H and uridine-^{3}H was not affected. Marks *et al.* (1971) found decreased incorporation of thymidine-^{3}H in human skin *in vitro* 2 hours after treatment with an aqueous extract, which suggests an effect on the S phase.

Hennings *et al.* (1969) showed that intraperitoneal injections in mice of a crude aqueous extract of mouse skin inhibits epidermal mitosis immediately and DNA synthesis (25–30%) within 9–12 hours. This suggests that epidermal chalone may inhibit DNA synthesis by inhibiting the synthesis of RNA and protein molecules. To elucidate this problem, Elgjo *et al.* (1971a) used water extracts from mouse skin treated with actinomycin D to inhibit DNA-dependent RNA synthesis. While the chalone obtained from mouse skin treated with a topical application of acetone (as control) depressed the epidermal mitotic rate by 60%, extracts from skin treated for 3–6 hours with actinomycin produced similar mitotic inhibition. Extracts obtained after 13 hours showed insignificant inhibition, and the one prepared after 24 hours had no mitosis-inhibiting effect at all. However, all extracts showed 30% inhibition of DNA synthesis with a 10-hour delay. These results suggest that dividing cells can be inhibited at different phases of the cell cycle, and that the chalones may contain two or more different components.

According to current theories, the epidermal chalone is produced by differentiating cells and inhibits the mitotic rate of basal cells. Elgjo *et al.* (1971b) separated basal and differentiating cells and prepared an aqueous extract of each portion. Equal amounts (4 mg) of each extract injected intraperitoneally showed a striking difference. While the extract obtained from basal cells inhibited the mitotic rate 56–75%, the average mitotic inhibition was 22% with the differentiating cell extracts. In another experiment (Elgjo *et al.*, 1972), both extracts were tested for their ability to inhibit DNA synthesis, using autoradiographic techniques. The basal cell extract had no inhibitory effect on epidermal DNA synthesis, but the differentiating cell extracts showed a variable but consistently inhibiting (average 38%) effect. According to Elgjo *et al.* (1972), basal cells produce a factor which acts on the G_2 phase of the mitotic cycle and is responsible for the inhibition of cell division. Another factor is produced by the differentiating cells, which acts on late G_1 so the cells do not proceed into the S phase, thus inhibiting epidermal DNA synthesis. These results indicate that mitosis and DNA synthesis of the epidermis are regulated by at least two different local inhibitors acting on M and S phases of the cell cycle.

The mitosis-inhibiting factor, chalone or M factor, has been previously described. The fraction that inhibits DNA synthesis, or S factor, was isolated from pig skin by Marks (1971), by use of gel chromatography. The S factor has an apparent molecular weight of more than 10^5 daltons, contains 80–90% carbohydrate and a high proportion of sialic acid, and seems to be completely resistant to heating in neutral solution and to proteolytic digestion. It is not known if, or how, epinephrine and other glucocorticoid hormones interact with the G_1 inhibitor. Recently, Elgjo and Edgehill (1973) found that water extracts from rat dermis inhibited mouse epidermal cells in G_2, and that the G_2-inhibiting activity of crude skin extracts was reduced or lost when they were dissolved in fresh serum. The G_1 inhibitor was only slightly affected. Heated serum had no influence on either inhibitor.

An attractive theory is that chalones exert their effect in a manner similar to some hormones such as adrenalin, that is, by activating adenylcyclase in the cell membrane (Iversen, 1969).

Another tentative explanation at the molecular level was proposed by Duell *et al.* (1971). According to these investigators, epidermal mitosis seems to be inhibited by beta-adrenergic stimulation due to a rise in intraepidermal cyclic adenosine monophosphate (cAMP). Following this line of thought, Voorhees and Duell (1971) speculated that psoriatic lesions may be the result of a faulty system generating cAMP in the epidermis.

The possibility of using chalones in the treatment of neoplastic disorders has also been investigated. Bullough and Laurence (1968a,c) demonstrated that mitosis in rabbit Vx2 epidermal tumors can be controlled by the epidermal chalone. Bullough and Deol (1971) reported that the Hewitt keratinizing epidermal carcinoma of the mouse also responds to treatment with partially purified skin extracts. The treatment of a transplantable squamous cell carcinoma of the hamster with epidermal chalone (Elgjo and Hennings, 1971) produced depression of the mitotic rate of tumor cells of about 34% during the first 4 hours after treatment, and an inhibition of DNA synthesis of more than 80% 8 hours after chalone administration. Repeated injections produced no change in the size of the tumor. Laurence and Elgjo (1971) showed that the epidermal chalone, adrenalin or hydrocortisone, or a combination of all of them had no effect on the mitotic rate of tumor cells *in vitro*. In recent years extracts of normal epidermis have been tested in parallel with extracts obtained by maceration of the tumors themselves. Findings indicate that the epidermal chalone is present in these tumors which presumably synthesize it but in inadequate concentrations.

H. Spleen

As discussed previously (Lozzio, 1973), enlargement of the spleen occurs in the majority of humans and animals suffering from a malignant process. The proliferation of spleen cells is the result of cellular and humoral immune reactions against cancer cells. The important immunological functions of the spleen in tumor rejection have been the subject of numerous investigations (Gershon and Kondo, 1969; von Hoepke, 1952; Hilgert and Krištofová, 1967; Pollard and Bussell, 1953; Meltzer and Bartlett, 1972; Vaillier *et al.*, 1972; Woglom, 1919, 1929), and are not considered in this article. The occurrence of antimitotic spleen factors, other than antibodies and specific substances involved in immune reactions, has also been reported. Crude extracts of mammalian spleens produced *in vitro* lysis of certain human cancer cells and normal chicken fibroblasts (Ludwig and von Ries, 1935). The injection of an alcoholic-aqueous extract of beef spleen into normal mice diminished cell division in the small intestine, as denoted by a significant decrease in mitotic figures in the crypts of Lieberkuhn (Fardon *et al.*, 1948).

MacFarlane *et al.* (1948) mixed suspensions of dbrB tumor cells with a "deproteinized" extract of beef spleen at a temperature of 5°C and reported the disappearance of cell division and a reduction in the size of the nucleoli and heterochromatic segments. There was an increase in the number of vacuolated nuclei and the destruction of chromatic material, leading to pycnosis and hollow nuclear membranes.

Diller and Watson (1949), Diller *et al.* (1954), and Watson *et al.* (1947) reported that injection of calf's spleen extract produced degeneration of cells of sarcoma 37 in mice, and also of methylcholanthrene-induced tumors in A-strain mice. Cell changes consisted of vacuolization of the cytoplasm, nuclear pycnosis, or granulation and condensation of the chromatin. Undamaged cells became greatly enlarged, and polyploidy in these cells was unusually frequent. Normal tissue cells were unaffected, but there was mitotic stimulation in the blood-forming organs. These studies and those from several earlier works were reviewed by Diller (1955).

Katzberg (1952) reported that fragments of mouse spleen tissue produced a cytolytic factor when incubated in Parker's medium. The cytolysin was found to have a greater effect on sarcoma 180 than on normal cells.

The effect of a crude spleen extract of bovine origin on a chemically induced tumor has been studied by Cassano (1955). Fibromyomas of the uterus were induced by subcutaneous implantation of 20 mg of α-estradiol in guinea pigs. The incidence and size of fibromatous nodules was markedly decreased in intact and castrated animals 100 days after implantation of the hormone pellet in animals treated with a daily injection of the spleen extract. Treatment with the beef spleen preparation produced greater inhibiting action on the development of fibromyomas in intact than in castrated female guinea pigs.

A glycopeptide has been purified from bovine spleen by Araki and Sugihara (1969). A similar or identical substance was found in Ehrlich tumors (Sugihara and Araki, 1964a, 1966) and gastric cancer (Funaoka *et al.*, 1964). The substance with the greatest carcinostatic activity was precipitated with 70% ethanol and partially purified by gel filtration. Its apparent molecular weight was about 5000, it was heat-labile (100°C), and it contained no sialic or nucleic acids. The compound was characterized as a glycopeptide with a peptide chain of aspartic and glutamic acid, serine, and valine, but it was not crystallized. The addition of this glycopeptide to Ehrlich tumor cells produced "dilation and destruction of the mitochrondria, dilation of the endoplasmic reticulum and marked decrease of lactic dehydrogenase activity." Although this substance produced obvious cell damage, it was not clearly stated whether or not the glycopeptide prevented multiplication of Ehrlich carcinoma cells.

The treatment of human malignancies with spleen extracts originated in 1928, when injection of a beef spleen extract "into patients with chronic lymphatic leukemia caused immediate liquifaction of two large lymph nodes" (Watson, 1960).

During the following years several patients suffering from a variety of tumors were treated with the spleen extract. Fresh calf spleens were homogenized at the rate of 40 gm per liter of saline, and the tissue suspension was incubated at 42°C. About 400 ml of the supernatant was passed through a Seitz filter. Phenol (0.5%) was added as a preservative, and the crude extract stored up to 6 months at about 2°C (Watson, 1960). Two to five milliliters of this extract was given to the patients intravenously and subcutaneously twice a day from 2 months to 1 year. The treatment was given for as long as deemed appropriate. Fourteen patients suffering from various carcinomas were given the spleen extract alone. Some of these patients had tumor recurrence after surgery (e.g., breast carcinoma); in others the tumor was removed at the same time treatment was started; and in a few no surgery was performed. Eight of the fourteen cases were followed for an average period of 13 years without evidence of malignant growth (Watson, 1960). The majority of the malignant processes involved the mammary gland or the gastrointestinal or urogenital tract. Another group of 25 patients with different malignancies was treated with the spleen extract in combination with surgery and/or radiotherapy. Twenty-four of them lived 5 years or more, with an average survival time of nearly 10 years (Watson, 1966).

Amersbach *et al.* (1946) also reported on 21 cases of basal cell carcinoma treated by injection of spleen or liver extract. Fourteen cases showed complete regression, and only one case treated with spleen extract failed to regress.

Recently, an inhibitory cell factor was isolated from normal human and bovine spleens and from those of patients with various hematological diseases (Lozzio and Lozzio, 1973; Lozzio *et al.*, 1973a,b). Thus far, we have purified it nearly 1500 times, in terms of biological activity. Briefly, a 30% spleen homogenate was prepared in Ringer's solution, and the cells were disrupted by sonic vibration. The suspension was then heated to 80°C, and the precipitate discarded. The supernatant was added to a hollow fiber concentrator. Ninety percent of the starting volume was collected as a filtrate of less than 10,000 molecular weight. This filtrate was then passed through hollow fibers with a 5000-molecular-weight cutoff, and 85% of the volume was collected as a filtrate of less than 5000 molecular weight. This portion was next subjected to membrane ultrafiltration (500-molecular-weight cutoff) and concentrated to 10% of the starting volume. This concentrate (< 5000 > 500 molecular weight) was recovered and precipitated with 80% ethanol. After lyophilization, this precipitate was subjected to column (2.6 × 100 cm) chromatography on Sephadex G-25, using 0.01 *M* ammonium bicarbonate (pH 8.6) as eluent. The volume eluted between 325 and 472 ml was pooled and rechromatographed twice as indicated above. The active compound

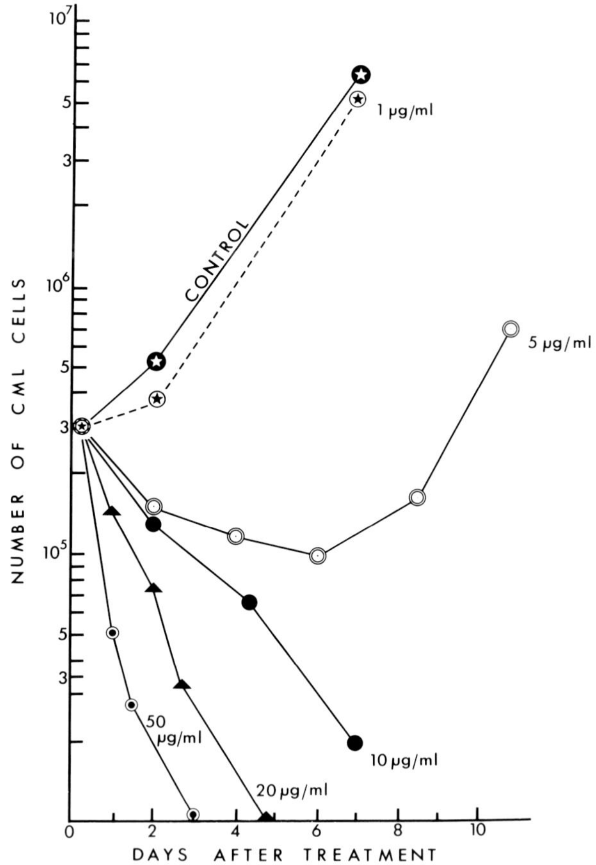

FIG. 2. Cytotoxicity of a low-molecular-weight peptide isolated from human and bovine spleens on CML cells containing the Philadelphia chromosome.

was eluted in two consecutive tubes with a ratio $V_e/V_0 = 1.78$ (356 ml) and 1.84 (367 ml) V_0 was equal to 200 ml. Although the exact composition of the mitotic inhibitor has not been worked out as yet, analyses made indicate that it is a low-molecular-weight (1000) peptide.

Its cytotoxic activity was tested on human chronic myelogenous leukemic (CML) cells having the Philadelphia chromosome as a marker. This cell line has been maintained in culture for more than 3 years (Lozzio and Lozzio, 1973). The cytotoxicity of the low-molecular-weight peptide to CML is illustrated in Fig. 2. The peptide was added in various concentrations to 20-ml cultures of CML cells containing an initial inoculum of 3.5×10^5 cells. The addition of 1 μg/ml of the culture medium produced no effect on the growth of CML compared with control cultures. A reversible cytotoxic effect was observed when 5 μg/ml was added. The effect was irreversibly cytotoxic when 10 μg/ml or more was added to CML cells. As expected, a progressive increase in the peptide concentration resulted in a more effective cell-killing effect in a gradually shorter period of time. Other studies are underway to characterize the peptide, as well as its biological activity and relevance as a cell homeostatic regulator. Since a similar cytotoxic factor has been found in human sera and urine, we think this peptide may be a hormonelike substance which may not be produced solely in the spleen.

I. Other Tissues

The search for growth regulatory factors has also been carried out using a variety of human and animal tissue extracts. The presence of growth inhibitors in nearly all bovine organs has been investigated by Zicha *et al.* (1947). Extracts from calf organs were found to be markedly cytolytic to neoplastic cells. The amount of antimitotic compound in bovine organs diminished with the age of the animal, that of the oldest having little or no cytotoxic effect on cancer cells. None of the extracts prepared was cytolytic to normal cells.

A systematic investigation of the presence of growth inhibitory substances in most bovine and porcine organs was made by Bardos *et al.* (1968), using a variety of methods for the extraction and partial purification of some of these compounds. Out of 1140 fractions isolated, 14 showed positive *in vivo* antitumor activity against sarcoma 180, adenocarcinoma 755, and Ehrlich ascites tumor. The active fractions were obtained from plasma, erythrocytes, bone marrow, thymus, pancreas, and prostate. Significant cell culture cytotoxicity to cells of a human epidermoid carcinoma of the nasopharynx (KB cell line) was found in eight fractions from the liver, one from the lung, and two from the pineal gland. Some of the inhibitors appeared to be proteins or lipids, and others were nonidentified low-molecular-weight compounds.

Incubation of Yoshida sarcoma ascites cells for several hours with various normal rat tissue homogenates produced marked inhibition of tumor growth when they were transplanted intraperitoneally into rats (Druckery *et al.*, 1958). Druckery and co-workers (1959) showed that homogenates of various tissues in rats, especially spleen and lung, suppressed the growth of transplanted Yoshida sarcoma, Walker 256 carcinoma, T sarcoma, and DS carcinosarcoma. Further studies by Hartmann (1959) confirmed and extended the findings of Druckery and co-workers (1958, 1959) and established that the factor was protein in nature, most probably associated with the nuclear fraction of lung tissue. The factor was found in lungs and spleens of rats and mice, but not in livers. Surprisingly, extracts from rabbit lung, prepared in the same manner, had no inhibitory effect on transplantable tumors.

The injection of a tissue suspension prepared by homogenization of a whole mouse and crude extracts from mouse spleen, pancreas, thymus, stomach, thyroid gland, fatty tissue, intestine, kidney, and liver partially protected C3H mice bearing a transplantable leukemia (Olsen, 1963).

A 65% inhibition of the growth and "takes" of transplantable 180 and M-1 sarcomas was obtained by giving tumor-bearing mice a 10% brain tissue emulsion (Timoshechkina, 1963).

A thermolabile macromolecule of unknown chemical composition was found in extracts of liver, kidney, and spleen of human origin (Nilsson and Philipson, 1968). This compound inhibited RNA and protein synthesis and produced cell death. The tissue extracts were active against human diploid fibroblasts, and KB and HeLa cells.

Recently, Lord *et al.* (1974) tested the specificity of growth inhibition of mature blood cell extracts on their respective progenitor cells by measuring the effects on the structuredness of the cytoplasmic matrix by the technique of fluorescence polarization (Cercek and Cercek, 1972; Cercek *et al.*, 1973). Saline extracts of bovine lymphocytes, rat granulocytes, and rat erythrocytes were partially purified by ultrafiltration. Each extract was fractionated into three molecular-weight ranges: 500 to 1000, 1000 to 10,000, and 30,000 to 50,000. The fractions were tested against proliferative populations of human lymphoid cells, mouse granulocytic cells, and mouse erythroid cells, and in all cases complete specificity was found. Also, the activity was found in only one molecular-weight fraction, with the exception of the erythrocyte extract. In each case, with the possible exception of the erythrocyte extract, the active fractions corresponded in molecular weight to those reported for the lymphocytic chalone, granulocytic chalone, and erythrocytic chalone. Furthermore, the tests were successful across species barriers, another of the criteria for chalones.

III. Serum Inhibitors of Cell Growth

Carrel and Ebeling (1921, 1922a,b, 1923a,b) reported that the inhibitory effect of serum on chick fibroblasts can be attributed to the antagonistic action of growth-activating and growth-inhibiting substances, the growth-inhibiting substances exerting a more pronounced effect. According to these investigators, the inhibiting substances resist heating to 65°C and remain in the serum with the albumin fraction. The increased inhibitory action of serum on homologous fibroblasts in old age is partly due to the decrease in the activating substances and to the enhanced activity of the growth-inhibiting principle. Lumsden and Kohn-Speyer (1929) reported that the fresh serum of several species contained heat-labile specific cytotoxins (heterotoxins) which were effective against cultures of Jensen's sarcoma of rats and normal cells of mice, but no other details were given. Simms and Stillman (1937b) found that a euglobulin fraction of chicken serum caused degenerative changes in cultures of chick fibroblasts.

Sacerdote de Lustig and Lyonnet (1946) and Norris and Majnarich (1948b) reported that the serum of cancer patients enhanced the growth of adenocarcinomas of the rectum and breast in humans, chondrosarcoma, squamous cell carcinoma, and neoplastic thyroid gland in culture, while normal serum had an inhibitory effect. Norris and Majnarich (1948a, 1949a) reported that two types of factors affecting cell proliferation have been observed in human sera. One accelerates the rate of normal cell proliferation, and the other inhibits mitosis of normal cells. Normal serum contains, predominantly, factors that accelerate the rate of normal cell division, while sera from patients with neoplastic diseases inhibit normal cell proliferation, probably because of an excess of inhibiting substances.

A thermolabile, complementlike, cytotoxic factor has been found in human serum by various investigators. Chang (1947) described a serum factor toxic to homologous spermatocytes with properties similar to those of complement. Penttinen and Saxén (1957), Penttinen *et al.* (1958), and Saxén and Penttinen (1956) showed that thermolabile substances present in human sera caused clumping of cultured HeLa cells. Cell aggregation was abolished by heating or by treating the serum with trypsin and hyaluronidase.

Bolande and Todd (1958) have described a factor in pooled human sera, which was toxic to human malignant fibroblasts (U 12) from the uterus and HeLa cells, but was innocuous to normal human fibroblasts. The activity was ascribed to the complement components C′3, C′4, and probably C′2. Bolande and McClain (1960) reported the presence of a heat-labile cytotoxic factor in human serum effective against Ehrlich ascites tumor and sarcoma 180 cells. To a certain extent this factor has some properties similar to those of the heterotoxins mentioned earlier. These observations were in close agreement with those of Willheim *et al.* (1957, 1959), who have reported previously essentially similar results using normal human sera and Ehrlich tumor cells.

Chang *et al.* (1959) reported the presence of cytotoxic activity in normal human sera and plasma against normal (conjuctival cells) and malignant (HeLa cells) human tissues. The cytotoxicity was associated with Cohn fractions II and III. On further fractionation the cytotoxicity was found in subfractions II and III. Separations by the cold-ethanol method yielded three main fractions: II-1,2,3 (mainly γ-globulins), III-O (rich in lipids), and III (with proteolytic enzymes). None of these fractions was found to be cytotoxic when tested separately. Cytotoxic activity was again demonstrable when the three fractions were recombined.

Bolande (1960) found that pooled normal human serum or sera from cancer patients had a marked cytotoxic effect on six strains of atypical mammalian cells. However, such an effect was not demonstrable on three normal human cell strains and on two of rabbit fibroblasts. The adsorption of the serum onto a large number of cells eliminated the cytotoxicity, and this was associated mainly with inactivation of C′2 and C′4 complement components. The fact that heat-inactivated human serum could not be reactivated with fresh guinea pig serum indicates that the toxic substances were not related to hemolytic activity of the complement. Since the extraction of lipids from serum eliminates its activity, there is a strong possibility that the activity may be associated with the lipoprotein fraction. Terasaki *et al.* (1961) also reported on the toxicity of heterologous sera. The activity was lost by heating or by storage and could not be fully restored by the addition of fresh homologous or autologous serum.

Fedoroff (1956, 1958), Fedoroff and Cook (1959), and Federoff and

Doerr (1962) showed that the serum of schizophrenic patients was toxic to transformed fibroblasts (L cells, Pomerat's strain FF of human cells, and mouse subcutaneous tissue cells), but had no effect on HeLa cells in human or horse serum. This toxicity has been shown to be due to the presence of at least two nondialyzable substances in the serum. One of these was inactivated by heating at 56°C for 30 minutes, whereas the other was not. The presence of the heat-stable cytotoxic substance in human serum was demonstrated by adsorbing it from the serum onto L cells at 4°C without affecting the heat-labile substance. The heat-stable component of the cytotoxic system may be regarded as an antibody.

Kuru *et al.* (1959) demonstrated that the thermostable factor is destroyed by heating at 65°C for 30 minutes. By fractionation on zone electrophoresis, it was concluded that this factor is present in the γ-globulin fraction. The factor apparently could be adsorbed by Ehrlich ascites tumor cells. It was also shown by Landy *et al.* (1960) that normal human serum depleted of the complement components lost its ability to destroy mouse sarcoma 37 cells. They indicated that the fixation of complement by a preformed antigen–antibody complex destroyed the activity. Removal of divalent calcium and magnesium ions by EDTA caused a similar effect. The activity, however, was restored when free Ca^{2+} and Mg^{2+} were added, and when sera depleted of C′3 and C′4 were mixed. From these experiments it was concluded that complement was one of the factors responsible for heterotoxicity. The other factor, which was heat-labile, was shown to be adsorbed by tumor and normal mouse tissue. This factor could be eluted from the tissues after reaction with human serum. From these experiments, Landy and associates have concluded that serum heterotoxicity is dependent on an antibody complement system.

Ginsburg *et al.* (1961) described a heat-stable factor in normal human serum, which was cytotoxic to Lanschutz ascites tumor cells. This cytotoxic action required the presence of human or rabbit complement, but guinea pig complement was ineffective. The active principle was found in the β-globulin fraction and has characteristics similar to C′4 of human complement. Normal sera from guinea pigs were shown to produce regression of certain lymphomas and lymphosarcomas in mice and rats by Kidd (1953a,b) and Ainis *et al.* (1958). The toxic factor was thermostable, absent in newborn guinea pig sera, and distinct from hemagglutinins. Kwak *et al.* (1963) also reported a cytotoxic factor in guinea pig serum which was stable at 56°C and active even after 21 days at 4°C. Their results indicated that the tumor inhibitory factor in guinea pig serum was independent of antibody and complement and was active against malignant cells without the participation of other host defense mechanisms.

Wilkins (1962) showed that an ether-extractable, saline-insoluble fraction of human serum influences mitotic activity in regenerating rat liver. The same serum fraction obtained from patients with malignant brain tumors stimulates the mitotic activity of regenerating rat liver. However, a similar serum factor from individuals with normal brains and that from sera of patients with benign brain lesions inhibited the mitotic activity of regenerating rat liver. Holmberg (1962) found that interstitial fluid from Walker carcinoma of rat and solid Ehrlich-Landschutz hyperdiploid tumor of mice contain a dialyzable factor deleterious to cultured strain-L cells. Normal interstitial fluid was not cytotoxic (see also Section II,G). Watts (1963) reported a similar dialyzable factor in the interstitial fluid from a squamous cell kidney carcinoma (SCK1) cytotoxic to homologous tumor cells grown in cultures. This cytotoxic factor was apparently associated with a polypeptide fraction with an estimated molecular weight low enough (3000–10,000) to pass quite freely into the general circulation. This fact was confirmed by Watts (1963), who reported that sera from SCK1 tumor-bearing rats also contained a factor, presumably the same, toxic to autologous and homologous tumor cells in culture.

Nicolau *et al.* (1963) reported a cytotoxic effect of serum from patients suffering from visceral cancer on HeLa cells, while normal sera were completely inactive against these cells. Complement as well as heterophilic antibodies did not appear to be indispensable for this action.

Sera from women in the second half of a normal pregnancy inhibited cell proliferation (Norris and Majnarich, 1949b; Penttinen and Saxén, 1962). Rejnek *et al.* (1963) have related this inhibitory action to the occurrence of abnormal immunoelectrophoretic characteristics of α_1 lipoprotein. It is also possible, however, that the high serum levels of glucocorticoid hormones present at this stage of pregnancy can produce mitotic inhibition.

Bias *et al.* (1973) reported cytotoxicity of normal human sera to acute lymphocytic leukemic cells. This cytotoxicity was complement-dependent, was removed by adsorption on tumor cells, and resided in the IgM fraction of the serum.

De Luca and collaborators (1964) reported that the Cohn fraction IV-1 prepared from bovine plasma and certain batches of human serum yields an extract which exhibits irreversible cytotoxic activity against cells in culture. Although the exact chemical nature of this factor is not known, it is likely that its biological activity is associated with proteins. Its lability toward heat, acid pH, and urea, and its nondialyzable nature, have been shown. Work by De Luca *et al.* (1966) clarified the chemical nature of this factor. The cytotoxic activity was due to two components. One was a heat-labile, nondialyzable material assumed to be a protein, and the other was lipoid in nature. The active lipids appear to be free fatty acids. Paul (1973) presented evidence that this toxic serum factor was rather specific for malignant cells. The toxic serum factor was detectable in sera of rat, mouse, and guinea pig, and some batches of horse serum after acidification to pH 2. The toxic factor in rat serum was purified on DEAE-cellulose columns, and its molecular weight was 80,000 to 100,000. A concentration of 40–400 μg/ml of the column-purified preparation was needed in order to observe a cytotoxic effect.

Tritsch and Grahl-Nielsen (1969) isolated two biologically active peptides from tryptic digests of the α chain of human hemoglobin. Both peptides had unique amino acid sequences, and when the smaller of the two, Val-Leu-Ser-Pro-Ala-Asp-Lys, was synthesized from L-amino acids, it was identical to the peptide isolated. Both the heptapeptide and the nonacosapeptide were cytotoxic to cell line RPM1 no. 2402 which originated from a carcinoma of the Syrian hamster. The two peptides were equipotent, and levels of 10^{-5}–10^{-7} *M* reduced the viable cell number in suspension culture without a lag period. The toxicity of the two peptides was counteracted by the addition of serum protein.

Stjernholm (1974) isolated two protein fractions from human and animal sera. One fraction (a chalone) prevented the mitosis of human lymphocytes *in vitro,* whereas the other (an antichalone) promoted cell division. Increasing concentrations of the chalone produced a progressive inhibition of mitosis; it also prevented the stimulation of lymphocytes exposed to phytohemagglutinin or pokeweed mitogen. In addition to the inhibition of DNA synthesis, the chalone depressed RNA and protein synthesis. The fraction was specific for lymphocytes, as both lymphocytes from peripheral blood and a human lymphoblastoid cell line responded to the chalone, but human polymorphoneuclear leukocytes and guinea pig peritoneal macrophages did not. This lymphocyte chalone may be confined to mammals, as sera from guinea pig, rat, dog, and chimpanzee all contained the factor, while sera from rattlesnake, water moccasin, chicken, and turtle showed a complete absence of chalone activity. The mitosis-promoting factor or antichalone neutralized or inhibited the chalone action when added to lymphocyte cultures.

A study of patients with chronic lymphocytic leukemia revealed three groups with normal, low, or no chalone activity in the sera. Patients with normal activity had white blood cell counts of less than 50,000, while the other two groups had high counts ranging from 70,000 to 275,000.

IV. Summary

The material in this section is summarized in Table I.

The limitation of the mitotic activity of many mammalian cells in high-density cultures (see Section II,A) cannot be explained only on the basis of available nutrients in the medium. It appears that cultured cells release substances which in turn inhibit cell division when the appropriate concentration in the medium is reached. Thus specific and nonspecific inhibitors of cultured cell growth have been reported. Some normal cells appear to produce an inhibitor of the proliferation of oncogenic virus-containing cells which are unable to synthesize a similar antimitotic substance. The growth of normal fibroblasts may be controlled by contact inhibition, protein factors present in the serum added to the medium, and attachment to rigid surfaces (anchorage-depending growth).

Extracts from mice and chick embryos have been found to suppress cell growth *in vivo* and *in vitro*, respectively. The administration of mouse embryonic and placental extracts inhibited the growth of 70% of spontaneous and transplanted tumors. Since the treatment was mainly effective on carcinomas and not on sarcomas, the extract appeared to have some tissue specificity. Low-molecular-weight inhibitors of normal cells have also been obtained from chick embryos. Some of the compounds isolated inhibited protein and RNA synthesis, but their chemical compositions are not yet known. It is possible that one of the antimitotic substances studied was a nucleic acid derivative. Crude extracts from human placenta inhibited the growth of some transplantable tumors. A protein of relatively low molecular weight (70,000) was recently isolated from human amniotic and chorionic membranes. It inhibits DNA synthesis and division of several normal and malignant cells (see Section II,B).

Specific and nonspecific mitotic inhibitors have been partially purified from mammalian and amphibian kidneys. It is interesting to note that the majority of inhibitory substances extracted from the kidneys have a lipoid composition. The existence of a chalone, probably protein in nature, has been reported in amphibian kidneys. It may control growth and differentiation from the early stages of the pronephros, as well as kidney cell renewal in adult life (see Section II,C).

A variety of inhibitors has been isolated from liver tissue. Some of the reported inhibitors fulfill the criteria for a liver chalone, while others are distinctly different in their biological properties. In the majority of cases in which chemical composition has been studied, the inhibitors have been found to be proteins or polypeptides. The enzyme arginase is a constituent of normal liver and has been found to be inhibitory to cells grown in tissue culture by virtue of arginine depletion in the culture medium. It is not known how many of the reported protein inhibitors were arginase, but it is clear that there are other protein inhibitors in liver tissue that are definitely not arginase. Ethanolamine, another inhibitor of cell growth, has been found in the liver. Practically all the substances inhibit cell growth, mitosis, and DNA synthesis. Some have also been found to inhibit RNA and protein synthesis. Numerous growth inhibitors isolated from the liver were active against various tumor cells both *in vitro* and *in vivo* (see Section II,D).

Some tissues of tumor-bearing animals and/or the tumor itself have been found to contain mitotic inhibitors. The substances, isolated from a variety of tumors, have been partially characterized as lipids, proteins, and polypeptides. One small polypeptide present in various human and animal tumor fluids, has been isolated and crystallized. In most instances the compound extracted from one tumor inhibited the growth of other unrelated malignant processes. Chalones have been prepared from cancer tissues of rodents. The growth of skin carcinomas, melanomas, and lymphomas, and the proliferation of myelogenous leukemic cells, was inhibited by the respective chalone. Inhibitors of the growth of various experimental tumors have been detected in a variety of unrelated tissues from embryological, anatomical, and functional standpoints. None of them has ever been purified, and their relevance to abnormal growth remains obscure (see Section II,E).

The results obtained with striated muscle extracts indicate that unknown components of muscle tissue are cytotoxic to malignant cells *in vitro* and *in vivo* (see Section II,F). Complete regression or inhibition of transplantable carcinomas and sarcomas was observed by treatment with muscle extracts. The antimitotic substances contained in crude muscle extracts appeared to be more cytotoxic to malignant than to normal cells. Similarly, a complex of protein with mucopolysaccharide and/or nucleic acid derivatives, obtained from connective tissue, had greater cytotoxic activity on human cancer cells of mesenchymal origin than on homologous normal cells. Two antimitotic nucleoproteins, one with adenine as the base and the other with hypoxanthine as the base, have been isolated from beef cornea and rabbit muscle. Both substances depressed the incorporation of phosphorous into DNA and RNA of regenerating rat liver and inhibited the increase in sulfhydryl groups in proteins prior to cleavage of sea urchin eggs.

The nature and mechanism of action of the epidermal chalone can be summarized as follows. It is a tissue-specific, but not species-specific, diffusible substance which depresses mitotic activity and DNA synthesis of skin cells *in vitro* and *in vivo*. To express itself, the chalone requires the presence of adrenaline and hydrocortisone which probably act on a chalone antagonist. The epidermal chalone consists of two fairly well-characterized factors which operate at different phases of the cell cycle. The M (mitosis) factor is a substance produced by basal epidermal skin, which acts on the G_2 phase of the cell cycle, inhibiting mitosis. The S (synthesis) factor is produced by the differentiating cells and acts on late G_1 phase, inhibiting DNA synthesis. Chemically, the M factor is probably a protein or glycoprotein with a molecular weight of 20,000 to 40,000, stable at low pH, and resistant to pepsin but destroyed by trypsin. The S factor has an apparent molecular weight of 100,000, is rich in carbohydrates and sialic acid, and appears to be resistant to heat and proteolytic digestion (see Section II,G).

The inhibitory effect of mammalian spleen extracts on homologous neoplastic growth has been claimed for many years by clinicians and basic researchers (see Section II,H). Whether or not factors produced in the spleen are able to suppress malignant growth *in vivo* remains open to question. However, it has been demonstrated that purified spleen products inhibit proliferation of cancer cells *in vitro*. It is possible that more than one substance cytotoxic to cancer cells are produced in the spleen. The most recent data suggest that the active fraction may be a peptide. Most of the results of the treatment of human malignancies with spleen extracts are impressive. It is puzzling that this line of research has not been continued by other investigators, improving the technique of extraction of the spleen factor to obtain a more effective product. As far as we are aware, there is no publication confirming or denying the beneficial effects of the administration of spleen extracts to humans suffering from various types of carcinomas, leukemias, and lymphomas. Although preliminary evidence has been presented that a glycopeptide extracted from human spleen primarily inhibits RNA synthesis, the mechanism of action of the spleen factor(s) on cancer cells is not yet known.

Antimitotic substances have been isolated from several other tissues during the past few years. Most recently, crude extracts of mature blood cells have been shown to have a specific inhibitory effect on their respective progenitor cells. Although cell-specific, these extracts are not species-specific, suggesting that they may act as chalones.

The cytotoxic effect of sera is presumably due to the presence of at least two nondialyzable components. One of them is presumably a protein, and the other is most likely lipoid in nature (free fatty acids). Healthy people as well as those with various diseases may have cytotoxic serum. Normal human serum was found to have marked cytotoxicity toward atypical mammalian cells (some derived from malignant tumors and others derived from spontaneously transformed cultures of normal cells), however, it was noncytotoxic for normal human or animal cells. Sera from patients suffering from neoplastic diseases and that from women in the later stages of pregnancy were strongly inhibitory to normal cell proliferation. Sera from schizophrenic patients was cytotoxic to fibroblastlike cells, while it was innocuous against epitheliumlike cells. Sera from normal and tumor-bearing animals have been found to have a cytotoxicity similar to that of human sera. A lymphocytic chalone and its antichalone appear to be present in normal and leukemic human sera (see Section III).

V. Concluding Remarks

The current knowledge of endogenous mitotic inhibitors of mammalian cells is still very limited and provides a somewhat confusing picture. Except for the polypeptide crystallized by Holmberg (1968), those purified by Tritsch and Grahl-Nielsen (1969), and the low peptide isolated in our laboratory (Lozzio and Lozzio, 1973; Lozzio *et al.*, 1973a), the inhibitors considered in this review are a very heterogenous group of compounds. They have been isolated from cultured cells, many normal mammalian tissues, a variety of tumors, and serum. In instances in which the chemical composition was partially studied, the mitotic inhibitors were identified as proteins, lipids, nucleic acid derivatives, mucopolysaccharides, or a combination of two or more of these. Both tissue-specific (chalone) and nonspecific inhibitors have been reported, and there is no doubt that antimitotic substances are normal constituents of mammalian tissues. Unfortunately, many of the reports deal with very crude preparations in

TABLE I: ENDOGENOUS INHIBITORS OF

Source	Inhibitor	Nature	Molecular weight (approx.)
Culture medium from Syrian hamster kidney cells (BHK 21)	Anomin	Unknown	1000–2000
Culture medium from Chinese hamster cells	Growth inhibitor	Unknown	>13,000
Human fibroblasts and culture medium of same	Fibroblast chalone	Unknown	30,000–50,000
Culture medium of human lymphocytes	DNA synthesis inhibitor	Unknown	?
Chick embryo	Growth inhibitor	"Minor" nucleic acid derivatives (two fractions)	<12,000
Human amniotic and chorionic membranes	DNA synthesis inhibitor	Protein	70,000
Sea urchin ovaries	Cleavage retarding factor	Polynucleotide	?
Kidneys, liver, and myocardium	Mitotic inhibitor	Phospholipid	?
Xenopus laevis kidneys	"Chalone"?	Unknown	?
Rat liver and serum	Hepatic chalone	Protein?	?
Rabbit liver	Hepatic chalone	Polypeptide	Dialyzable (<12,000)
Normal rat liver	DNA synthesis inhibitor	Proteinlike	?
Adult rat liver	RNA synthesis inhibitor	Protein	125,000
Human liver	DNA and RNA synthesis inhibitor	Unknown	?
Normal rat liver	Growth inhibitor	Unknown	Dialyzable (<12,000)
Bovine liver	Growth inhibitor	Protein	?

GROWTH AND/OR MITOTIC ACTIVITY

Effects	Reference
Nonspecific inhibition of cell division; inhibition of protein synthesis? Addition of serum to the medium neutralizes the inhibitor's effect	Bürk, 1966, 1967
Limits exponential growth of Chinese hamster cells; acts synergistically with adrenaline	Froese, 1971
Inhibits thymidine-^{3}H uptake of human WI-38 fibroblasts	Houck *et al.*, 1972
Suppresses DNA synthesis and strongly inhibits incorporation of uridine-^{3}H by normal human lymphocytes stimulated by PHA	Smith *et al.*, 1970
Inhibits growth of homologous explants of normal liver cells at less than 1×10^{-6} *M*	Coogan *et al.*, 1968, 1969
Inhibits DNA synthesis and cell division of human lymphocytes, fibroblasts and epidermic cells	Baden, 1973
Retards cleavage of sea urchin eggs; coexists with a dinucleotide accelerator	Menkin, 1959
Marked antimitotic activity for kidney and liver tissues, completely suppresses cell division when given intraperitoneally to rats in large doses	McJunkin and Henry, 1935
Inhibits mitosis in cultured larval pronephros but not in epidermal cells; cortisone and adrenaline enhance the inhibitory effect on pronephros	Chopra and Simnett, 1969
Specifically decreases mitotic index of hepatocytes. Found in intact rat liver but absent in the liver of partially hepatectomized rats	Saetren, 1956; Stich and Florian, 1958
Inhibition of DNA synthesis, tissue specific for liver cells	Verly *et al.*, 1971
Inhibits DNA synthesis of intact cells	Otsuka, 1967
Reversible inhibition of uptake of RNA and protein precursors	Frank, 1968
Weakly inhibits penetration of DNA precursors and strongly inhibits phosphorylation and incorporation of thymine into DNA. Impairs phosphorylation of UMP in RNA synthesis	Nilsson, 1970
Retards growth and causes marked decrease in cell numbers in cultures of rat ascites hepatoma cells and Yoshida sarcoma cells	Suzuki, 1959
Inhibits growth of rat ascites hepatoma cells (AH-130) and Yoshida sarcoma cells	Hori and Ukita, 1962

(Continued)

TABLE I

Source	Inhibitor	Nature	Molecular weight (approx.)
Ox liver	Mitotic inhibitors (2)	Polypeptides or low MW compounds attached to polypeptides	Dialyzable
Necrotic mouse sarcoma 180	Growth inhibitor	Lipid	?
Normal mouse liver and spleen			
Various human and animal tumor fluids	Cytotoxic polypeptide	Polypeptide	1000–2000
Tendons, muscles, and aorta of chick, cow, and dog	Growth inhibitor	Protein complex with mucopolysaccharides and/or nucleic acids	High
Beef cornea and rabbit muscle	Cornin	Nucleoprotein	?
Epidermal cells	Epidermal chalone M factor	Basic protein–glycoprotein or mixture of proteins, glycoproteins, and/or polysaccharides	30,000–40,000
Bovine spleen, Ehrlich tumors, gastric cancer	Carcinostatic factor	Glycopeptide	5000
Human and bovine spleens	Splenic peptide	Peptide or glycopeptide	1000?
Lungs and spleen of rats and mice	Growth inhibitor	Protein	?
Pig skin	Epidermal chalone S factor	80–90% carbohydrate + sialic acid	100,000
Bovine plasma and some human serum, rat serum, mouse serum, some horse serum	Cytotoxic factor	Two components—one protein and one lipid	80,000–100,000
Human and animal serum	Lymphocytic chalone	Protein	?
α-Chain of human hemoglobin	Cytotoxic peptides	(2) Peptides	1,200 3,500
Interstitial fluid from kidney carcinoma	Cytotoxic factor	Polypeptide	3,000–10,000

(*Continued*)

Effects	Reference
Both completely arrest mitotic activity of HeLa cells and one also inhibits mitotic activity of Chang liver cells	Oftebro *et al.*, 1963
Inhibits growth of mammary gland tumor, sarcoma 180, and Ehrlich ascites carcinoma	Miller and Kimsey, 1967
Retards growth of L cells, Chang liver cells, HeLa cells, and MB64E malignant lymphoid cells. Higher concentrations cause cell death. Acts only in S phase	Holmberg, 1968
Inhibits growth of human cancer cells of mesenchymal origin and to a lesser extent, normal fibroblasts	Simms and Stillman, 1937a; Parshley, 1965; Parshley and Mandl, 1963; Parshley *et al.*, 1965
Decreases incorporation of ^{32}P into nucleic acids and inhibits DNA synthesis in sea urchin eggs. Also depresses incorporation of ^{32}P into DNA and rRNA of regenerating rat liver	Nisida *et al.*, 1964; Nisida and Murakami, 1965a,b
Tissue-specific inhibition of mitosis of epidermal cells. Both adrenaline and hydrocortisone required *in vitro* but not *in vivo*	Bullough *et al.*, 1964
Damages mitochondria and endoplasmic reticulum and decreases lactic dehydrogenase activity of Ehrlich ascites cells	Araki and Sugihara, 1969; Sugihara and Araki, 1964a, 1966; Funaoka *et al.*, 1964
Cytotoxic for chronic myelogenous leukemia and multiple myeloma cells. Reversible in small concentrations, irreversible in larger amounts	Lozzio and Lozzio, 1973; Lozzio *et al.*, 1973a,b
Inhibits *in vivo* growth of Yoshida sarcoma cells when given intraperitoneally to rats	Druckery *et al.*, 1958, 1959; Hartmann, 1959
Inhibits DNA synthesis of epidermal cells	Marks *et al.*, 1971
Irreversibly cytotoxic to cells in culture; apparently specific for malignant cells	De Luca *et al.*, 1964, 1966; Paul, 1973
Inhibits DNA synthesis, depresses RNA and protein synthesis; specific for lymphocytes	Stjernholm, 1974
Cytotoxic for cell line RPMI #2402 originating from a carcinoma of Syrian hamsters. Toxicity neutralized by the addition of serum protein	Tritsch and Grahl-Nielsen, 1969
Cytotoxic to homologous tumor cells grown in culture	Watts, 1963

which the chemical composition is questionable or unknown. Other investigators have attempted to purify the inhibitors, but they have only partially succeeded. Most of the preparations that are still being tested are mixtures of amino acids, polypeptides, nucleotides, nucleosides, and carbohydrates. With the recent advances in biochemistry, it should be possible to apply modern techniques to the isolation and characterization of these compounds. The work to be done does not appear to be more cumbersome than that done in the past to isolate well-known hormones.

Except for the inhibitors from the kidneys, which appear to be lipoid in nature, we think that the great majority of mitotic inhibitors reported are polypeptides or glycopeptides of low molecular weight, which can attach and easily penetrate the cell, thus becoming part of a larger-molecular-weight intracellular complex.

To determine whether or not nucleosides, nucleotides, and nucleotide–peptide complexes can function as endogenous mitotic inhibitors may be a profitable line of research. At this point one asks the question: Are these low-molecular-weight compounds, found in tissues and fluids, the result of protein degradation or split fragments of nucleic acids? This question and many others will remain unanswered until a purified product(s) is available for chemical and metabolic studies. It is imperative that purified preparations be made available for precise determination of the mechanisms of action and for possible therapeutic use. The latter point is especially important, as it has been demonstrated that many of the inhibitors have significant antitumor activity. There are several reports in the literature of successful treatment of human carcinomas, leukemias, and lymphomas with spleen extracts. Surprisingly, this line of research has not yet been pursued further, perhaps because "the efficacy of splenic extracts in clinical application has been affirmed and denied through the years" (Diller, 1955).

Although research on chalones has been pushed forward in recent years, none of them has been purified enough as yet, and their exact chemical composition remains unknown. Thus the studies on chalones, as well as on many other regulators of cell division, are in a primitive stage, as indicated by the heterogeneous extracts used and the variable results obtained. In view of recent results (reviewed by Lozzio, 1973) indicating remission of an experimental leukemia with a granulocytic chalone, we felt it appropriate to reproduce a figure from an article by A. H. Roffo who in 1937 achieved the first tumor regression with a tissue extract (see Fig. 1). Thirty-seven years have elapsed since Roffo's experiments, and we have improved very little the purity of the majority of mitotic inhibitors now being tested. We have found that some mitotic inhibitors may be tissue-specific, but the results are essentially similar to those reported nearly 40 years ago.

The problem of specific and nonspecific inhibitors of cell division is a debatable matter. Chalones are by definition endogenous (noncytotoxic) inhibitors of cell division. Since the characteristics of inhibition of many cells and tissues by various crude extracts are quite similar, it is possible that a single substance may be responsible for the antimitotic action. For example, it is possible that compounds of low molecular weight may form complexes with proteins of various tissues, thus appearing as specific inhibitors of a particular tissue, when in fact they function as general regulators of cell homeostasis. If this were the case, studies oriented toward finding the source of the natural regulator(s) of cell growth (hormone?) would be of great importance.

The regulatory mechanisms of cell growth in cultures are poorly known. A distinction must be made between the limits of growth of cells in suspension cultures and cell growth that depends entirely on anchorage to rigid surfaces, which produces a monolayer. In our opinion the growth of cells in liquid media appears to be primarily controlled by inhibitors released into the medium, which act through a negative feedback mechanism on essential cell metabolic pathways (e.g., protein or nucleic acid syntheses). Therefore cells growing free in crowded conditions would produce their own mitotic inhibitors which may or may not be specific. When the inhibitor reaches an appropriate concentration in the medium, the cells stop dividing in spite of all necessary nutrients being available to them. A contact inhibition phenomenon in high-density suspension cultures seems less likely to affect cell growth than in monolayer cultures in which each cell is in contact with several other cells. It is precisely the physical contact that primarily would produce a change in the cell membrane structure and perhaps intercellular transfer of soluble substances (growth-limiting messengers), resulting in subsequent alteration of intracellular metabolic processes necessary for the growth of the monolayer. In this context we stress that inhibitors produced by cells growing in suspension could also attach and/or alter the cell membrane, thus triggering the release of lysosomal enzymes, endonucleases, and so on, which in turn would stop cell division and eventually cause cell death.

The relevance of *in vitro* studies on cell growth to the hemeostatic system operating *in vivo* is questionable, even though the changes at higher cell densities are obviously more likely to be related to the *in vivo* conditions.

It is hoped that this review will acquaint researchers with the accomplishments in this field and will provide the impetus for further investigation. The study of endogenous mitotic inhibitors could become one of the most fruitful areas for further research, because elucidation of the biochemical mechanisms controlling normal cellular proliferation is fundamental for an understanding and possible control of malignant growth.

Acknowledgments

Aided by research grants ET-64A, CI-87B, and CI-87C from the American Cancer Society, Inc. The authors thank Mrs. Vanessa Faircloth for her help in the preparation of this article.

References

Ainis, H., Kurtz, H. M., Kramer, P. I., Weimer, H. E., Ryan, R. M., and Jameson, E. (1958). *Cancer Res.* **18**, 1309.

Amersbach, J. C., Walter, E. M., and Sperti, G. S. (1946). *Arch. Dermatol. Symp.* **54**, 119.

Araki, F., and Sugihara, Y. (1969). *Hiroshima J. Med. Sci.* **18**, 253.

Aujard, C., Chang, E., and Frayssinet, C. (1973). *Exp. Cell Res.* **78**, 476.

Baden, H. P. (1973). *J. Nat. Cancer Inst.* **50**, 43.

Baden, H. P., and Sviokla, S. (1968). *Exp. Cell Res.* **50**, 644.

Bardos, T. J., Gordon, H. L., Chmielewicz, Z. F., Kutz, R. L., and Nadkarni, M. V. (1968). *Cancer Res.* **28**, 1620.

Bellanger, F., and Harel, L. (1969). *C. R. Acad. Sci.* **269**, 113.

Bias, W. B., Santos, G. W., Burke, P. J., Mullins, G. M., and Humphrey, R. L. (1973). *Transplant. Proc.* **5**, 949.

Bolande, R. P. (1960). *Lab. Invest.* **9**, 475.

Bolande, R. P., and McClain, J. P. (1960). *Proc. Soc. Exp. Biol. Med.* **103**, 345.

Bolande, R. P., and Todd, E. W. (1958). *AMA Arch. Pathol.* **66**, 720.

Brues, A. M., Jackson, E. B., and Aub, J. C. (1936). *Proc. Soc. Exp. Biol. Med.* **34**, 270.

Brues, A. M., Subbarow, Y., Jackson, E. B., and Aub, J. C. (1940). *J. Exp. Med.* **7**, 423.

Bullough, W. S. (1962). *Nature (London)* **193**, 520.

Bullough, W. S. (1965). *Cancer Res.* **25**, 1683.

Bullough, W. S. (1972a). *Brit. J. Dermatol.* **87**, 187.

Bullough, W. S. (1972b). *Brit. J. Derm.* **87**, 347.

Bullough, W. S., and Deol, J. U. R. (1971). *Eur. J. Cancer* **7**, 425.

Bullough, W. S., and Deol, J. U. R. (1972). *Brit. J. Dermatol.* **86**, 586.

Bullough, W. S., and Laurence, E. B. (1960a). *Proc. Roy. Soc., Ser. B* **151**, 517.

Bullough, W. S., and Laurence, E. B. (1960b). *Exp. Cell Res.* **21**, 394.

Bullough, W. S., and Laurence, E. B. (1964a). *Exp. Cell Res.* **33**, 176.

Bullough, W. S., and Laurence, E. B. (1964b). *Exp. Cell Res.* **35**, 629.

Bullough, W. S., and Laurence, E. B. (1966). *Nature (London)* **210**, 715.

Bullough, W. S., and Laurence, E. B. (1968a). *Nature (London)* **220**, 134.

Bullough, W. S., and Laurence, E. B. (1968b). *Nature (London)* **220,** 137.
Bullough, W. S., and Laurence, E. B. (1968c). *Eur. J. Cancer* **4,** 587.
Bullough, W. S., and Laurence, E. B. (1968d). *Eur. J. Cancer* **4,** 607.
Bullough, W. S., and Laurence, E. B. (1970a). *Cell Tissue Kinet.* **3,** 291.
Bullough, W. S., and Laurence, E. B. (1970b). *Eur. J. Cancer* **6,** 525.
Bullough, W. S., Hewett, C. L., and Laurence, E. B. (1964). *Exp. Cell Res.* **36,** 192.
Bullough, W. S., Laurence, E. B., Iversen, O. H., and Elgjo, K. (1967). *Nature (London)* **214,** 578.
Bürk, R. R. (1966). *Nature (London)* **212,** 1261.
Bürk, R. R. (1967). *Wistar Inst. Symp. Monogr.* **7,** 39.
Califano, A. (1962). *Boll. Soc. Ital. Biol. Sper.* **38,** 639.
Carrel, A. (1930). *Proc. Inst. Med. Chicago* **8,** 62.
Carrel, A., and Ebeling, A. H. (1921). *J. Exp. Med.* **34,** 599.
Carrel, A., and Ebeling, A. H. (1922a). *J. Exp. Med.* **35,** 647.
Carrel, A., and Ebeling, A. H. (1922b). *J. Exp. Med.* **36,** 399.
Carrel, A., and Ebeling, A. H. (1923a). *J. Exp. Med.* **37,** 653.
Carrel, A., and Ebeling, A. H. (1923b). *J. Exp. Med.* **38,** 419.
Cassano, F. (1955). *Folia Endocrinol.* **9,** 747.
Cercek, L., and Cercek, B. (1972). *Int. J. Radiat. Biol.* **21,** 445.
Cercek, L., Cercek, B., and Ockey, C. H. (1973). *Biophysik* **10,** 187.
Cevese, P. G., and Ferro, A. (1951). *Riv. Anat. Patol. Oncol.* **4,** 78.
Chang, M. C. (1947). *J. Gen. Physiol.* **30,** 321.
Chang, R. S., Pennell, R. B., Keller, W., Wheaton, L., and Liepens, H. (1959). *Proc. Soc. Exp. Biol. Med.* **102,** 213.
Chany, E., and Frayssinet, C. (1971). *C. R. Acad. Sci.* **272,** 2644.
Chopra, D. P., and Flaxman, B. A. (1973). *J. Nat. Cancer Inst.* **50,** 281.
Chopra, D. P., and Simnett, J. D. (1969). *Exp. Cell Res.* **58,** 319.
Chopra, D. P., Yu, R., and Flaxman, B. A. (1972). *J. Invest. Dermatol.* **59,** 207.
Coogan, P. S., Friese, J., Moganini, H., and Hass, G. M. (1968). *Lab. Invest.* **18,** 139.
Coogan, P. S., Friese, J., Moganini, H., Berardi, R., and Hass, G. M. (1969). *Lab. Invest.* **20,** 371.
De Luca, C., Rakowski, P. F., and Tritsch, G. L. (1964). *Biochim. Biophys. Acta* **86,** 346.
De Luca, C., Carruthers, C., and Tritsch, G. L. (1966). *Exp. Cell Res.* **43,** 451.
Diller, I. C. (1955). *Antiseptic* **52,** 95.
Diller, I. C., and Watson, G. F. (1949). *Growth* **13,** 27.
Diller, I. C., Watson, G. F., and Fisher, M. (1954). *Biol. Bull.* **107,** 350.
Doljanski, L., Hoffman, R., and Tenebaum, E. (1944). *Growth* **8,** 13.
Druckery, H., Schmahl, D., and Rajewsky, M. (1958). *Naturwissenschaften* **45,** 16.
Druckery, H., Schmahl, D., Steinhoff, D., Rajewsky, M., Bannasch, P., and Flaschentrager, T. H. (1959). *Z. Krebsforsch.* **63,** 28.
Duell, E. A., Voorhees, J. J., Kelsey, W. H., and Hayes, E. (1971). *Arch. Dermatol.* **104,** 601.
Dvorak, H. J., and Byram, J. W. (1930). *Proc. Soc. Exp. Biol. Med.* **27,** 967.
Elgjo, K. (1969). *Acta Pathol. Microbiol. Scand.* **76,** 31.
Elgjo, K., and Edgehill, W. (1973). *Virchows Arch., B* **13,** 14.
Elgjo, K., and Hennings, H. (1971). *Virchows Arch., B* **7,** 1.
Elgjo, K., Hennings, H., and Edgehill, W. (1971a). *Virchows Arch., B* **7,** 342.
Elgjo, K., Laerum, O. D., and Edgehill, W. (1971b). *Virchows Arch., B* **8,** 277.
Elgjo, K., Laerum, O. D., and Edgehill, W. (1972). *Virchows Arch., B* **10,** 229.

Engelhardt, D. L. (1971). *J. Cell. Physiol.* **78,** 333.
Fardon, J. C., Prince, J. E., and Berning, S. N. (1948). *Cancer Res.* **8,** 531.
Fedoroff, S. (1956). *J. Lab. Clin. Med.* **48,** 55.
Fedoroff, S. (1958). *Tex. Rep. Biol. Med.* **16,** 32.
Fedoroff, S., and Cook, B. (1959). *J. Exp. Med.* **109,** 615.
Fedoroff, S., and Doerr, J. (1962). *J. Nat. Cancer Inst.* **29,** 331.
Frank, W. (1968). *Hoppe-Seyler's Z. Physiol. Chem.* **349,** 237.
Frankfurt, O. S. (1971). *Exp. Cell Res.* **64,** 140.
Froese, G. (1971). *Exp. Cell Res.* **65,** 297.
Funaoka, H., Ezaki, H., Araki, F., and Sugihara, Y. (1964). *J. Radiat. Res.* **5,** 159.
García-Giralt, E., and Macieira-Coelho, A. (1969). *C. R. Acad. Sci.* **268,** 2316.
García-Giralt, E., Berumne, L., and Macieira-Coelho, A. (1970). *J. Nat. Cancer Inst.* **45,** 649.
Gershon, R. K., and Kondo, K. (1969). *J. Nat. Cancer Inst.* **43,** 545.
Ginsburg, I., Dishon, T., Bloch, M., and Gross, J. (1961). *Proc. Soc. Exp. Biol. Med.* **107,** 235.
Graf, L., and Rapport, M. M. (1961). *Cancro* **14,** 415.
Guidetti, E. (1953). *Neoplasie* **6,** 1.
Guidetti, E. (1956). *Acta Unio Int. Contra Cancrum* **12,** 365.
Guidetti, E. (1964). *Acta Unio Int. Contra Cancrum* **20,** 241.
Guidetti, E., and Castoldi, C. (1956). *Rev. Belge Pathol. Med. Exp.* **25,** 431.
Hartmann, E. (1959). *Brit. J. Cancer* **13,** 693.
Heaton, T. B. (1926). *J. Pathol. Bacteriol.* **29,** 293.
Heaton, T. B. (1929). *J. Pathol. Bacteriol.* **32,** 565.
Heilbrunn, L. V., Wilson, W. L., and Harding, D. (1951). *J. Nat. Cancer Inst.* **11,** 1287.
Henderson, I. W. D. (1970). *J. Surg. Oncol.* **2,** 221.
Hennings, H., Elgjo, K., and Iversen, O. H. (1969). *Virchows Arch., B* **4,** 45.
Herbut, P. A., and Kraemer, W. H. (1960). *Amer. J. Pathol.* **36,** 105.
Hilgert, I., and Krištofová, H. (1967). *Folia Biol. (Prague)* **13,** 110.
Holmberg, B. (1962). *Nature (London)* **195,** 45.
Holmberg, B. (1968). *Eur. J. Cancer* **4,** 263.
Hondius-Boldingh, W., and Laurence, E. B. (1968). *Eur. J. Biochem.* **5,** 191.
Hori, M., and Ukita, I. (1962). *J. Biochem. (Tokyo)* **51,** 322.
Houck, J. C., Weil, R. L., and Sharma, V. K. (1972). *Nature (London), New Biol.* **240,** 210.
Iversen, O. H. (1969). *Homeostatic Regulators, Ciba Found. Symp.* p. 29.
Iversen, O. H., Aandahl, E., and Elgjo, K. (1965). *Acta Pathol. Microbiol. Scand.* **64,** 506.
Kagen, L. J., and Linder, S. (1972). *Proc. Soc. Exp. Biol. Med.* **140,** 1325.
Kaliss, N. (1952). *Cancer Res.* **12,** 379.
Kaliss, N., and Snell, G. D. (1951). *Cancer Res.* **11,** 122.
Katzberg, A. A. (1952). *Growth* **16,** 15.
Kidd, J. G. (1953a). *J. Exp. Med.* **98,** 565.
Kidd, J. G. (1953b). *J. Exp. Med.* **98,** 583.
Kuru, G., Kosaki, H., Itouzi, H., Isibasi, H., and Tokunaga, A. (1959). *Acta Unio Int. Contra Cancrum* **15,** 165.
Kwak, K. S., Kurtz, H. M., Jameson, E., and Bradley, B. (1963). *Proc. Amer. Ass. Cancer Res.* **4,** 36.
Landy, M., Michael, J. G., Trapani, R. J., Achinstein, B., Woods, M. W., and Shear, M. J. (1960). *Cancer Res.* **20,** 1279.

Laurence, E. B., and Elgjo, K. (1971). *Virchows Arch., B* **7**, 8.
Laurence, E. B., and Randers-Hansen, E. (1971). *Virchows Arch., B* **9**, 271.
Laurence, E. B., and Randers-Hansen, E. (1972). *Virchows Arch., B* **11**, 34.
Laurence, E. B., Randers-Hansen, E., Christophers, E., and Rytömaa, T. (1972). *Rev. Eur. Etud. Clin. Biol.* **17**, 133.
Lord, B. I., Cercek, L., Cercek, B., Shah, G. P., Dexter, T. M., and Lajtha, L. G. (1974). *Brit. J. Cancer* **29**, 168.
Lozzio, B. B. (1973). *Exp. Hematol.* **1**, 309.
Lozzio, B. B., Lozzio, C. B., and Bamberger, E. G. (1973a). *Experientia* **29**, 87.
Lozzio, B. B., Lozzio, C. B., and Bamberger, E. G. (1973b). Proc. Symp. Pathophysiology of the Spleen. *Clujul Med.* (Romania) **46**, 373.
Lozzio, C. B., and Lozzio, B. B. (1973). *J. Nat. Cancer Inst.* **50**, 535.
Ludwig, F., and von Ries, J. (1935). *Schweiz. Med. Wochenschr.* **65**, 5.
Lumsden, T., and Kohn-Speyer, A. C. (1929). *J. Pathol. Bacteriol.* **32**, 185.
MacFarlane, E. W., Schmock, N. G., and Nadeau, L. V. (1948). *Cancer Res.* **8**, 438.
McJunkin, F. A., and Hartman, C. D. (1933). *Amer. J. Pathol.* **9**, 739.
McJunkin, F. A., and Henry, J. W. (1935). *Amer. J. Pathol.* **11**, 353.
McJunkin, F. A., and Matsui, T. (1931). *Arch. Pathol.* **12**, 794.
Marks, F. (1971). *Hoppe-Seyler's Z. Physiol. Chem.* **352**, 1273.
Marks, F., Fukui, K., and Halprin, K. (1971). *Brit. J. Dermatol.* **84**, 453.
Marrs, J. M., and Voorhees, J. J. (1971a). *J. Invest. Dermatol.* **56**, 174.
Marrs, J. M., and Voorhees, J. J. (1971b). *J. Invest. Dermatol.* **56**, 353.
Matsumara, M., and Ishida, Y. (1954). *Gann* **45**, 109.
Meltzer, M. S., and Bartlett, G. L. (1972). *J. Nat. Cancer Inst.* **49**, 1439.
Menkin, V. (1959). *J. Exp. Zool.* **140**, 441.
Miller, A. J., and Kimsey, L. S. (1967). *Cancer* **20**, 471.
Miyamoto, M., and Terayama, H. (1971). *Biochim. Biophys. Acta* **228**, 324.
Mohr, V., Althoff, J., Kinzel, V., Suss, R., and Volm, M. (1968). *Nature (London)* **220**, 138.
Moskalewski, S. (1971). *Przegl. Dermatol.* **63**, 348.
Murphy, J. B., and Sturm, E. (1933). *Science* **77**, 631.
Murphy, J. B., and Sturm, E. (1934a). *J. Exp. Med.* **60**, 293.
Murphy, J. B., and Sturm, E. (1934b). *J. Exp. Med.* **60**, 305.
Nakahara, W., and Fukuoka, F. (1961). *Gann* **52**, 197.
Nakahara, W., Fukuoka, F., and Naora, H. (1962). *Gann* **53**, 1.
Nicolau, S. G., Badanoiu, A. L., and Gavrilescu, M. (1963). *Acta Unio Int. Contra Cancrum* **19**, 183.
Nilsson, G. (1970). *Exp. Cell Res.* **59**, 207.
Nilsson, G., and Philipson, L. (1968). *Exp. Cell Res.* **51**, 275.
Nisida, I., and Murakami, T. H. (1965a). *Acta Med. Okayama* **19**, 1.
Nisida, I., and Murakami, T. H. (1965b). *Acta Med. Okayama* **19**, 11.
Nisida, I., Murakami, T. H., and Kanso, H. (1964). *Symp. Soc Cell Chem.* **14**, 57.
Norris, E. R., and Majnarich, J. J. (1948a). *Amer. J. Physiol.* **153**, 483.
Norris, E. R., and Majnarich, J. J. (1948b). *Amer. J. Physiol.* **153**, 492.
Norris, E. R., and Majnarich, J. J. (1949a). *Proc. Soc. Exp. Biol. Med.* **70**, 229.
Norris, E. R., and Majnarich, J. J. (1949b). *Proc. Soc. Exp. Biol. Med.* **70**, 663.
Oftebro, R., Laland, P., Dedichen, J., Laland, S., and Thorsdalen, N. (1963). *Brit. J. Cancer* **17**, 189.
Olsen, K. (1963). *Lancet* **1**, 507.
Otsuka, H. (1967). *Wistar Inst. Symp. Monogr.* **7**, 3.

Otsuka, H., and Terayama, H. (1966). *Biochim. Biophys. Acta* **123**, 274.
Parshley, M. S. (1965). *Cancer Res.* **25**, 387.
Parshley, M. S., and Mandl, I. (1963). *Proc. Amer. Ass. Cancer Res.* **4**, 50.
Parshley, M. S., and Mandl, I. (1965). *Nature (London)* **208**, 800.
Parshley, M. S., Blackwood, C. E., and Mandl, I. (1965). *Proc. Amer. Ass. Cancer Res.* **6**, 50.
Paul, D. (1973). *Nature (London), New Biol.* **242**, 186.
Penttinen, K., and Saxén, E. (1957). *Ann. Med. Exp. Biol. Fenn.* **35**, 335.
Penttinen, K., and Saxén, E. (1962). *Acta Pathol. Microbiol. Scand.* **154**, 148.
Penttinen, K., Saxén, E., Saxén, L., Toivonen, S., and Vainio, T. (1958). *Ann. Med. Exp. Biol. Fenn.* **36**, 27.
Pollard, M., and Bussell, R. (1953). *Tex. Rep. Biol. Med.* **11**, 48.
Rejnek, J., Bednařík, T., Řeřábková, E., and Doležal, A. (1963). *Clin. Chim. Acta* **8**, 108.
Roffo, A. H. (1926). *Bol. Inst. Med. Exp. Estud. Trat. Cancer, Buenos Aires* **2**, 978.
Roffo, A. H. (1927a). *Bol. Inst. Med. Exp. Estud. Trat. Cancer, Buenos Aires* **3**, 5.
Roffo, A. H. (1927b). *Bol. Inst. Med. Exp. Estud. Trat. Cancer, Buenos Aires* **3**, 328.
Roffo, A. H. (1937a). *Prensa Med. Argent.* **24**, 2048.
Roffo, A. H. (1937b). *Bull. Ass. Fr. Etude Cancer* **26**, 602.
Roffo, A. H. (1937c). *Bol. Inst. Med. Exp. Estud. Trat. Cancer, Buenos Aires* **14**, 257.
Roffo, A. H. (1938). *Lancet* **2**, 184.
Roffo, A. H., and Correa, L. M. (1929). *Bol. Inst. Med. Exp. Estud. Trat. Cancer, Buenos Aires* **5**, 200.
Roffo, A. H., and García-Velloso, L. (1927). *Bol. Inst. Med. Exp. Estud. Trat. Cancer, Buenos Aires* **3**, 531.
Roffo, A. H., and Lopez-Ramirez, R. (1930). *Bol. Inst. Med. Exp. Estud. Trat. Cancer, Buenos Aires* **7**, 1225.
Roffo, A. H., and Villanueva, J. (1927a). *Prensa Med. Argent.* **13**, 1017.
Roffo, A. H., and Villanueva, J. (1927b). *Bol. Inst. Med. Exp. Estud. Trat. Cancer, Buenos Aires* **3**, 92.
Roffo, A. H., and Villanueva, J. (1930). *Bol. Inst. Med. Exp. Estud. Trat. Cancer, Buenos Aires* **5**, 449.
Rytömaa, T., and Kiviniemi, K. (1968). *Eur. J. Cancer* **4**, 595.
Rytömaa, T., and Kiviniemi, K. (1970). *Eur. J. Cancer* **6**, 401.
Sacerdote de Lustig, E., and Lyonnet, J. (1946). *Rev. Soc. Argent. Biol.* **22**, 487.
Saetren, H. (1956). *Exp. Cell Res.* **11**, 229.
Sato, T., and Grob, D. (1966). *Fed. Proc., Fed. Amer. Soc. Exp. Biol.* **25**, 292.
Saxén, E., and Penttinen, K. (1956). *Ann. Med. Exp. Biol. Fenn.* **34**, 396.
Scaife, J. F. (1970). *Experientia* **26**, 1071.
Shibuya, T., Inaba, M., and Kawaguchi, A. (1935). *Trans. Japan Pathol. Soc.* **25**, 412.
Shiokawa, K., and Yamada, K. (1967). *Develop. Biol.* **16**, 398.
Simms, H. S., and Stillman, N. P. (1937a). *J. Gen. Physiol.* **20**, 603.
Simms, H. S., and Stillman, N. P. (1937b). *Arch. Pathol.* **23**, 316.
Simnett, J. D., and Chopra, D. P. (1969). *Nature (London)* **222**, 1189.
Smith, R. T., Bauscher, J. A. C., and Adler, W. (1970). *Amer. J. Pathol.* **60**, 495.
Stich, H. F., and Florian, M. L. (1958). *Can. J. Biochem. Physiol.* **36**, 855.
Stjernholm, R. L. (1974). *J. Reticuloendothel. Soc.* **15**, 25a.
Stoker, M. (1969). *Homeostatic Regulators, Ciba Found. Symp.* p. 264.
Strittmatter, C. F. (1968). *Proc. Soc. Exp. Biol. Med.* **128**, 366.
Strong, L. C. (1968). *Int. Ser. Monogr. Pure Appl. Biol. Div.: Mod. Trends Physiol. Sci.* **31**, 195.

Sugihara, Y., and Araki, F. (1963). *Gann* **54,** 1.
Sugihara, Y., and Araki, F. (1964a). *Hiroshima J. Med. Sci.* **13,** 65.
Sugihara, Y., and Araki, F. (1964b). *Hiroshima J. Med. Sci.* **13,** 299.
Sugihara, Y., and Araki, F. (1966). *Gann* **57,** 287.
Suzuki, S. (1959). *Jap. J. Exp. Med.* **29,** 341.
Terasaki, P. I., Esail, M. L., Cannon, J. A., and Longmire, W. P. (1961). *J. Immunol.* **87,** 383.
Timoshechkina, M. E. (1963). *Vop. Onkol.* **9,** 98.
Tritsch, G. L., and Grahl-Nielsen, G. (1969). *Biochemistry* **8,** 1816.
Vaillier, D., Donner, M., and Burg, C. (1972). *Experientia* **28,** 1352.
Van Snick, P., and Brachet, J. (1971). *Arch. Biol.* **82,** 173.
Verly, W. G., Deschamps, Y., Pushpathodam, J., and Desrosiers, M. (1971). *Can. J. Biochem.* **49,** 1376.
Voaden, M. J. (1968). *Exp. Eye Res.* **7,** 326.
von Hoepke, H. (1952). *Z. Krebsforsch.* **58,** 378.
Voorhees, J. J., and Duell, E. A. (1971). *Proc. Int. Symp. Psoriasis, 1971* p. 301.
Walton, A. J. (1914). *J. Exp. Med.* **20,** 554.
Watson, G. F. (1960). *Acta Unio Int. Contra Cancrum* **16,** 1220.
Watson, G. F. (1966). *Growth* **30,** 491.
Watson, G. F., Diller, I. C., and Ludwick, N. V. (1947). *Science* **106,** 348.
Watts, J. (1963). *Nature (London)* **197,** 196.
Wayss, K., Mattern, J., and Volm, M. (1973). *Naturwissenschaften* **60,** 354.
Weiss, P., and Kavanau, J. L. (1957). *J. Gen. Physiol.* **41,** 1.
Werner, H. (1944). *Nature (London)* **155,** 827.
Werner, H. (1945). *Proc. Soc. Exp. Biol. Med.* **59,** 128.
Wilkins, R. H. (1962). *Surg. Forum* **13,** 421.
Willheim, R., Ivy, A. C., and Janecek, H. M. (1957). *J. Exp. Med. Surg.* **15,** 300.
Willheim, R., Revici, E., and Auber, M. (1959). *Fed. Proc., Fed. Amer. Soc. Exp. Biol.* **18,** 604.
Woglom, W. H. (1919). *J. Cancer Res.* **4,** 281.
Woglom, W. H. (1929). *Cancer Res.* **4,** 281.
Zicha, O., Urbanek, G., and Zicha, K. (1947). *Cas. Lek. Cesk.* **86,** 1341.

Ultrastructure of Mammalian Chromosome Aberrations

B. R. BRINKLEY AND WALTER N. HITTELMAN

Division of Cell Biology, Department of Human Biological Chemistry and Genetics, The University of Texas Medical Branch, Galveston, Texas, and Department of Developmental Therapeutics, The University of Texas, M. D. Anderson Hospital and Tumor Institute, Houston, Texas

I. Introduction

Chromosome aberrations are induced by a variety of agents, including radiation (Wolff, 1961), chemicals (Shaw, 1970), viruses (Nichols, 1970), temperature changes (Hampel and Levan, 1964; Dewey *et al.*, 1971), and mycoplasms (Paton *et al.*, 1965). However, relatively little is known about the molecular basis of such damage and how it is expressed within the architecture of the chromosome. The serious deficiencies in our knowledge of chromosome aberrations are due to many factors, not the least of which is a general ignorance of eukaryotic chromosome structure. In addition, morphological evaluation of chromosome damage has been largely confined to light microscope observations. Although this approach has been effective in identifying and classifying the various types of aberrations expressed at metaphase, the limitations of resolution by light optics are obvious. Structures associated with damaged regions smaller than 0.1–0.2 μm are invisible. Since most clastogens (Shaw, 1970) interact directly or indirectly with DNA or nucleoprotein to bring about breaks and rearrangements which are later expressed in metaphase chromosomes, much greater resolution is necessary to properly evaluate damage. Progress made in recent years now permits direct observation of chromosome aberrations by both transmission and scanning electron microscopy. With the improved resolution afforded by these instruments, it should be theoretically possible to evaluate the molecular basis of chromosome damage and provide a smoother correlation of damage at the DNA and protein level with aberrations visible in the metaphase chromosome. This article presents a review of the progress made in the analysis of chromosome damage by electron microscopy, and attempts to correlate these findings with information available from light microscope studies at one end of the spectrum and molecular level of organization at the other. The article considers several questions: First, what can be said about chromosome aberrations such as breaks, gaps, exchanges, and related phenomena at the light microscope level? Second, what new information is provided by ultrastructural analysis of chromosome damage? Third, from this and other information, what is the target in the chromosome for the initial lesion produced by damaging agents or clastogens? And finally, what changes occur during the condensation of chromatin to transform the molecular lesion into a cytologically visible aberration? No attempt is made to review exhaustively the voluminous literature on chromosome aberrations. The publications of Evans (1962), Wolff (1961), Kihlman (1971), and Comings (1974) provide excellent review of this subject and serve as resources for topics discussed in this article.

II. Light Microscope Observations and Terminology

Chromosomal aberrations have been categorized according to their structural configuration in the metaphase chromosome observed by light microscopy (Fig. 1). For the purposes of this article, the following definitions are in order (see Evans, 1962, for greater details).

A. Chromosome-Type Aberrations

Generally, this is an aberration that affects both chromatids at the same locus (Fig. 1c, e, and h). If the damage occurs prior to DNA replication and is not repaired, the aberration is expressed as a *break* in both chromatids at metaphase. An identical lesion, an isolocus

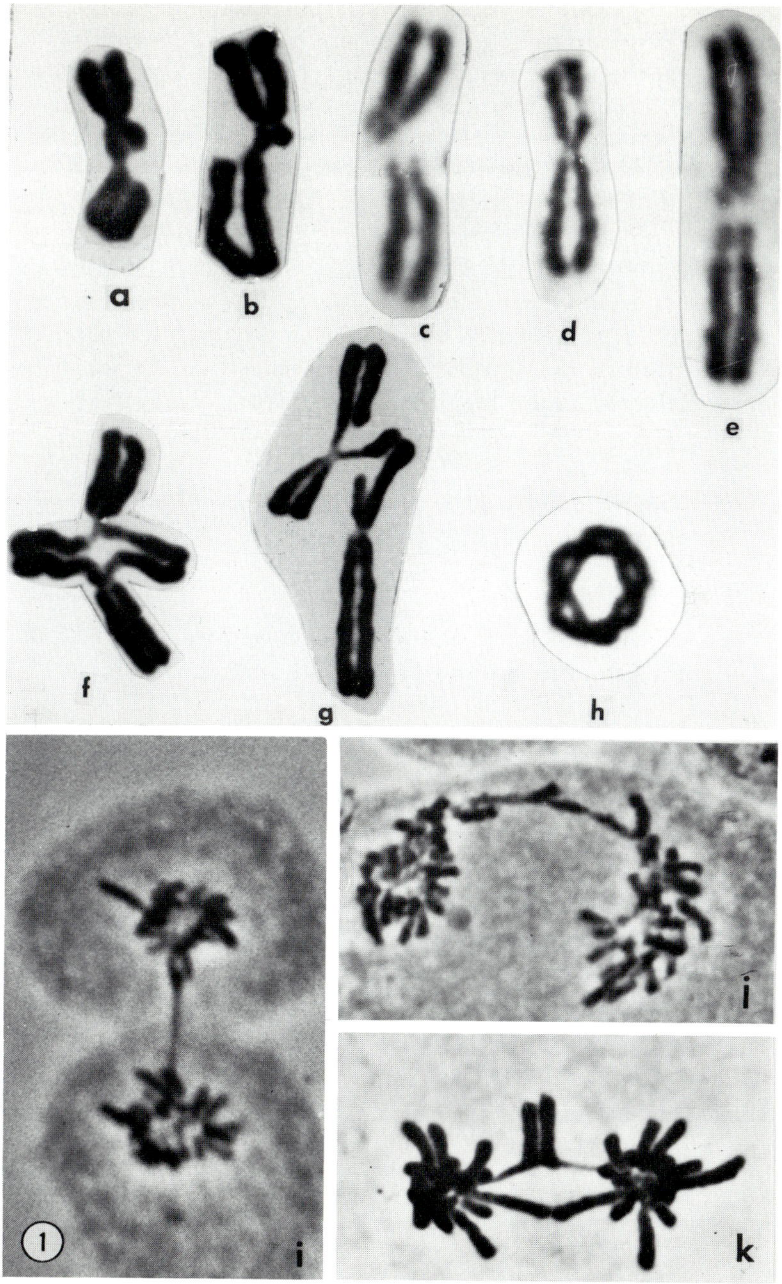

FIG. 1. Types of chromosome aberrations scored by light microscopy. a, Chromatid break; b, unaligned chromatid break; c, isochromatid break; d, chromatid gap; e, isochromatid gap; f and g, exchanges; h, ring; i, chromosome bridge; j, stickiness; k, side-arm bridge. (From Brinkley and Shaw, 1970.)

break, may result from two independent chromatid breaks occurring, by chance, at the same locus. Although chromosome breaks are identical to isochromatid aberrations, the two can be identified experimentally on the basis of the time in the cell cycle when the cell is exposed to the breaking agent (see Comings, 1974).

B. Chromatid-Type Aberrations

This is an aberration that affects only one of a pair of chromatids (Fig. 1a, b, d, f, and g). Although important exceptions have been reported (Hsu *et al.*, 1962; Wolff, 1969), chromatid aberrations generally occur during or after DNA synthesis, when the chromosome becomes a double target for breaking agents.

C. Gaps

A gap is expressed as a severe attenuation or achromatic region along the chromatid arm (Fig. 1d and e).

D. Exchanges

This is a structural rearrangement of chromosomes in which two breaks from different chromosomes (interchange) or from two different loci within the same chromosome (intrachange) interact (Fig. 1f, g, and h). Exchanges may result in dicentric bridges at anaphase (Fig. 1i).

E. Subchromatid Aberrations

Although these types of aberrations are controversial, several investigators have described them as aberrations involving subunits within the chromatid (Fig. 1k). Adherents to the subchromatid hypothesis of chromosome damage generally support a bineme or polyneme model of chromosome structure.

F. Chromosome Stickiness

This is the adherence of chromosomal segments at anaphase by means other than breaks or exchanges (Fig. 1j).

G. Damage to Specialized Chromosome Regions

In addition to the more "classic" types of aberrations described above, many drugs and physical agents can induce damage to specialized chromosome regions such as the kinetochore, the nucleolar organizer, and the telomeres, as is discussed in subsequent sections.

TABLE I
COMPARATIVE ASPECTS OF CHROMOSOME DAMAGE OBSERVED WITH LIGHT AND ELECTRON MICROSCOPY

Type of damage	Light microscope	Electron microscope
Breaks		
Aligned	Apparent structural discontinuity; broken ends aligned	True structural discontinuity; broken ends aligned; sister chromatid connections True structural continuity
Unaligned	Apparent structural discontinuity; broken ends unaligned	True structural discontinuity; broken ends unaligned; sister chromatid connections True structural continuity
Gaps		
Chromatid	Achromatic lesions with apparent structural continuity	True structural continuity: 50- to 80-Å fibrils, 200- to 250-Å fibrils, 500- to 800-Å fibrils True structural discontinuity
Isochromatid	Achromatic with apparent structural continuity; may appear like exaggerated secondary constrictions or nucleolar organizer regions	True structural continuity, but not like nucleolar organizers
Bridges		
Chromosome	Dicentric with chromosome stretched across metaphase plate; acentric fragments often present	Same as light microscope except stretched regions often double, revealing "half-chromatids"
Side arm	Stickiness at sharply bent regions of anaphase arms; apparent structural continuity between associated arms	Same as light microscope except true structural continuity of subchromatid fibers
Stickiness	Adherence of chromosome arms at anaphase; no acentric fragments	Thin chromatin exchanges connecting various chromosome arms at metaphase and anaphase

III. Electron Microscope Observations

In recent years improved methods of cell preparation for electron microscopy have facilitated ultrastructural studies of most types of chromosome damage. The first ultrastructural analysis of irradiation damage was reported by Bloom and Leider (1962). Although the quality of morphological preservation afforded by the procedures available at the time of their study was less than ideal, they were

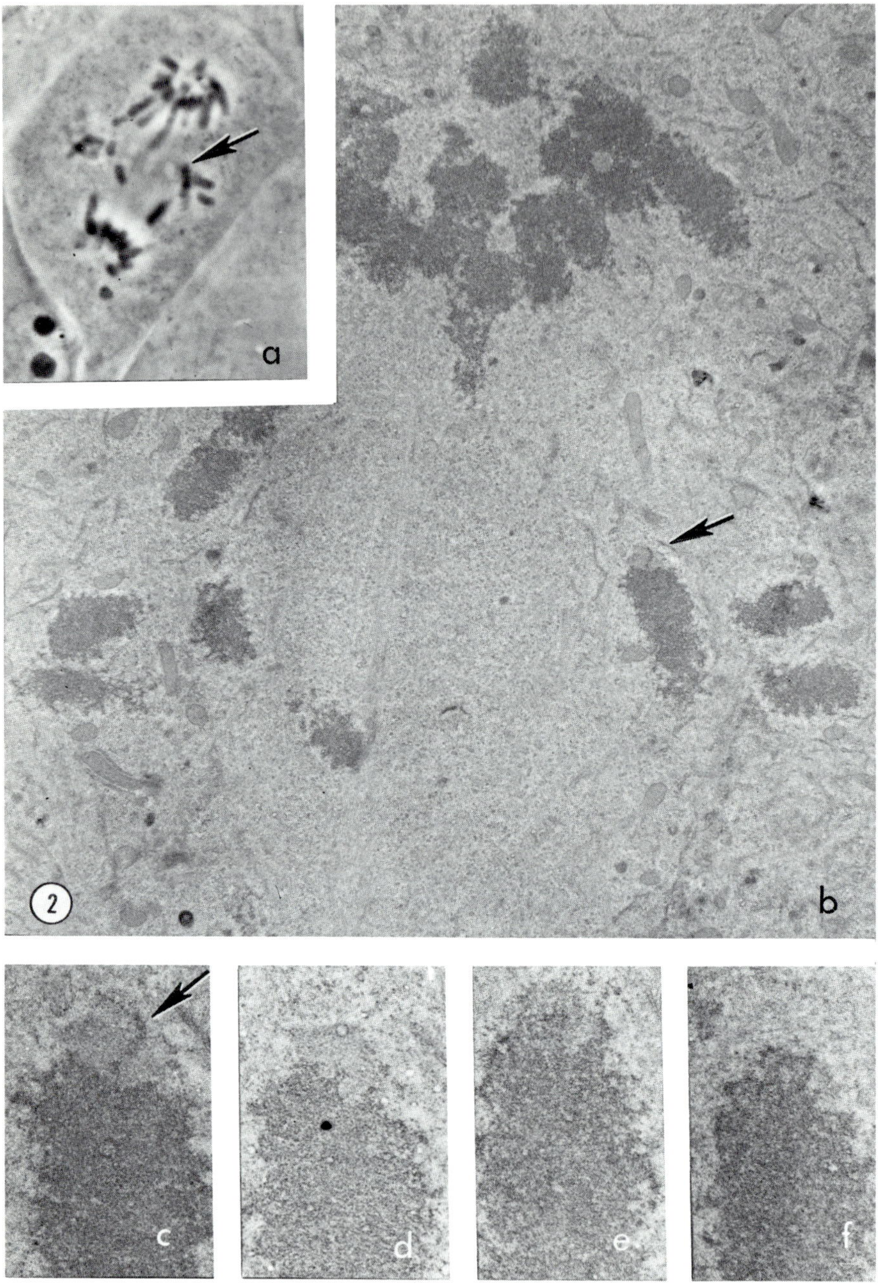
a
2
b
c
d
e
f

able to observe the phenomenon of the "paling" or loss of contrast in chromosomes due to ultraviolet irradiation. The paling was thought to be due to loss of DNA in the irradiated regions (Bloom and Özarslan, 1965). The first detailed ultrastructure studies of cultured mammalian cells using modern methods of fixation and embedding were carried out by Humphrey and Brinkley (1969). These investigators utilized a flat embedding procedure (Brinkley *et al.*, 1967) to preselect rat kangaroo (PtK_1) cells by phase microscopy with typical radiation-induced chromosome damage, and subsequently sectioned and examined the same cells in the electron microscope. A similar approach was used by these investigators to study subchromatid aberrations in rat kangaroo chromosomes (Brinkley and Humphrey, 1969). Brinkley and Shaw (1970) extended investigations to include studies of damage to human chromosomes, which either occurred spontaneously or was induced by x rays, mitomycin C, or thymidine-^3H. The latter study utilized both serial ultrathin sections of Epon-embedded materials, and air-dried chromosomes which were examined by both light and electron microscopy (Shaw *et al.*, 1972). More recently, several other investigations of chromosome aberrations by transmission and scanning electron microscopy have been reported (Scheid and Traut, 1970, 1971; Brogger, 1971; Golomb *et al.*, 1971; Yu, 1971; Wahren *et al.*, 1972). These studies provided additional information concerning breaks, gaps or achromatic lesions, exchanges, and bridges as described in the following sections. A comparison of the features of chromosome aberrations viewed by light and electron microscopy is given in Table I.

A. Breaks

Chromatid breaks viewed by the electron microscope in unsectioned, unstained chromosomes are demonstrated in Figs. 3 to 6. Ultrathin sections of broken fragments of rat kangaroo chromosomes are shown in Fig. 2a–f. In both types of preparations the broken region is

---

FIG. 2. Acentric fragments in anaphase cells of rat kangaroo fibroblast fixed 6 hours after radiation with 250 rads. (a) Phase-contrast micrograph of the cell in Epon. Arrow points to acentric fragment which can be identified as part of the X chromosome in the electron micrograph. ×1190. (b) Thin section of cell shown in (a). Arrow points to acentric fragment of X chromosome with break through the nucleolar organizer. ×8500. (c–f) Serial sections through the X chromosome fragment. ×25,500. Arrow in (c) points to dense axial element which extends through the nucleolar organizer. Compare with intact X chromosome shown in Fig. 26. (From Humphrey and Brinkley, 1969.)

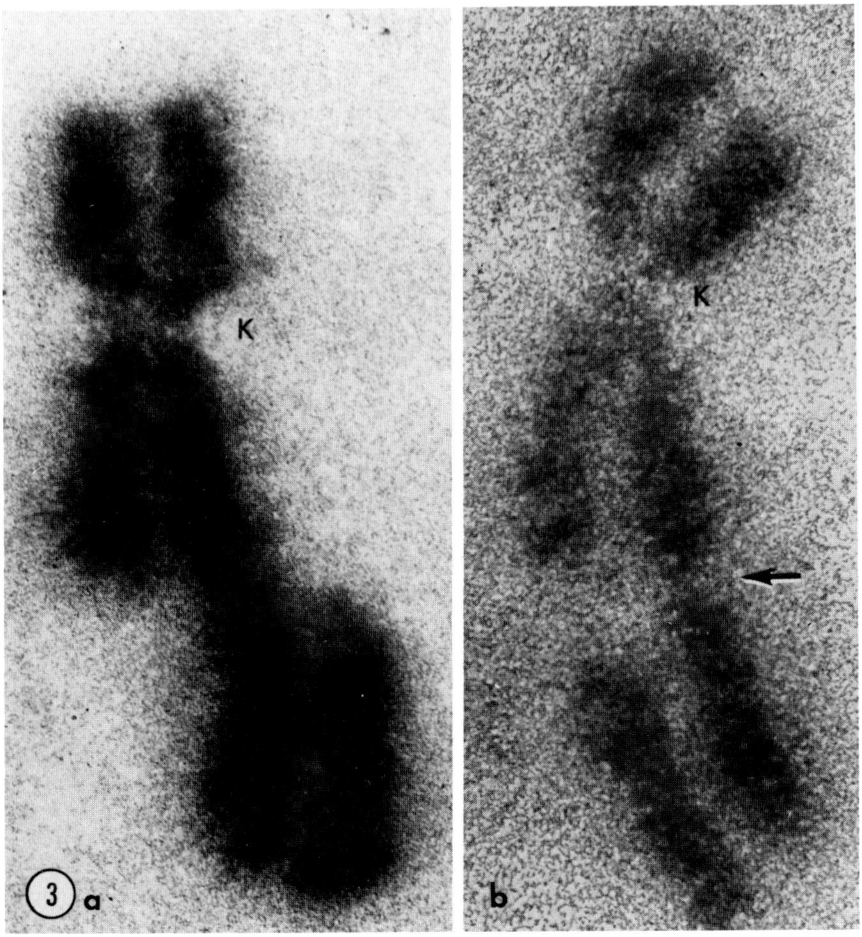

FIG. 3. Chromatid breaks observed in whole-mount, acetic acid–ethanol-fixed preparations. (a) Unaligned break. (b) Aligned chromatid break. Arrow points to apparent damage in sister chromatid at the same locus K, Kinetochore. ×7000. (From Brinkley and Shaw, 1970.)

characterized by an absence of interconnecting fibrils between the two broken parts. Moreover, the broken ends appear to be no different in fine structure than the normal ends or telomeres of chromosomes typically seen in the electron microscope. A notable exception is when the break occurs in the region of the nucleolar organizer as shown in Figs. 2a–f and 10a–b. In these types of breaks, one may see the electron-dense axial filament extending through the secondary constriction and numerous lightly stained fibrous loops which sur-

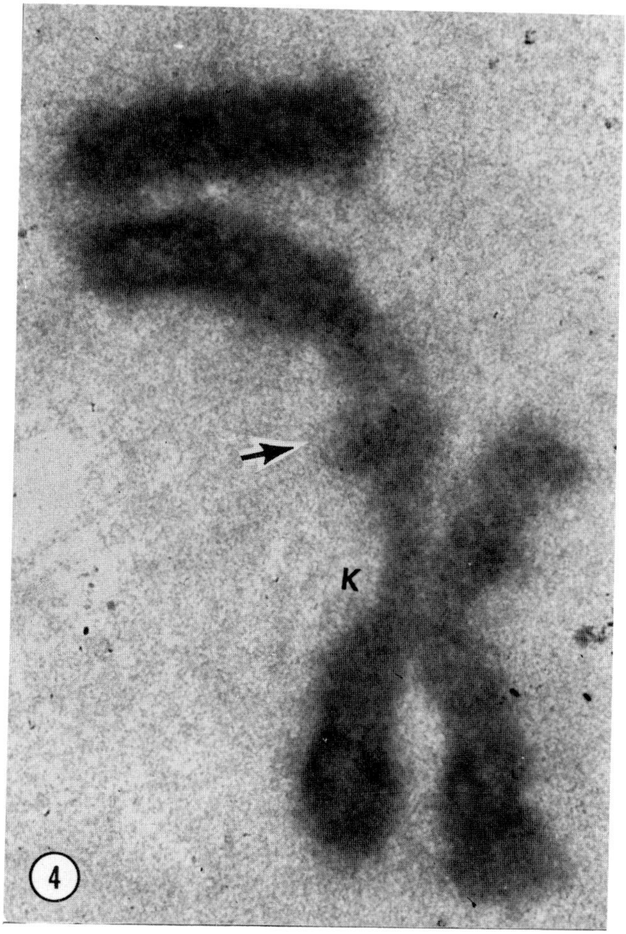

FIG. 4. Chromatid break in chromosome C from human lymphocyte. Arrow points to damaged area in sister chromatid K, Kinetochore. ×7000. (From Brinkley and Shaw, 1970.)

round the axial element. A break in this region does not appear to disrupt the organization seen in undamaged chromosomes (Hsu *et al.*, 1967), and each fragment contains a normal-appearing piece of secondary constriction. This observation suggests that small broken fragments of the secondary constriction become differentiated at metaphase, exactly as they would if they were on undamaged chromosomes. It is not unusual that damage to a nucleolar organizer has little effect on its organization and fine structure at metaphase. In-

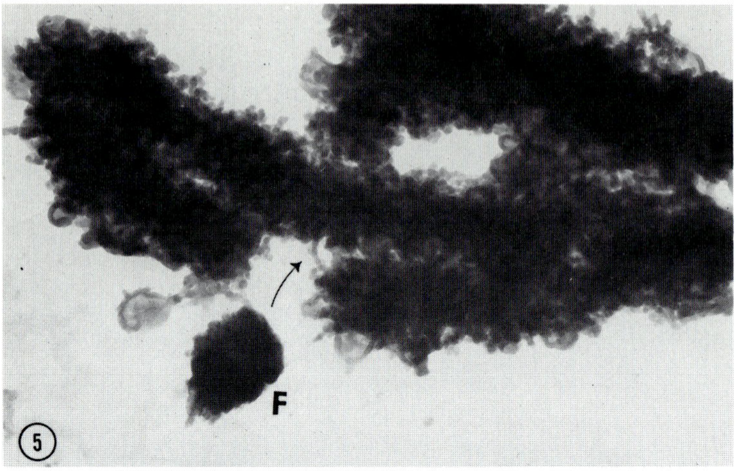
F
5

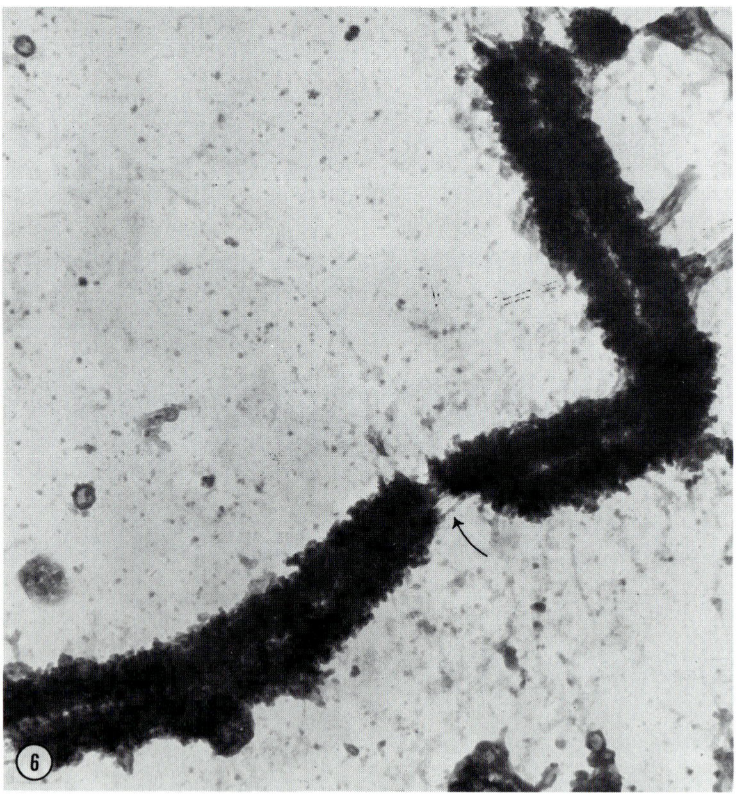
6

deed, the classic experiments of McClintock (1934) indicated that a partially deleted nucleolar organizer produced by radiation was capable of continued, although somewhat reduced, nucleolus formation.

In view of the fact that broken ends of chromosomes are unstable and sticky and tend to restitute (heal) or combine with other broken ends, it is not too surprising that they fail to display ultrastructural features that distinguish them from normal telomeres. Also, a break in the DNA molecule is likely to be masked by the associated chromatin proteins and the folding of the DNP fibrils into a metaphase chromosome.

Chromatid breaks are easily scored by light microscopy when the broken region is unaligned with the undamaged regions of the chromatid. When such preparations are examined by electron microscopy (Figs. 3–6), they appear similar to the broken regions described above; that is, there is no distinguishing feature of the broken end of the chromosome that might indicate the ultrastructural basis of the aberration. However, it is rather difficult, if not impossible, to score for chromatid breaks when the broken fragment remains aligned with the undamaged regions of the chromatid. Such regions are frequently scored as gaps. However, some aligned breaks represent true discontinuities along the chromatid when examined by electron microscopy. Thus it appears that some chromatid breaks remain aligned and can easily be confused with gaps (see Fig. 7a–c). One interesting feature noted in many aligned and unaligned chromatid breaks, as well as gaps, is the presence of an atypical structure in the sister chromatid at approximately the same locus. As shown in Figs. 3b, 4, 6, 15, and 16, the sister chromatid may display small attenuations or slightly swollen bands at the locus adjacent to the chromatid break. Whether or not these adjacent regions represent areas that were repaired or restored cannot be determined from these micrographs. Frequently, chromatids display numerous interchromatid fibrils or connections which appear to cross-link the parallel metaphase chromatids (Fig. 19). Similar structures have also been described by others (Abuelo and Moore, 1969). Normally, the in-

FIG. 5. Chromatid break with small deletion in isolated chromosomes of Chinese hamster. The small fragment (F) nearby may be the deleted segment from the chromatid. Note chromatin fibrils (arrow) at the broken ends. ×27,500.

FIG. 6. A broken or deleted segment from an isolated chromosome of Chinese hamster. Note chromatin fibrils which appear to connect the broken ends with the sister chromatid. A gap is apparent in the sister chromatid at the same locus. ×19,500.

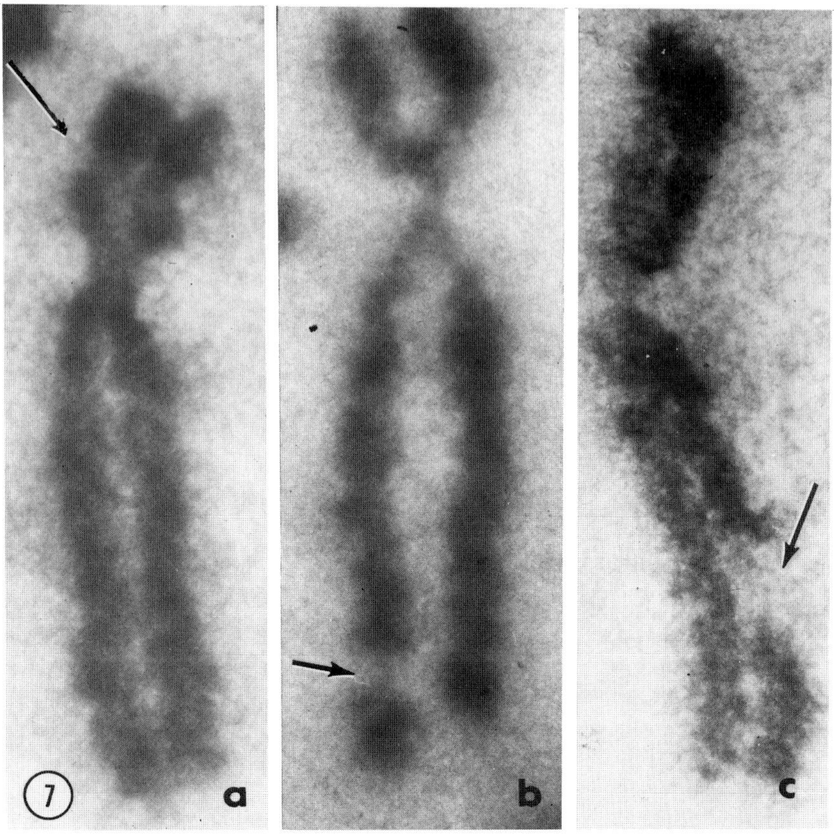

FIG. 7. Chromatid gaps (arrows) from human lymphocyte chromosome. The gap in (c) may have been scored as a break by light microscopy; however, thin chromatin fibrils are seen extending across the gap. ×6000. (From Brinkley and Shaw, 1970.)

terchromatid fibrils remain attached until early anaphase, but they may become detached in chromosomes that have been "over-colchicinized." It is apparent that the interchromatid fibrils play an important role in maintaining the association of broken fragments with undamaged chromatids. Figure 2 is a phase and electron micrograph of a cultured rat kangaroo cell which received 500 rads of x radiation 8 hours prior to fixation. Although the cell is in late anaphase or telophase, numerous chromosome and chromatid fragments are still present at the metaphase plate. Apparently, the broken fragments remain attached to the sister chromatids via interchromatid fibrils during prophase and prometaphase. Thus, in conventional light microscope preparations of colchicine-blocked cells, the broken chromatid fragments appear as being either aligned or unaligned. The broken fragments apparently arrive on the metaphase plate with the undamaged chromosomes, but are released at anaphase and remain aggregated at the metaphase plate if they lack a kinetochore or centromere. Of course, it is known from time-lapse cinematography studies that acentric fragments may undergo erratic movements during anaphase and telophase but are usually excluded from the daughter nucleus at telophase.

Recent improvements in methods for isolating mammalian metaphase chromosomes (Wray and Stubblefield, 1970) have permitted electron microscope analysis of intact, unfixed, unstained chromosomes in which high-molecular-weight DNA is retained (Wray *et al.*, 1972). In preliminary experiments we have used the above procedure to isolate chromosomes from cells that have been treated with mitomycin C to induce aberrations. Such chromosomes displayed numerous aberrations including breaks, gaps, and exchanges which were observable by both light and electron microscopy. Chromosomes from control cells isolated by identical procedures rarely showed aberrations aside from the usual preparative damage in the form of occasional stretching and distortion of the chromatids. As shown in Figs. 5 and 6, chromatid breaks in isolated chromosomes appear as a complete disruption of chromatin along the chromatid. The broken ends were characterized by 200- to 250-Å chromatin fibrils which were either "looped" out and reinserted into the broken end or inserted into the adjacent sister chromatid (see arrows in Figs. 5 and 6). The latter may have been interchromatid fibrils which are normally seen on metaphase chromosomes (see Table I and p. 53). In general, the ultrastructure of breaks in isolated chromosomes was not greatly different from that seen in sectioned or acetic acid–ethanol-fixed, air-dried preparations.

In summary, it can be concluded that the broken regions of chromosomes and chromatids, when viewed in the transmission electron microscope, appear greatly similar to the ends or telomeres of normal undamaged chromosomes. Apparently, during prophase normal condensation takes place and the broken regions undergo condensation much like the undamaged regions of the chromosomes. A notable exception to this observation is when the break occurs in a specialized region of the chromosomes such as the nucleolar organizer. The latter displays the typical organization seen in undamaged chromosomes, even if only a small part of it is present. Ultrastructural studies suggest that chromatid breaks may be either aligned or unaligned. In aligned breaks there is no evidence for the existence of attachments across the broken region. Apparently, the broken fragment is kept in register with the other region of the chromatid by interchromatid fibers which serve as cross-bridges between sister chromatids.

B. Gaps or Achromatic Lesions

When cells are exposed to x rays, various drugs, and viruses, chromosomal aberrations are frequently expressed as a slight attenuation in one or both sister chromatids. When observed by light microscopy such gaps may appear as complete breaks or display a thin connection across the lesion. Whether or not gaps represent losses of chromosomal material or alterations in chromosome structure (packing) remains to be determined (see further discussion in Section 6). They may be repaired in subsequent division cycles, or expressed as isolocus gaps or true breaks. When typical gaps are viewed in the electron microscope, they also appear as either clean breaks with no material extending across the lesion or as attenuations with chromatin strands of various thickness extending across the region (Humphrey and Brinkley, 1969; Brinkley and Shaw, 1970; Scheid and Traut, 1970, 1971; Brogger, 1971).

Chromatid gaps of varying intensities are shown in Figs. 7–18. In Fig. 7a, a weak isolocus gap is seen on the short arm of a B-group chromosome from a human lymphocyte. It is even doubtful that such minor attenuation would be noted by light microscopy. A more typical gap can be seen in the terminal region of the long arm of a B-group chromosome in Fig. 7b. Note that the length of the chromatid with the gap exceeds the length of its sister chromatid by approximately the same length as the gap itself. Such preparations suggest that the gap may represent an expanded region of the chromatid as the result of an error in folding or packing. Chromatids with gaps may also be shorter than their unaffected counterparts. Thus Brogger (1971) measured 67 spontaneous unilocus gaps and breaks in human chromosomes and found that the length of the affected chromatid was on the average 8.9% shorter than the undamaged sister chromatid. It is concluded that gaps may be due to differences in folding of the chromatin threads or actual loss of chromosomal material. Unfortunately, ultrastructural studies have thus far failed to resolve this question (see additional comments in Section 6). Artifacts due to the processing of chromosomes for light or electron microscopy may account for differences in gap morphology (Brogger, 1971). It is well known that isolocus gaps and constrictions are easily broken during slide preparation (Evans, 1961). Chromatid gaps can be "converted"

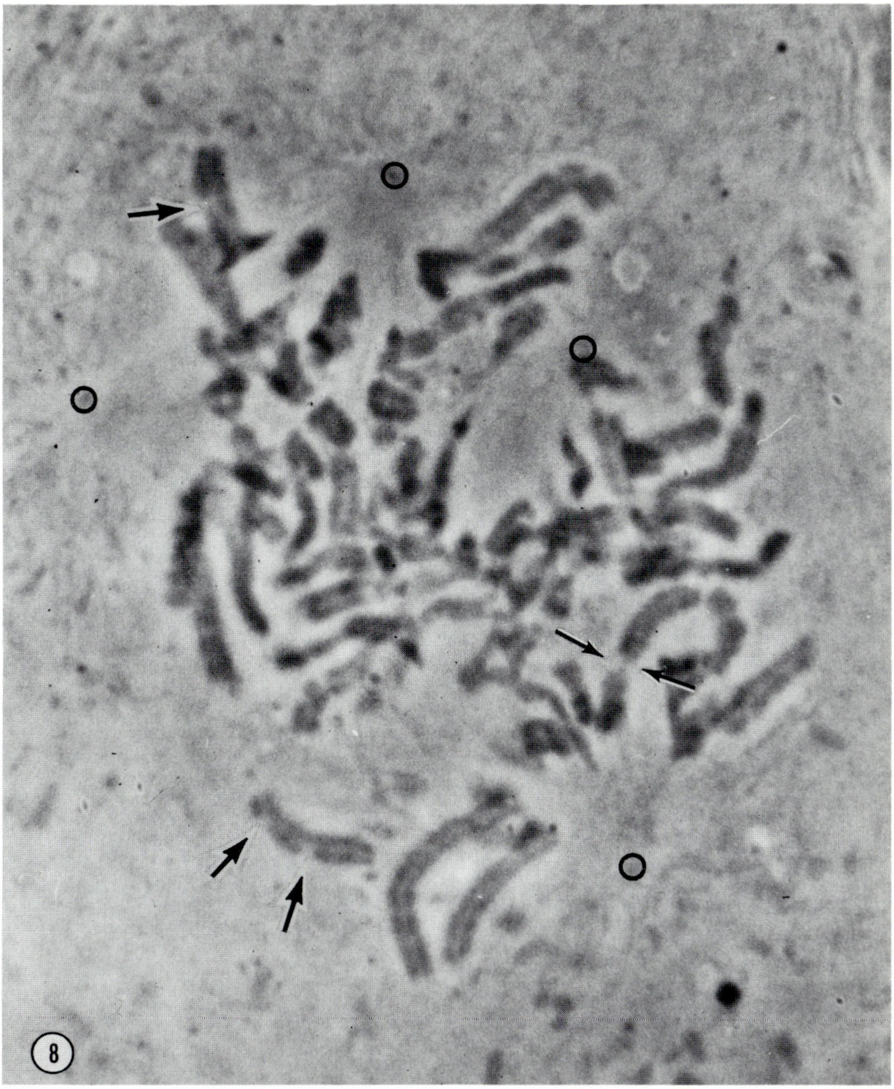

FIG. 8. A near-tetraploid rat kangaroo cell treated with mitomycin C. The cell is also tetrapolar (circles indicate centriole positions). Both chromatid gaps (large arrows) and isochromatid gaps (small arrows) are seen in the preparation which was fixed in 3% glutaraldehyde, postfixed in 1% osmium, and flat-embedded in Epon. Photograph was made through the plastic resin prior to ultrathin sectioning. ×2125. (From Brinkley and Shaw, 1970.)

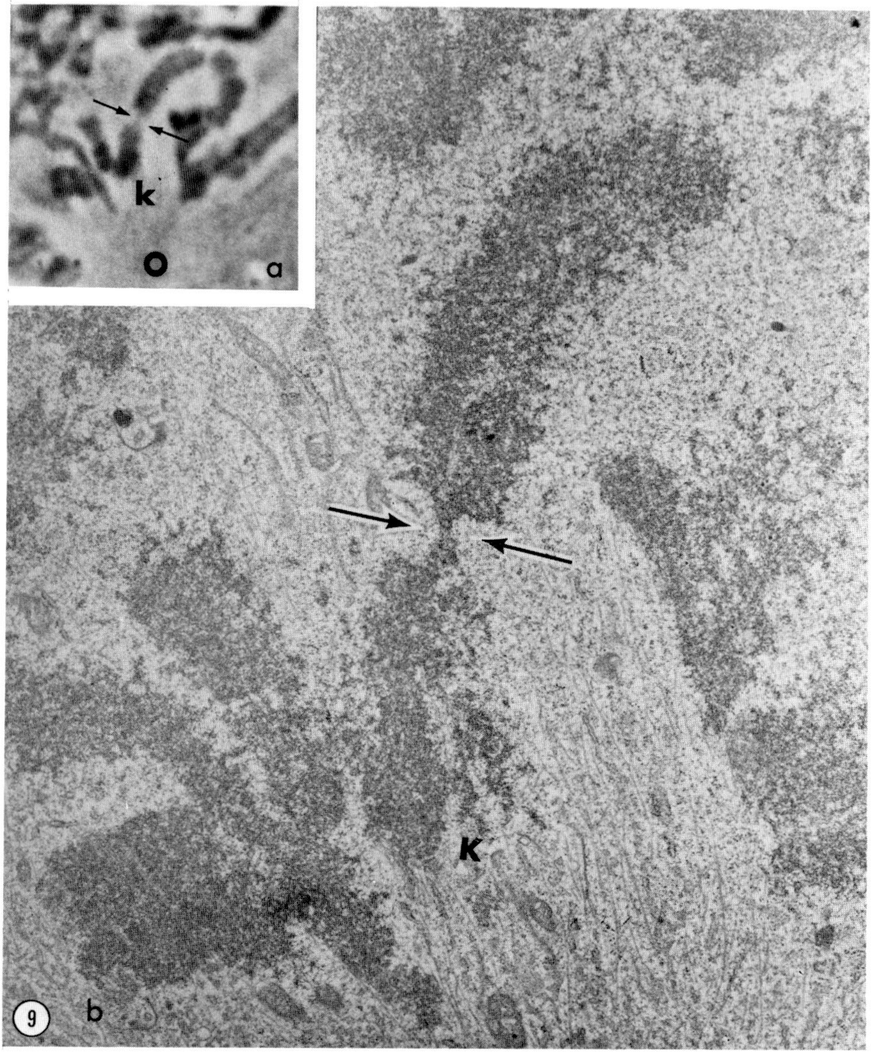

FIG. 9. Light (a) and electron (b) micrographs of an isolocus gap (arrows) in chromosome of rat kangaroo fibroblasts treated with MC. Circle in (a) indicates centriole position; k, kinetochore. In (b), K, kinetochore. (a) ×1875. (b) ×9000. (From Brinkley and Shaw, 1970.)

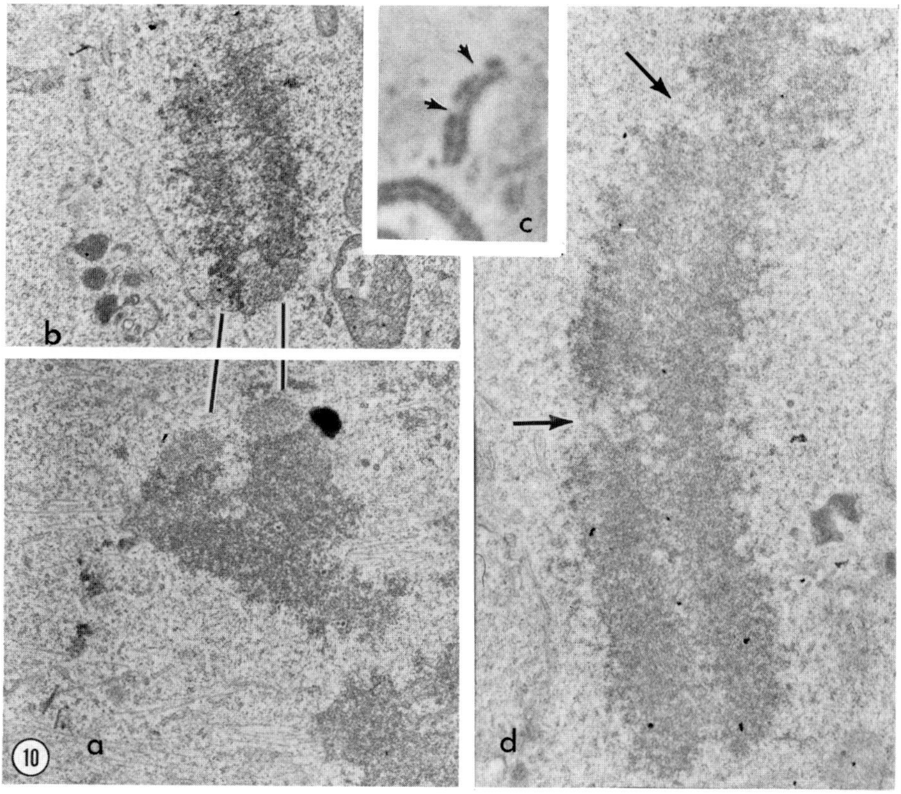

FIG. 10. X chromosome of rat kangaroo cell damaged by mitomycin C as viewed by electron microscopy (a, b, and d) and light microscopy (c). Chromatid gaps (arrows) are present on the acentric fragment. The black and white bars in (a) and (b) connect the broken ends and define the locus of the break as occurring through the nucleolar organizer. (a) ×15,000. (b) ×15,000. (c) ×1875. (d) ×15,000. (From Brinkley and Shaw, 1970.)

to breaks during examination in the electron microscope if the beam intensity is too high (P. S. Baur and B. R. Brinkley, unpublished observation).

Electron microscope analysis of gap morphology by serial ultrathin sections has provided new insight into their structure. By using a flat embedding procedure which allows preselection of a particular chromosomal lesion, Brinkley *et al.* (1967) sectioned gapped regions of chromosomes from cells that had been fixed in glutaraldehyde, postfixed in osmium tetroxide, and embedded in Epon. It is less likely that chromosomes from such preparations are as greatly distorted as

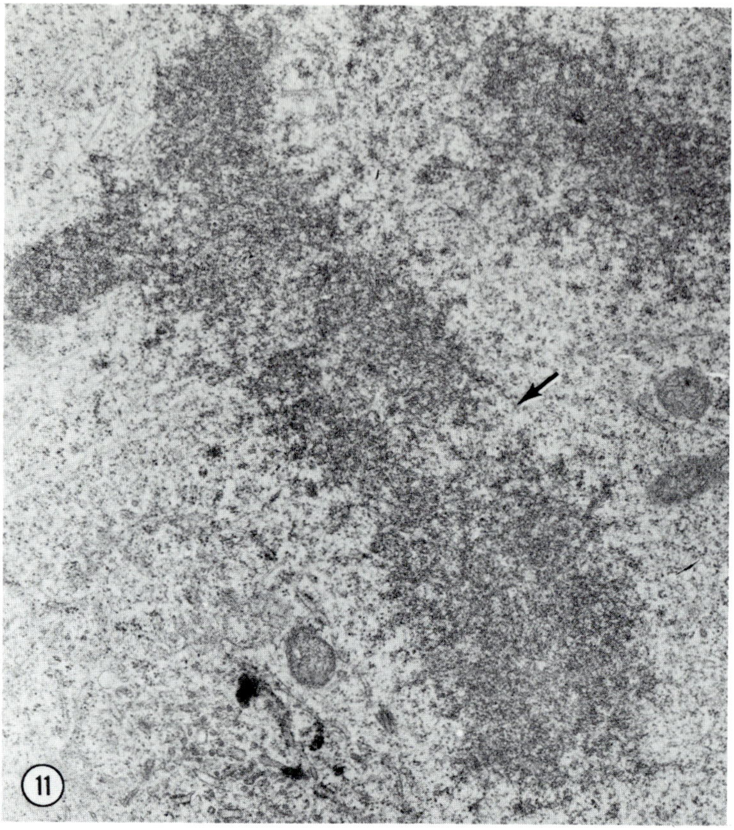

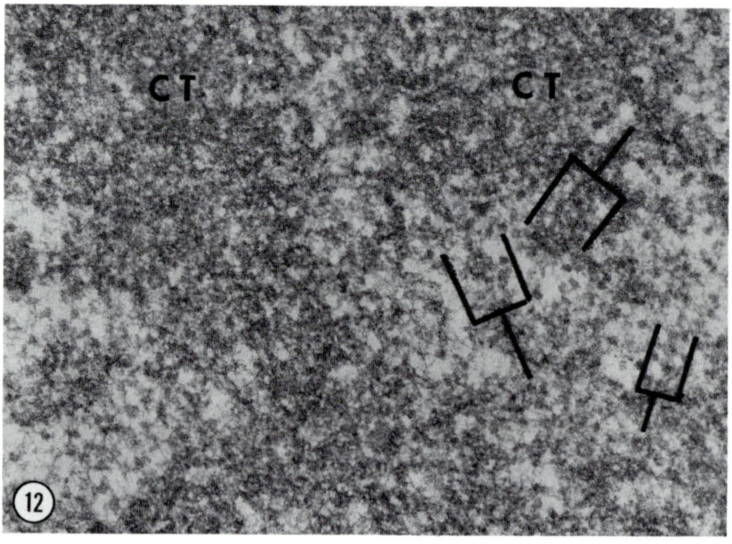
CT
CT

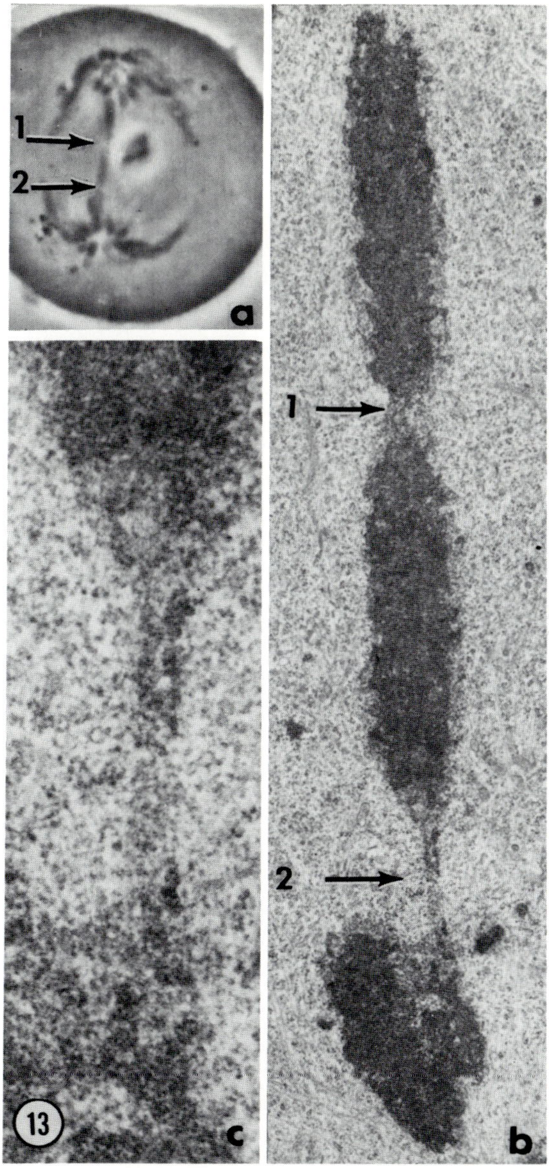

FIG. 11. Electron micrograph of a chromatid gap (arrow) from chromosome shown in upper left-hand corner of Fig. 8. ×17,000. (From Brinkley and Shaw, 1970.)

FIG. 12. Higher magnification of a portion of the chromosome shown in Fig. 11. Three general classes of chromosome fibrils can be seen in the gap (forks). CT, Chromatid. ×52,700. (From Brinkley and Shaw, 1970.)

FIG. 13. Light (a) and electron (b and c) micrographs of an anaphase bridge showing two gaps (arrows 1 and 2). The gap indicated by arrow 2 appears bipartite in the electron micrograph. (a) ×1400. (b) ×20,000. (c) ×70,000. (From Humphrey and Brinkley, 1969.)

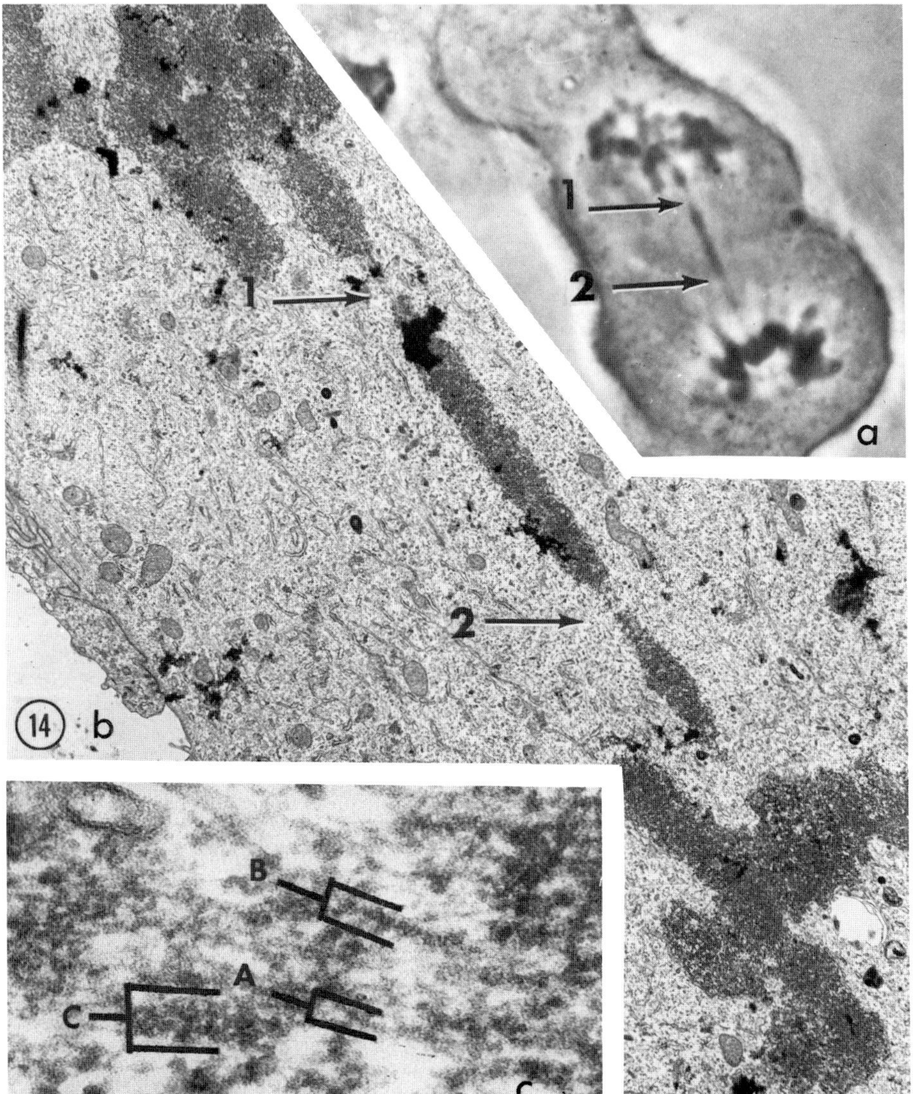

FIG. 14. An anaphase bridge as observed by both phase (a) and electron (b and c) microscopy. Arrows 1 and 2 point to gaps or attenuations in the bridge. At higher magnification the gaps consist of fibrils ranging in size from 50 to 70 Å (A); 200 to 250 Å (B); and 500 to 800 Å (C). (a) ×1600. (b) ×9000. (c) ×137,000. (From Humphrey and Brinkley, 1969.)

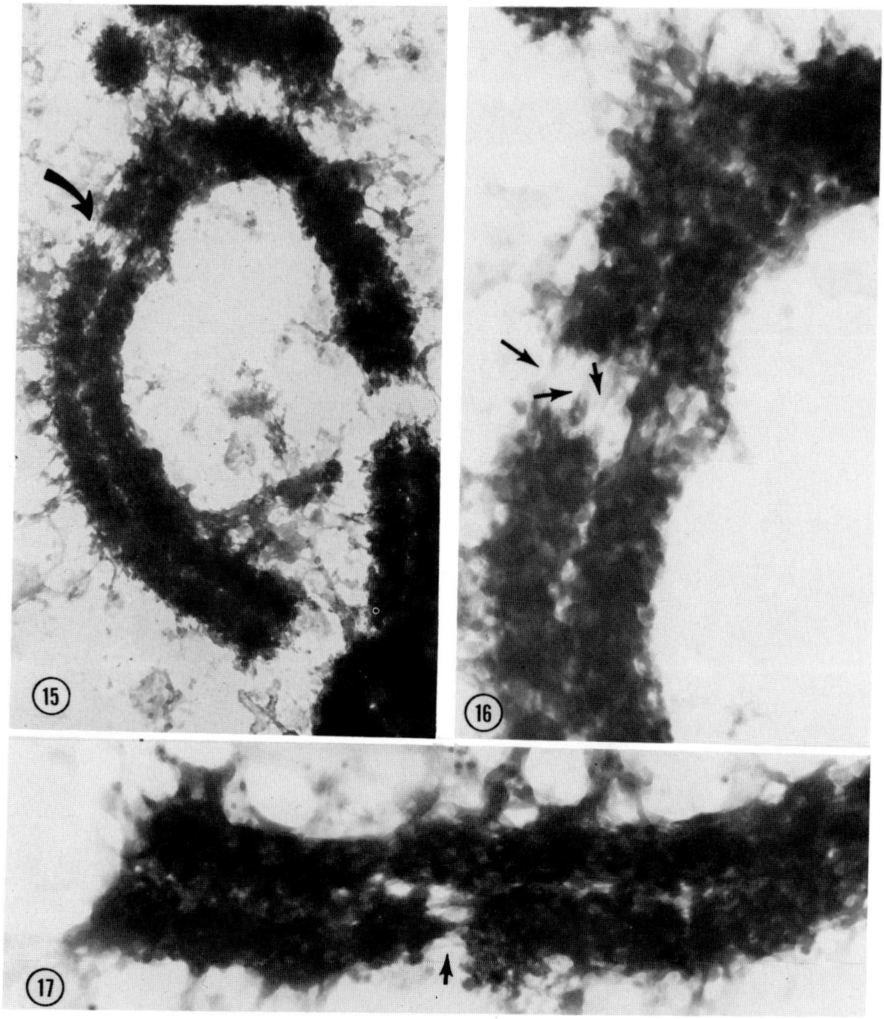

FIG. 15. Isolated chromosome of CHO cell treated with MC. Note gap in one chromatid (arrow). ×15,000.

FIG. 16. At a higher magnification thick and thin fibrils (arrows) can be seen extending across the gap. The adjacent sister chromatid also appears to be damaged, but less severely, at the same locus. ×30,000.

FIG. 17. Chromatid gap in isolated chromosome similar to that in Fig. 16, but note that adjacent sister chromatid appears undamaged. Note thick and thin fibrils extending across gap (arrow). ×30,000.

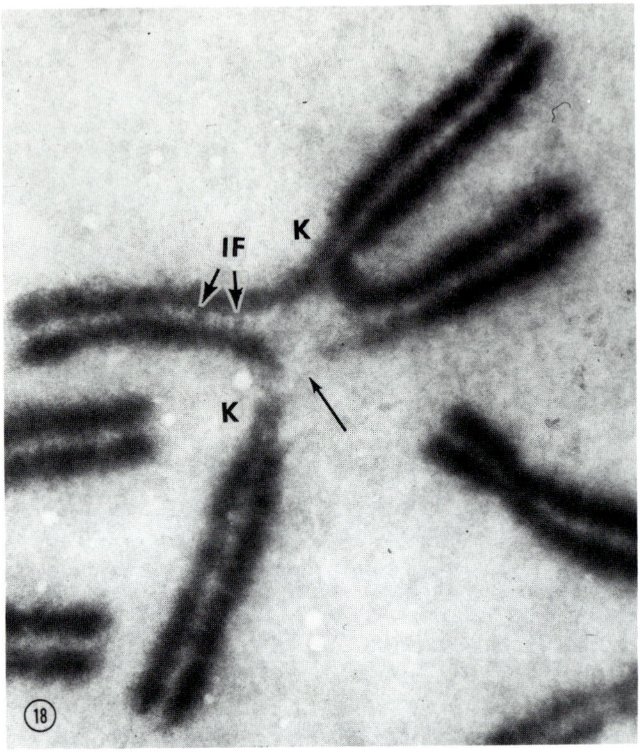
IF
K
K
18

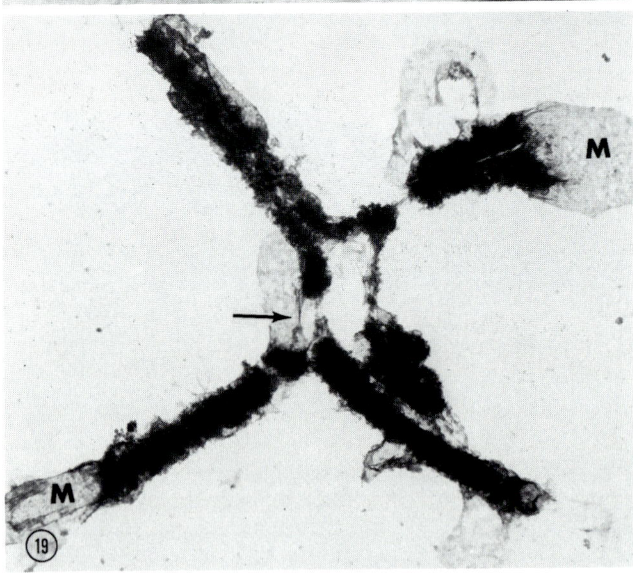
M
M
19

those examined after air- or critical-point-drying. In such preparations we have always found chromatin threads extending across the gap. As shown in Figs. 9–18, sections of gaps display several classes of fibrils. In Fig. 14 the gap is traversed by three general classes of microfibrils: class A fibrils which measured 50–80 Å in diameter, class B fibrils which measured 200–250 Å, and class C fibrils which measured 500–800 Å. In more attenuated gaps (Fig. 13), the chromatid often appeared to consist of a bipartite unit with two parallel chromatin threads extending across the gap. Similar observations were made by Scheid and Traut (1970) using ultraviolet microscopy. It should be pointed out that the gaps shown in Fig. 13 and 14 are from anaphase bridges of rat kangaroo cells that received 250 and 500 rads of irradiation 6 and 8 hours prior to fixation. Thus, although these lesions represent a form of achromatic lesion, they may not be the same as gaps seen on metaphase chromosomes arrested with colchicine. The latter are not under stress, as would be true of those seen on anaphase bridges. However, comparison of the gaps in Figs. 13 and 14 with isolated chromosomes arrested in metaphase with Colcemid (Figs. 15–17) indicates considerable similarities in the two types of lesions.

The chromosomes shown in Fig. 8 are from a near-tetraploid cell of the rat kangaroo (strain PtK_1), which was fixed in 3% glutaraldehyde, postfixed in osmium tetroxide, and flat-embedded in Epon. The cell was treated with 1.0 μg/ml mitomycin C (MC) 24 hours prior to fixation. Both chromatid gaps (large arrows) and isochromatid gaps (small arrows) are visible in the preparation. When these same chromosomes were viewed in thin sections, chromatin fibrils measuring 50–800 Å were observed extending across the gaps (Figs. 9–12).

Ultrastructural analysis of gaps or achromatic regions made on isolated chromosomes (Figs. 15–17) confirmed observations made by thin sections. In all gaps observed on isolated MC-damaged chromosomes, parallel chromatin fibrils ranging in diameter from 150 to 500 Å were seen to extend across the gap. As mentioned previously, the adjacent chromatid often displayed ultrastructural damage at the

---

FIG. 18. A quadriradial exchange between two number-1 human chromosomes. Arrows point to broken region. Interchromatid connections (IF) and kinetochore (K) region can be identified. ×4550. (From Brinkley and Shaw, 1970.)

FIG. 19. A quadriradial exchange observed in isolated, whole-mount chromosome of CHO cell. Note stretched region indicated by arrow. Chromatid appears to be double at arrow. Membrane fragments (M) are attached to the telomeres of the chromosome. ×4550.

same locus (Figs. 15–17). Although there were clearly fewer DNP fibrils within the gap, it was impossible to determine whether or not the region exhibited a loss of DNA or an aberration in the packing or condensation of chromatin.

C. Exchanges

Agents that break or cross-link DNA are known to cause frequent exchanges between arms of homologous and nonhomologous chromosomes. The effect of MC has been particularly well studied in regard to exchanges (Mertz, 1961; Cohen, 1963; Shaw and Cohen, 1965). Nowell (1964) and Shaw and Cohen (1965) have reported that MC caused nonrandom breaks and exchanges between human chromosomes 1, 9, and 16 which bear the secondary constriction, and that the point of exchange usually involved the secondary constriction. Brinkley and Shaw (1970) observed quadriradial configuration in whole-mount electron microscope preparations following MC treatment of human lymphocytes. More recently, McGill and Brinkley (1974) isolated quadriradial exchanges by the method of Wray and Stubblefield (1970). Such an exchange is shown in Figs. 18 and 19. In Figure 18 the exchange is between two number-1 chromosomes and is classified as "alternate symmetrical," or class 1, according to the classification of Shaw and Cohen. The presence of interchromatid connections argues convincingly for true exchanges rather than homologous association. Such exchange figures frequently displayed breaks or weak spots in the unexchanged region.

Breaks followed by proximal chromatid reunion result in dicentric chromosome formation, whereas distal chromatid healing leads to acentric fragments. Dicentric chromosomes produce characteristic anaphase bridges as shown in Figs. 1i, 13, and 14, whether the exchange is between sister chromatids (intrachange) or between chromatids of other chromosomes (interchange).

D. Subchromatid Aberrations

When cells are exposed to ionizing radiation or certain drugs, a specific chromosomal exchange is detected at anaphase, called a side-arm bridge (Fig. 1k). These are especially apparent in the anaphase of the first division (M_1) following radiation or drug treatment in G_2 or prophase. The unusual anaphase configuration is so named because it appears to involve chromosomal filaments smaller than either of the anaphase chromosome arms. This observation, along with the fact that such exchanges predominate when cells are radiated after the chromosomes have duplicated their DNA, has led to the conclusion that side-arm bridges represent subchromatid exchanges and reflect at least bineme strandedness in the chromatid arms (Nebel, 1932; Swanson, 1947; Crouse, 1954; La Cour and Rutishauser, 1954). Östergren and Wakonig (1955) were the first to challenge this interpretation. They treated cells with a breaking agent called coumarin and analyzed aberrations in the first (M_1) and second (M_2) metaphase. They reasoned that a half-chromatid aberration observed in M_1 should appear as a full chromatid break in M_2.

When the analysis was completed, they reported predominantly chromosome-type and not chromatid-type aberrations in M_2. Consequently, they proposed that side-arm configurations were in fact pseudochiasma resulting from stickiness of the chromosomal matrix and not subchromatid exchanges. A similar conclusion was reached by Kihlman and Hartley (1967) in *Vicia faba* (see also Kihlman, 1970). However, Peacock (1961) reported that subchromatid aberrations induced in M_1 of *V. faba* appeared as chromatid breaks in M_2. Peacock's work was supported by Heddle (1969), who found 8–17% chromatid aberrations at M_2 in *V. faba*.

Electron microscope observations of radiation-induced side-arm bridges are shown in Figs. 20 and 21. Although the light microscope usually fails to resolve chromatin strands extending across the bridge, electron microscope studies clearly showed chromatin connections across the bridge (Brinkley and Humphrey, 1969). Moreover, when the chromosomes were unstretched, the side-arm bridge was approximately one-half the diameter of the chromatid (Fig. 20). Unfortunately, the electron microscope observations failed to settle the question of the validity of subchromatid aberrations and whether or not they truly reflect chromosome strandedness. Indeed, alternate interpretations can be provided. Comings (1970, 1974) has disputed the evidence for the subchromatid exchange hypothesis on the basis of the expected mode of segregation of old and new DNA strands if two or more strands are present in the chromatid. He pointed out that, according to the multistranded model, subchromatid exchanges are the result of union of half-chromatid units. However, when DNA is duplicated, an old and new strand would attach to each side of the chromatid. When a lesion occurs in one strand prior to M_1, it should express as two half-chromatid breaks when observed in M_2. Double half-chromatid aberrations have not been reported. Comings therefore believes the half-chromatid lesion is just an exchange involving a single DNP fiber in the chromatid.

The question of the strandedness of the metaphase chromosome is an important one, and one which, thus far, has not been resolved by

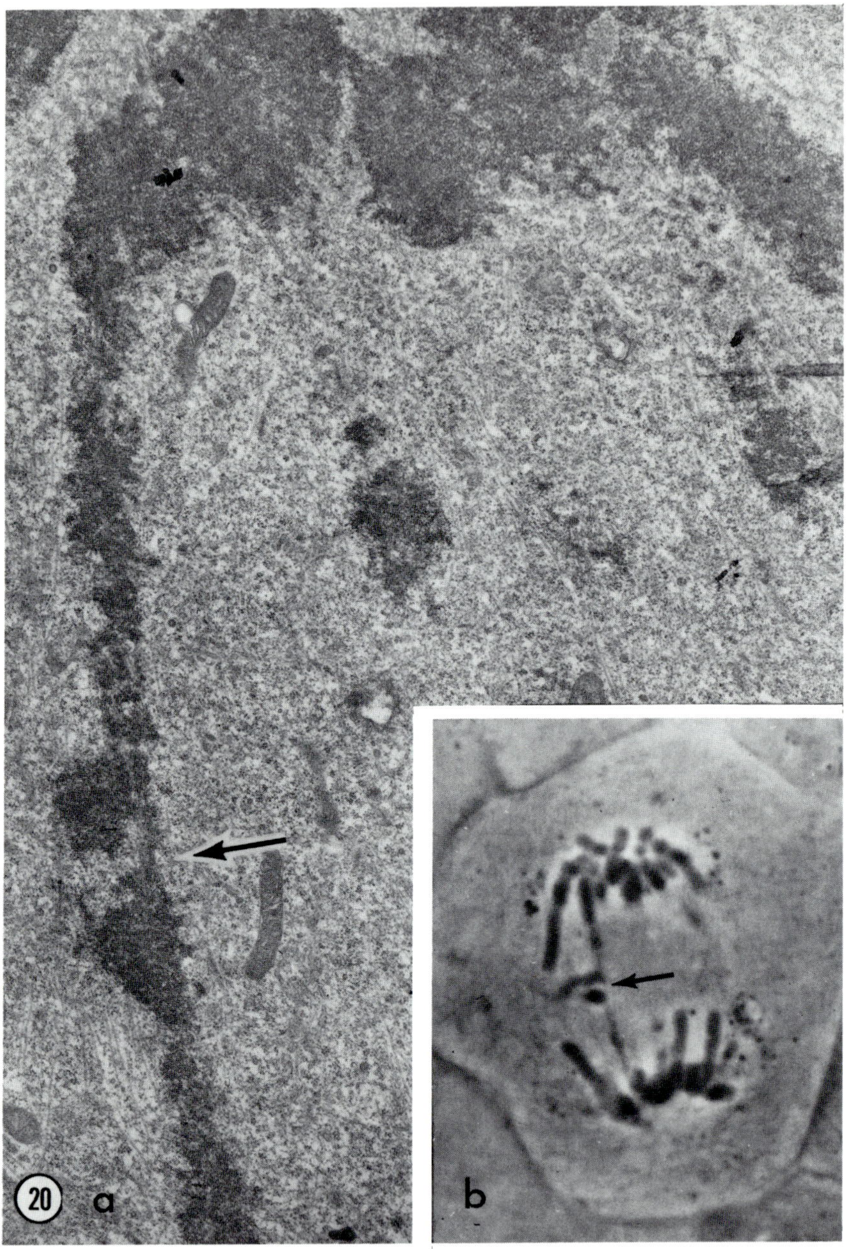

FIG. 20. Electron (a) and phase-contrast (b) micrographs of side-arm bridge in chromosomes of rat kangaroo (PtK_1) fibroblast. Arrows indicate subchromatid connections between anaphase arms. (a) ×12,000. (b) ×2000. (From Brinkley and Humphrey, 1969.)

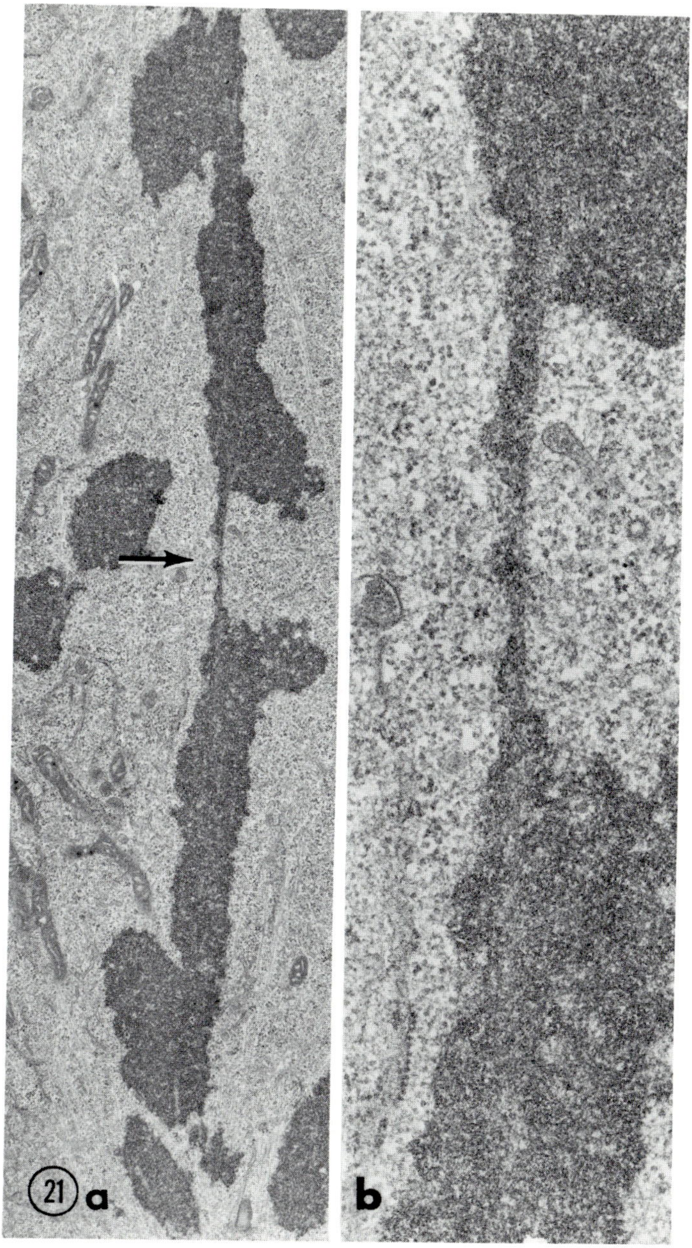

FIG. 21. (a) Side-arm bridge (arrow) of rat kangaroo fibroblast. Exchange is greatly attenuated. (b) Higher magnification of the bridge. (a) ×12,000. (b) ×48,000. (From Brinkley and Humphrey, 1969.)

any definitive experiments. For many years cytologists have described morphological substructures within chromatids which suggest that they are composed of more than a single DNP strand (Nebel, 1932; Manton, 1945; Giménez-Martín *et al.*, 1963; Brewen, 1964). These interpretations, which were made from fixed preparations, were strengthened by the elegant time-lapse cinemicrographs of Bajer (1965), which showed subchromatid structures in the large *Haemanthus* chromosomes of living cells. Also, certain treatments such as trypsin digestion lead to enhancement of half-chromatid structures (Trosko and Wolff, 1965; Wolfe and Martin, 1968; for review, see Wolff, 1969). For several years investigators have debated over the use of the transmission electron microscope for identification of strandedness in metaphase chromosomes. Stubblefield and Wray (1971) have presented micrographs which suggest that the chromatid may consist of more than a single strand of DNP extending down its length (also see discussion on p. 93). However, other investigators (DuPraw, 1965; Comings and Okada, 1972) interpret their electron micrographs of whole-mount water-spread chromosomes to indicate a single DNP fiber of approximately 200–250 Å, which is randomly folded along the chromosome arms. Unless present technology is improved considerably, the question of chromosome strandedness is not likely to be completely answered by transmission electron microscope analysis of whole-mount preparations. Frequently, images of whole-mount, water-spread chromosomes reflect serious artifacts of preparation, and the evidence for uninemy or polynemy is often in the eye of the beholder.

E. Chromosome Stickiness

The phenomenon of chromosome stickiness has been the subject of numerous cytological studies, but the molecular basis of this type of aberration remains largely unknown. The term is generally applied to those types of aberrations in which arms of chromosomes adhere to each other at anaphase. The anaphase bridge that forms in the case of stickiness appears to be different from the bridge produced by a dicentric chromosome as a result of the absence of acentric fragments in the neighboring cytoplasm (Fig. 1j). Moreover, anaphase chromosomes often attach at acute angles, and the bridges appear somewhat diffuse (McGill *et al.*, 1974). Aberrations arise when the chromosomes are torn apart during movement to the poles at anaphase.

Beadle (1932) first described stickiness in the chromosomes of *Zea mays* and identified the effect as being associated with a recessive allele located on chromosome 4. Chromosome stickiness is also a common feature of cells in pathological conditions (Swanson and Johnson, 1954; L. Y. F. Hsu *et al.*, 1970). Cultured cells may display varying degrees of chromosome stickiness during the normal course of mitosis. This observation, along with the fact that nucleoluslike material often persists in close proximity to chromosomes *in vitro*, led T. C. Hsu *et al.* (1965) to propose that stickiness may be due to RNA-containing particles which coat the surface of chromosomes. This problem has recently been reinvestigated in light of the behavior of Chinese hamster chromosomes after treatment with a drug, ethidium bromide (EB). EB is well known for its capacity to bind with DNA from a variety of sources (Waring, 1965; Angerer and Moudrianakis, 1973), resulting in conformational changes in the DNA molecule (Crawford and Waring, 1967). The binding of the drug to DNA in the nucleus of mammalian cells leads to several interesting abnormalities in metaphase chromosome structure as observed by Hsu *et al.* (1973), Unakul and Hsu (1973), and McGill *et al.* (1974). The last-mentioned investigators found that EB treatment produced an increase in the incidence of mitotic errors and chromosome exchanges which were evident at metaphase and anaphase of Chinese hamster cells. The aberrations were proportional to the duration of the treatment and the dose of the drug. The most characteristic feature of the anaphase bridge was the lack of acentric fragments, suggesting stickiness rather than a classic break-exchange effect. Although the attachments are best observed at anaphase (Figs. 1j and 22), even at metaphase thin chromatin exchanges were seen between chromatids and also connecting various chromosomes in the configuration (Fig. 23). These results were interpreted to indicate that EB intercalated with the DNA during interphase, and interfered with the normal chromosome condensation pattern at prophase, leading to entanglement of chromatin threads (McGill *et al.*, 1974). Such entanglement and "cross-linking," which occurred either between sister chromatids or adjacent chromosomes, could produce their sticky appearance at anaphase. In view of the intercalating nature of this drug, one wonders whether or not the stickiness is due to adherence of threads on the surface of the chromosome or to a physical break and exchange of individual threads. If EB causes physical exchanges of the chromatin, one would expect to see higher levels of chromosome aberrations in the subsequent metaphase following treatment. To test this, McGill *et al.* (1974) treated cells with EB for 3 and 6 hours and then allowed them to grow for a total of 24 hours in the absence of EB. When the cells were arrested in metaphase

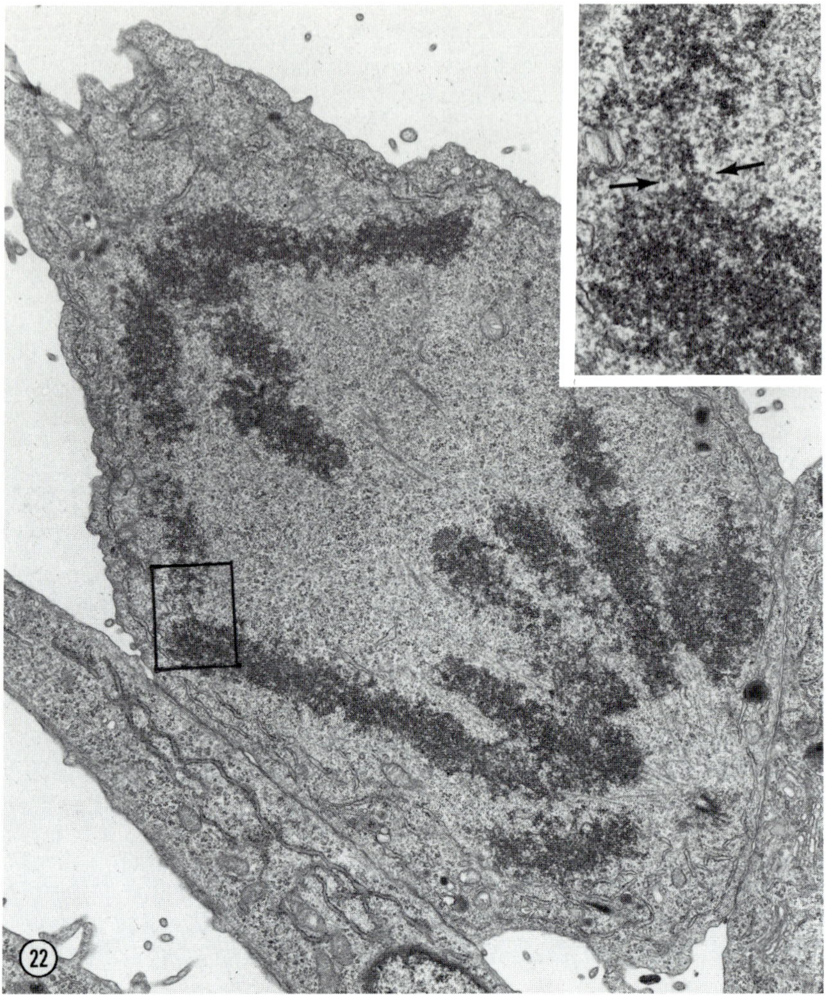

FIG. 22. Sticky anaphase bridge (arrows) of CHO cell treated with EB. Inset is an enlargement of the region indicated by the rectangle and shows apparent continuity of chromatin fibrils from opposite arms. ×9240. Inset: ×38,500. (Courtesy of M. McGill.)

and examined, a wide spectrum of chromosome damage was apparent, including chromatid and chromosome breaks, exchanges, dicentrics, and rings. Therefore, at least in the case of EB, chromosome stickiness is due in part to breaks and exchanges within the chromatin composing the chromatid fibers. As such, stickiness and subchromatid aberrations may be one and the same, even though cytological observations suggest that they are different (see Table I).

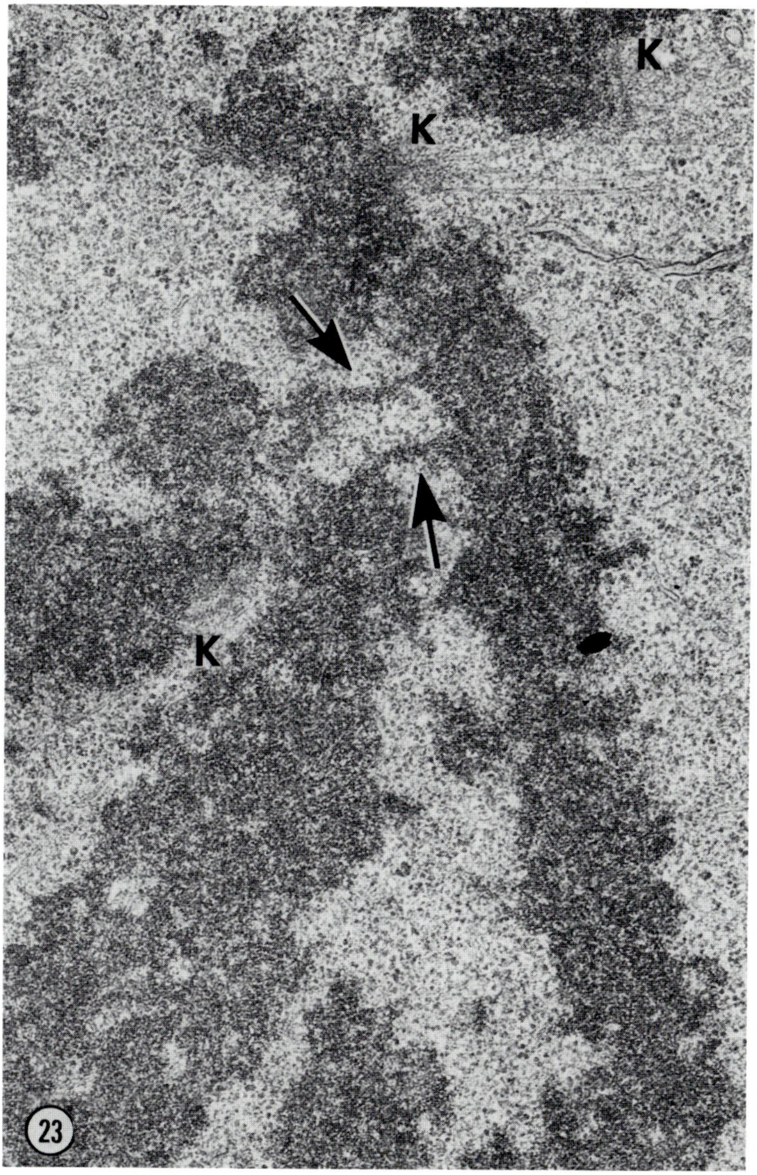

FIG. 23. Cross-arms connecting adjacent chromosomes (arrows) at metaphase following treatment of CHO cells with EB to produce stickiness K, Kinetochore. ×25,000. (From McGill *et al.*, 1974.)

F. Damage to Specialized Regions

Although transmission electron microscopy has been somewhat disappointing as a tool for evaluating metaphase chromosome strandedness in eukaryotic cells, it has provided important new dimensions in our knowledge of specialized chromosome regions such as the kinetochore, nucleolar organizer, and telomeres. Since the ultrastructure of these regions has already been described in previous publications, for example, the kinetochore (Robbins and Gonatas, 1964; Barnicot and Huxley, 1965; Brinkley and Stubblefield, 1966, 1970; Jokelainen, 1967; Comings and Okada, 1971; Roos, 1973), the nucleolar organizer and telomeres (Hsu *et al.*, 1967; Brinkley and Stubblefield, 1970), we will not cover their normal features in detail. However, since it is well known that such regions are often preferentially damaged by various clastogens, it is worthwhile to consider the ultrastructural basis of damage in these specialized regions.

1. *Kinetochore*

The kinetochore of mammalian chromosomes is a discrete structure consisting of three distinct layers. Brinkley and Stubblefield (1966) interpreted micrographs of kinetochores of Chinese hamster chromosomes as bipartite, "lampbrushlike" filaments which extend along the surface of the chromosome at the primary constriction (Fig. 24a). Serial sections through the kinetochore suggest that a pair of kinetochore filaments may be present on each chromatid. Studies representing different views of mammalian kinetochore structure have been published by Jokelainen (1967), Comings and Okada (1971), and Roos (1973). These investigators view the kinetochore as a multilayered disc or plaque which is roughly circular when viewed in a plane perpendicular to the metaphase spindle axis. Nowell (1964) reported that the kinetochore region was frequently damaged by MC. When rat kangaroo cells (strain PtK_1) were treated with 1.0 μg/ml MC for 24 hours, Brinkley and Shaw (1970) observed cells with apparent damage to the kinetochore. When these were subsequently examined by electron microscopy, disruption of the multilayered surface (axial elements according to the Brinkley and Stubblefield interpretation) was apparent (Fig. 24b). Although the kinetochore surface was pulled away from the chromosome, microtubules were still attached to it. In view of the accelerated research on the sites of microtubule assembly in cells, considerable new information on the structure and function of kinetochores is certain to be forthcoming. Such knowledge may be valuable in assessing the nature of aberrations in the kinetochore induced by drugs and physical agents.

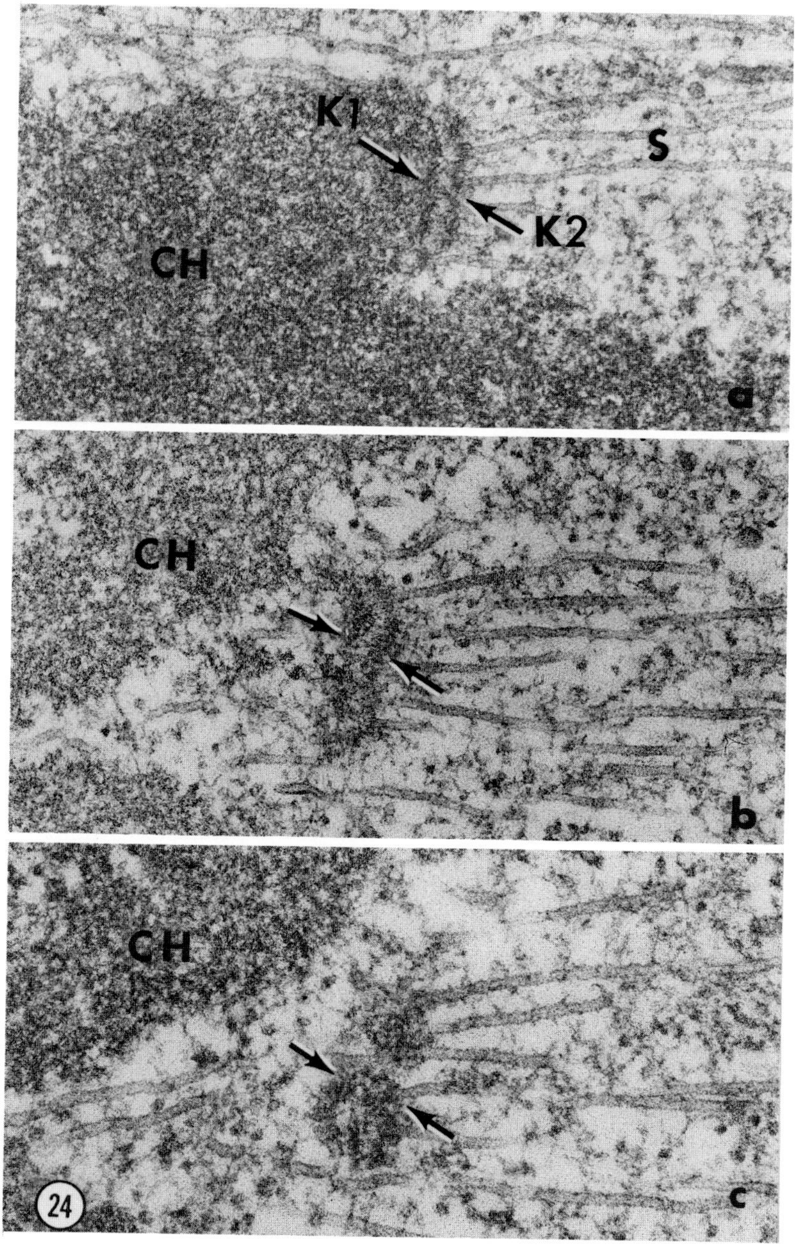

FIG. 24. (a) Normal kinetochore structure at metaphase showing spindle tubules (S) and axial elements (K1 and K2). ×50,000. (b and c) Serial sections of kinetochore damaged by MC treatment. The axial elements (K1 and K2) are separated from the chromosome (CH). Spindle tubules are still associated with the kinetochore. Such damage would likely lead to nondisjunction of the affected chromatid. ×60,000.

2. *Nucleolar Organizer*

Although the nucleolar organizer region has been elegantly characterized as the site for transcription of RNA at the molecular level, relatively little is known of its ultrastructure. Hsu *et al.* (1967) provided the first electron microscope interpretation of the region. Basically, it consists of an electron-dense chromatin core surrounded by less dense fibrils 50–70 Å in diameter. Although the nucleolar organizer appears as a constriction (secondary constriction) in typical light microscope images, Hsu *et al.* (1967) found that the width of the nucleolar organizer was the same as that of the chromatid arm. The similarity in ultrastructural organization of the nucleolar organizer to the kinetochore has been reviewed by Brinkley and Stubblefield (1970).

As shown in Figs. 25 and 26, the nucleolar organizer may become greatly attenuated following x irradiation (Humphrey and Brinkley, 1969). Such images suggest that the electron-dense core extending through the region may be stretched several times its actual length without breaking (compare with Fig. 27). The attenuation is best seen at anaphase and may be due to an exchange (side-arm bridge) between arms of chromosomes moving to opposite poles. Also, as already mentioned (p. 57) the characteristic morphology of the nucleolar organizer region is maintained at metaphase, even when a break passes through this region.

In an interesting series of experiments, Berns *et al.* (1969a,b, 1970) observed, by light microscopy, the effect of laser microbeam irradiation on selected chromosomes in living salamander lung cells *in vitro*. By focusing a small argon laser beam onto selected chromosome regions, specific lesions in the form of paling occurred in cells photosensitized with acridine orange. Such lesions were Feulgen-negative and remained on the chromosome throughout mitosis. Moreover, irradiation of special regions such as the nucleolar organizer led to the "turning off" of such regions, as indicated by a reduction in the number of nucleoli in subsequent daughter cells (Berns *et al.*, 1970). Although the mechanism of aberration induction by the laser beam is unknown, the lesions are expressed differently from those induced by other types or radiation. In more recent studies, Rattner and Berns (1973, 1974) extended their observations to the ultrastructural level. When rat kangaroo cells (strain PtK_2) were irradiated with the laser and examined by electron microscopy, the damaged area consisted of an electron-dense reticulum of fibrils and granules as shown in Fig. 28. Although much is yet to be learned about such aberrations, laser microirradiation is a potentially power-

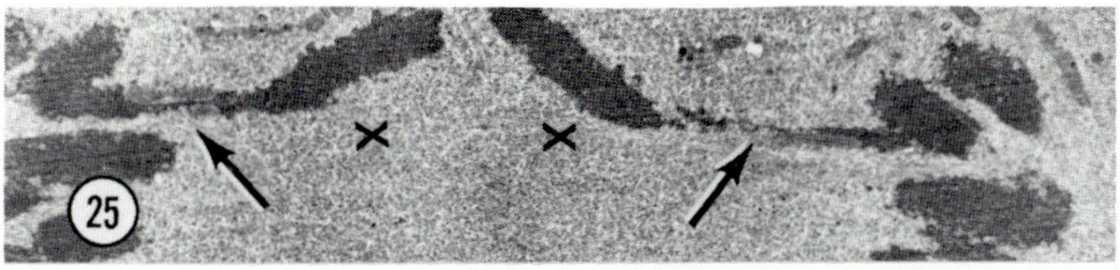

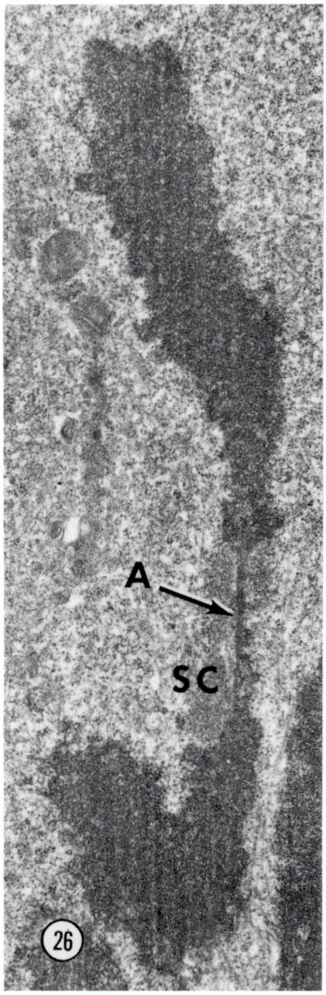

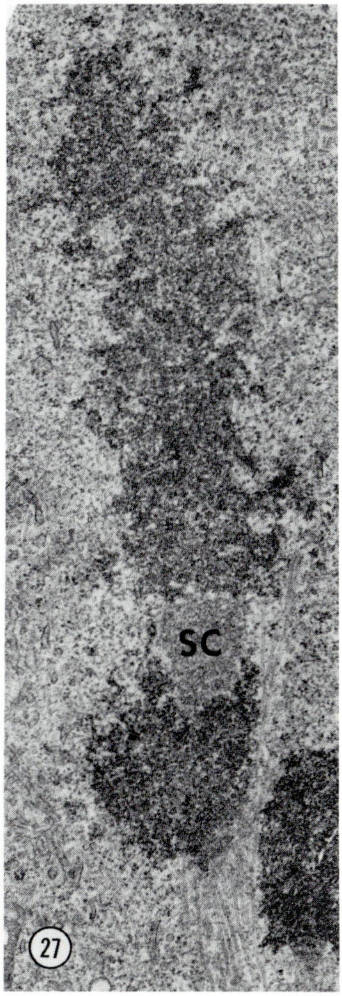

FIG. 25. Chromosomes of PtK_1 cells at anaphase following radiation with 250 rads. Arrows point to secondary constriction in daughter X chromosomes. ×6800. (From Humphrey and Brinkley, 1969.)

FIG. 26. Higher magnification of X chromosome showing dense axial element (A) and less dense fibrils in secondary constriction (SC). ×20,400. (From Humphrey and Brinkley, 1969.)

FIG. 27. Normal secondary constriction in control PtK_1 cell. ×17,000. (From Humphrey and Brinkley, 1969.)

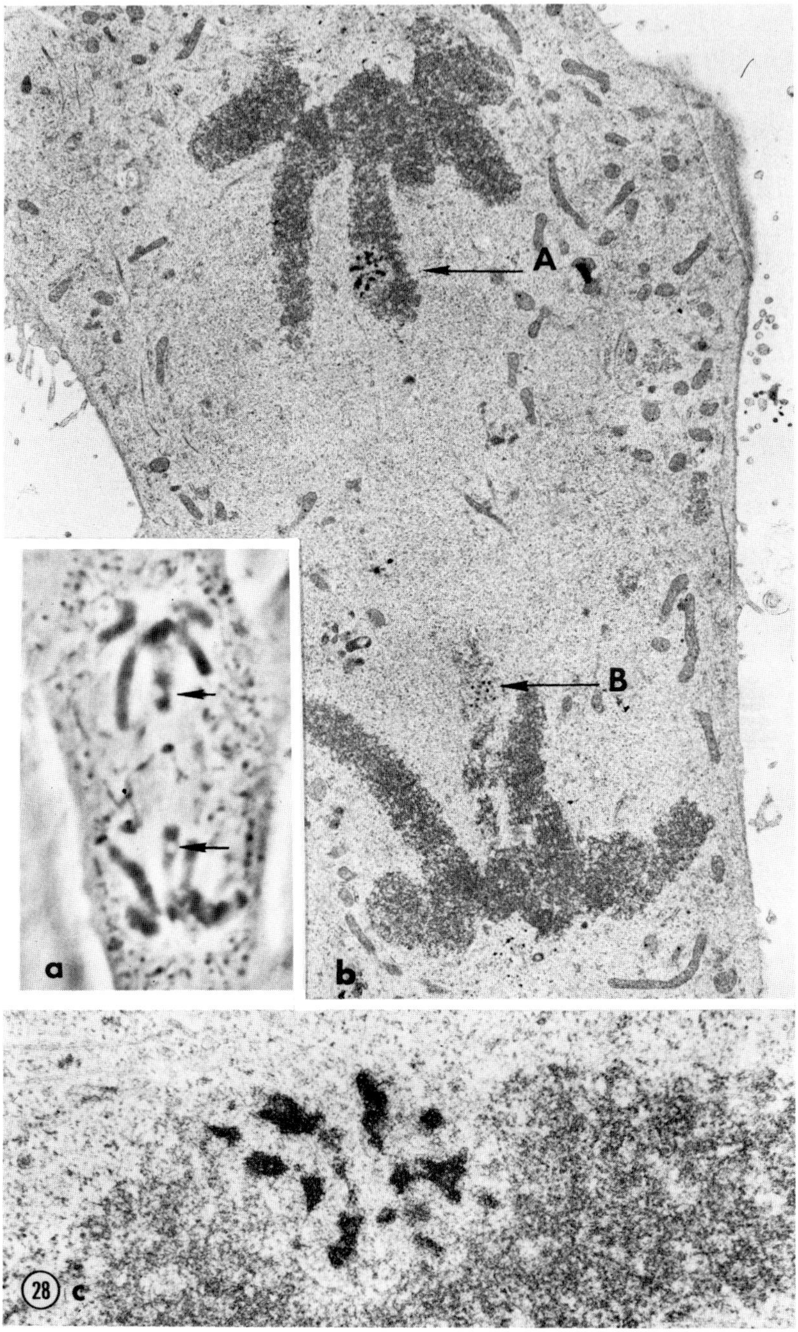
A
B
a
b
28
c

ful tool for inactivation of specific regions such as the nucleolar organizer, kinetochores, and other specialized chromosome areas.

IV. Target in the Chromosomes for Damage

A variety of natural and synthetic agents is known to produce damage to eukaryotic chromosomes. The biochemical nature of the initial lesion in the chromosome is poorly understood, yet considerable evidence suggests that damage to the DNA of the chromosome is necessary but not sufficient in the formation of a chromosome aberration. For example, there can be DNA breakage without chromosome breakage (Wray *et al.*, 1972). Nevertheless, most clastogens interact with the DNA, either directly or indirectly.

Some agents attack the chromosome directly in the formation of an aberration. Gall (1963) observed the breakage of lampbrush chromosomes from isolated newt oocytes by the sole addition of DNase to the culture. If, however, the chromosomes are not treated soon enough after isolation, they become resistant to breakage (W. N. Hittelman, personal observation).

Ionizing radiation, such as x rays, neutrons, and gamma rays, and bleomycin directly induce breaks both in the DNA (Lett *et al.*, 1967; Suzuki *et al.*, 1969; Kuo and Haidle, 1973) and in the chromosomes (Wolff, 1961; Paika and Krishan, 1973). Treatment of cells with ultraviolet light also induces chromosome lesions which later culminate in aberrations (Kirby-Smith and Craig, 1957; Humphrey *et al.*, 1963; Chu, 1965).

Chu (1965) showed with Chinese hamster cells that the ultraviolet-induced aberration frequency is wavelength dependent and reaches a broad maximum between 2400 and 2800 Å. This is interesting, because nucleic acids and proteins absorb well around 2650 and 2800 Å, respectively. Also, ultraviolet irradiation at 3130 Å has been found to produce more thymine dimers than at 2540 Å (Setlow, 1968). Chu suggested that DNA was the primary target, since pretreatment of the cells with bromodeoxyuridine (BUdR) increased their ultraviolet sensitivity for aberration formation.

FIG. 28. Damage to chromosomes of rat kangaroo cell (induced by argon laser—1000 $\mu J/\mu m^2$ energy). (a) Phase-contrast micrograph illustrating two sites of chromosome damage (arrows). (b) Electron micrograph of same cell showing central portion (A) lesion and electron-dense aggregates in peripheral (B) portion of the area. (c) Higher magnification of lesion area showing dense aggregated fibrils. (a) ×1155. (b) ×4620. (c) ×30,800. (From Rattner and Berns, 1974.)

Agents that are incorporated into the DNA of chromosomes also induce chromosome breaks. Radioactive deoxyribonucleosides, such as tritiated thymidine, have been shown to induce breaks both in the DNA (Cleaver *et al.*, 1972) and in the chromosome (Klevecz and Hsu, 1964). In fact, tritium, when incorporated into the DNA of cells, has a higher killing efficiency (related to the level of chromosome breakage) than tritium incorporated into RNA or protein (Burki and Okada, 1968).

Klevecz and Hsu observed chromosome breaks in Chinese hamster cells after thymidine-^{3}H, uridine-^{3}H, and x-ray treatment. Whereas x-ray and uridine-^{3}H breakage was randomly distributed, breaks induced by thymidine-^{3}H were nonrandom and corresponded to the time that a given chromosomal segment underwent DNA synthesis.

Alkylating agents, such as nitrogen mustard and maleic hydrazide, attack DNA directly (Lawley, 1966), and at the same time induce chromosome lesions (Evans and Scott, 1964, 1969). Similarly, agents known to interact with the DNA through intercalation, as in the case of daunomycin, are also known to induce chromosome aberrations (Vig *et al.*, 1970).

Metabolic agents used under conditions that specifically interfere with DNA synthesis (induce unbalanced growth) have also been shown to induce chromosome aberrations, while agents that interfere with protein or RNA synthesis do not seem to induce chromosome aberrations unless they act directly on the DNA at the same time (e.g., actinomycin D). For example, fluorodeoxyuridine (FUdR) and aminopterin induce chromosome aberrations unless enough thymidine is added at the same time to overcome thymidine starvation (Hsu *et al.*, 1965; Hittelman, 1973).

The evidence cited above implies a correlation between DNA breakage and the induction of chromosome aberrations. But breakage of DNA does not seem to be the only criterion for the production of a chromosome aberration.

Most clastogens induce both single- and double-strand breaks in DNA, as measured by the apparent lowering of DNA molecular weight on alkaline and neutral sucrose gradients. However, caution must be observed in these interpretations, since degradation of DNA measured on alkaline sucrose gradients might be a reflection of drug-induced, alkali-labile bonds rather than drug-induced strand breakage (Spataro and Kessel, 1973).

While several workers have shown that rejoining of single-strand breaks or alkali-labile bonds can occur after treatment by a variety of agents (e.g., Horikawa *et al.*, 1972), little conclusive information is available concerning the repair of double-strand breaks. Kaplan (1966) showed that there is no repair of double-strand breaks in the DNA of bacteria. However, Corry and Cole (1973) and Lange *et al.* (1973) have suggested that some double-strand repair does occur after irradiation of CHO cells at high biological doses. Double-strand break repair has also been reported for bleomycin (Saito and Andoh, 1973) and nitroquinoline 1-oxide (Andoh and Ide, 1972). In all cases, however, it is difficult to distinguish DNA repair from nucleoprotein repair (see discussion by Saito and Andoh, 1973).

In any event, Veatch and Okada (1969) have reported that there are 60 double-strand breaks per D_{37} dose of irradiation (the dose required to produce an average of one lethal event per cell). Since chromosome aberrations seem to be the primary cause of cell death after x irradiation (Dewey *et al.*, 1970; Bhambhani *et al.*, 1973), and one or two chromosome breaks are formed per D_{37} dose, there seem to be many more double-strand breaks in the DNA than there are apparent chromosome breaks. This again suggests that a chromosome aberration is something more than a double-strand break in the DNA.

V. Transition from Lesions to Aberrations

Several models for the formation of the initial chromosome lesion have been presented through the years. In fact, different agents are believed to induce chromosomal lesions (i.e., double-strand breaks in the DNA) in different ways, and these differences are expressed in the timing of aberration formation and the types of aberrations produced. A new technique for the visualization of chromosome aberrations has been recently developed and is useful in detecting these differences.

Johnson and Rao (1970) discovered that, when interphase cells are fused with metaphase cells with the aid of Sendai virus, the chromatin of the interphase nucleus prematurely condenses into discrete chromosomes (prematurely condensed chromosomes, PCC) whose morphology reflects the position of the interphase cell in the cell cycle at the time of fusion (Fig. 29a). This technique allows immediate visualization of drug-induced chromosome damage and eliminates the need to wait for the cell to reach mitosis in order to detect damage. When Chinese hamster cells were treated with x rays or bleomycin, aberrations (gaps, breaks, and exchanges) were detectable within minutes in G_2 PCC (Hittelman and Rao, 1974a,c). Aber-

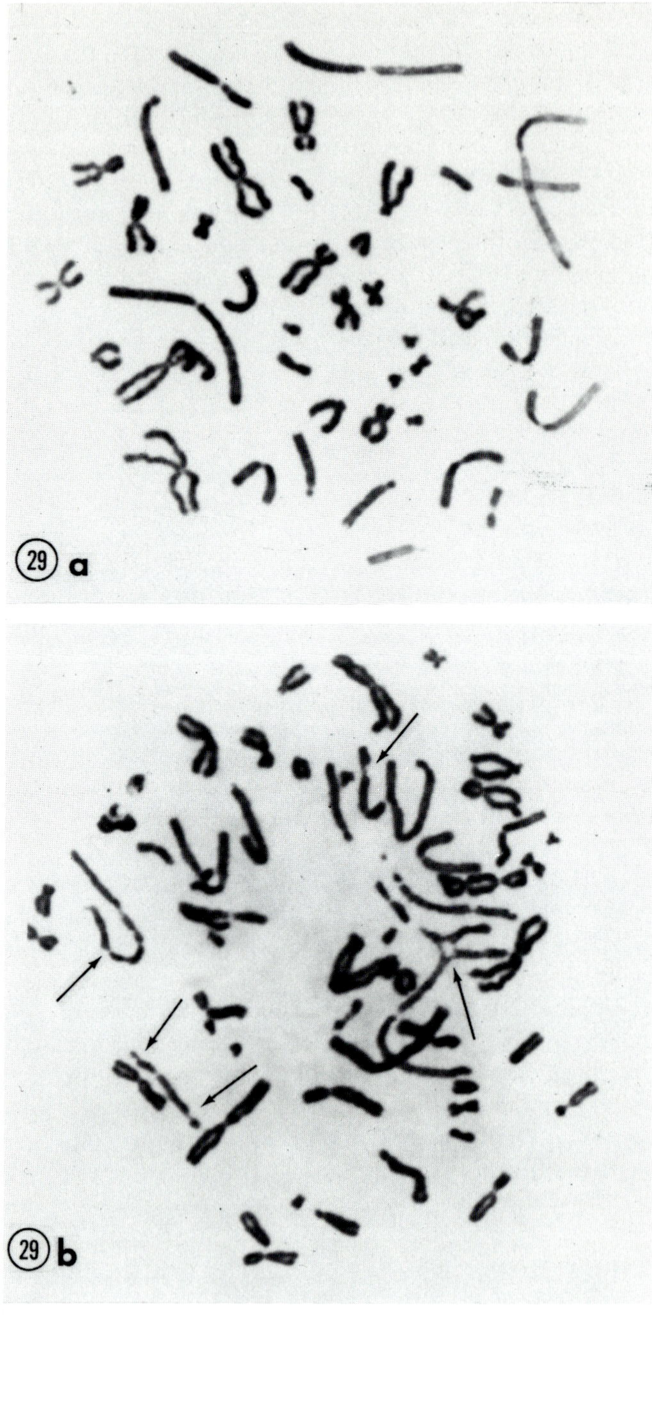
29 a
29 b

rations in G_2 PCC are shown in Fig. 29b. However, when the cells were treated with ultraviolet light or alkylating agents, only gaps could be detected immediately in the G_2 PCC (Hittelman and Rao, 1974b). These and other data (see Kihlman, 1971) suggest that chromosome lesions are produced in different ways.

Two major theories exist for the formation of chromosome lesions by ionizing radiation. One school of thought suggests that ionizing radiation creates direct breaks in the DNA, which may be repaired or misrepaired, leading to an aberration. Another school (Revell, 1959) suggests that ionizing radiation creates a primary lesion in the DNA, which can heal unless it interacts with another primary lesion through an exchange process, in which case it can lead to an aberration visible at metaphase. This controversy has been recently reviewed (Comings, 1974) and is not discussed further here. In either case a double-strand break or exchange is believed to be involved in the appearance of an aberration.

Some chemical agents and ultraviolet light require passage through S phase in order to form an aberration, usually of the chromatid type. Cells treated in G_2 with alkylating agents, for example, do not show chromosome damage (breaks and exchanges) until the second mitosis. Maleic hydrazide and nitrogen mustard (Evans and Scott, 1964, 1969) are cases in point. *Vicia* bean root tip cells treated in G_1 and S show chromatid aberrations at the first mitosis, while cells treated in G_2 fail to show aberrations (chromatid type) until the second mitosis. Furthermore, in the case of nitrogen mustard, 95% of the aberrations in the first cell cycle are located at or near regions of heterochromatin, whereas the localization is less marked for aberrations observed in the second posttreatment cell cycle. It was proposed that nitrogen mustard produces chromosomal lesions (repairable) independent of cell phase, and that chromosome structural changes are produced only as a result of the misreplication of DNA at the sites of alkylation. The high frequency of intrachange and localization of aberrations in heterochromatin was thought to be a consequence of simultaneous replication of DNA strands involved in an exchange aberration.

More recently, a similar mechanism was postulated (Brogger and Johansen, 1972) for MC-induced aberrations. Bender *et al.* (1973)

FIG. 29. (a) PCC of an untreated CHO cell from G_2 phase. Note that the PCC are slender and are less condensed than the mitotic chromosomes. (b) G_2 PCC of a CHO cell treated with 217.5 rads of x rays. Gaps, breaks, and an exchange are indicated by arrows.

also suggested that unrepaired single-chain gaps, arising possibly from incomplete excision repair or synthesis of defective template chains, might become double-strand breaks if a single-strand-specific DNase cleaves the remaining strand. In these cases single-strand lesions must become double through DNA synthesis, misrepair (Dubinin and Soyfer, 1969), or DNase action before an aberration can be observed at metaphase.

A third mechanism has been proposed for DNA and chromosome breakage, which is based upon the effects of some inhibitors of DNA synthesis (Coyle and Strauss, 1970; Hittelman, 1973). Interference of DNA synthesis by hydroxyurea, FUdR, or aminopterin, for example, might allow the transformation of single-strand lesions (associated with DNA replication) into double-strand breaks (as a result of nuclease action on an exposed, decondensed region of the replication DNA). After release from inhibition the double-strand breaks could then lead to chromatid aberrations observed at metaphase. This third mechanism of aberration formation is unique in that few if any exchanges are produced. A similar DNase-specific mechanism for the formation of aberrations has been proposed for virus-induced aberrations (ZurHausen, 1973). In this case, however, the endonuclease is suspected to be coded by viral genes, since inactivation of the viral genome reduces the virus capacity to induce aberrations.

These models for the breakage of DNA should be tempered with some information regarding the relationship between repair (and/or misrepair) of DNA and induction of chromosome aberrations. Wolff (1969) showed that repair of chromosome breakage after irradiation is dependent on protein synthesis. At the same time, Gautschi *et al.* (1973) showed that repair replication does not require concomitant protein synthesis. Wolff and Scott (1969) and Shaeffer *et al.* (1971) have reported that repair of radiation-induced chromosome breaks and the formation of exchanges can take place in the absence of unscheduled synthesis (thought to be representative of dark repair described by Painter, 1970).

Xeroderma pigmentosum (XP) cells show a severely reduced capacity to repair 4-nitroquinoline 1-oxide (4NQO) damage but can repair methyl methane sulfonate (MMS) damage. Sasaki (1973) has reported that XP cells are highly susceptible to chromosome breakage by 4NQO but not by MMS. This suggests that faulty or incomplete repair of DNA damage is responsible for the formation of chromosome damage. Similarly, in experiments with Yoshida cell lines with differential sensitivity to killing by sulfur mustard but the same sensitivity to x rays, Scott *et al.* (1974) found that while drug and radiation resistance was perhaps mediated through a reduced amount of induced chromosome damage, it was not quantitatively related to the cell's capacity for DNA repair replication.

However, Winans *et al.* (1972) suggest from x-ray survival experiments with Chinese hamster ovary cells that there is competition between fixation and repair of damage. They reported that potentially lethal lesions are either rapidly converted at 37°C into lethal lesions (chromosome aberrations) or are repaired, while at 20°C fixation processes are inhibited while repair continues. Also, hypertonic treatment of S-phase cells blocked repair of both sublethal and potentially lethal damage. Thus the transition of a lesion to an aberration is dependent on the state of the chromatin at the time of treatment.

The PCC technique has been recently utilized in the direct measurement of repair of chromosome damage. By waiting various periods of time after clastogenic treatment before fusion, one can directly measure chromosome repair in interphase cells. Hittelman and Rao (1974a,c) found that 30–50% of x-ray- and bleomycin-induced breaks and gaps can be repaired within an hour, but exchanges, once formed, are not repaired. This method now allows a more direct study of the effect of inhibitors of DNA repair on chromosome repair.

Chromosome aberrations reflect the state of the chromatin at the time of treatment in other ways, too. First, little damage is immediately observed in cells irradiated in mitosis when the chromosomes are condensed. If, however, mitotic cells are irradiated and allowed to proceed to the next mitosis, many aberrations are observed. It has been suggested that the condensed structure of the chromosomes at metaphase conceals the damage. This notion has recently been tested by Hittelman and Rao (1974a,b), again utilizing the PCC technique.

Since the formation of PCC during fusion is a time-dependent function, one can treat cells with x rays at different stages of chromosome condensation during the fusion and measure the resultant aberration frequency in the PCC. The chromosome aberration frequencies observed in G_2 PCC decreased when the G_2 chromosomes were irradiated during later stages of condensation. Thus, even though the G_2 cells had sustained chromosome damage, the aberrations were hidden later on in the fusion when the interphase chromatin was more condensed. In fact, in cells irradiated just before fusion, one could still observe a few aberrations in the induced PCC but not in the chromosomes from the mitotic cell. This was thought to be a consequence of differences in the degree of chromosome condensation.

It is well known that the types of chromosome aberrations induced by ionizing radiation and bleomycin are determined by the stage of the cell cycle at which the cell was treated. As reviewed by Evans (1962) and Wolff (1961), chromosome-type aberrations are produced in early G_1, and chromatid-type aberrations are produced in late G_1, S, and G_2. The late G_1 transition from chromosome-type to chromatid-type aberrations after irradiation is interesting, because it suggests that the chromosome becomes a double structure even before DNA synthesis. This has been interpreted as evidence for a multistranded chromosome (Wolff and Luippold, 1964). However, it is possible that this simply represents a loosening of the chromosome structure in preparation for DNA synthesis, as can be observed in studies of actinomycin binding to chromatin (Pederson, 1972). In line with this interpretation, chromatid-type aberrations are induced by uridine-^{3}H in G_0 and G_1 cells (Lindahl-Kiessling *et al.*, 1970). This suggests that single-strand breaks might be induced in an open section of DNA during RNA synthesis and, if resistant to repair, are translated into chromatid-type breaks during DNA synthesis. A similar model might explain the G_1–S transition for types of aberrations induced by ionizing radiation.

Several investigators have reported localization of chromosome breaks and exchanges in the heterochromatin after treatment with a variety of agents including x rays (Evans and Bigger, 1961), nitrogen mustard (Evans and Scott, 1969), hydroxyurea (Somers and Hsu, 1962), MC (Shaw and Cohen, 1965), neutrons and gamma rays (Natarajan and Ahnström, 1970), and tritiated uridine (Natarajan and Sharma, 1971), just to name a few. With regard to banded regions, MC preferentially induces interchange aberrations in constitutive heterochromatic regions (Morad *et al.*, 1973), while breaks appear to be localized in the R bands. The distribution of different types of aberrations among the chromosomes is thought to reflect the spatial arrangement of the chromosomes in the cell at the time of the formation of the structural aberration (Natarajan *et al.*, 1974; Gatti *et al.*, 1973).

Natarajan and Ahnström (1969) compared the relative frequencies of induced chromosome exchanges for a variety of agents in the root meristematic cells of *Nigella damascena* and in *V. faba*. While both species have nearly equal total chromosome length, *Nigella* exhibits little heterochromatin when compared to *V. faba*. In line with this very few interchanges were induced in *Nigella* when compared to

Vicia. This evidence indicated that the sensitivity of cells to induced aberration formation was related to the amount of heterochromatin in the cell. It is not known, however, why heterochromatin should render the cell sensitive to aberration production. Hsu (1974) has proposed the "bodyguard hypothesis" which states that the constitutive heterochromatin is used by the cell to protect the euchromatic genes from damage by providing an absorbing barrier on the outer surface of the nucleus. By decondensing the heterochromatin with a hypotonic solution pretreatment before MC treatment, Hsu shifted the aberration burden toward the euchromatic regions.

It was mentioned earlier that the production of chromosome aberrations is very closely related to the dispersion of chromatin in the cell at the time of treatment (Dewey *et al.*, 1972). The aberration frequency (and radiosensitivity) of cells varies with the stage in cell cycle (Sinclair and Morton, 1966; Dewey *et al.*, 1970), and qualitatively parallels the dispersion of chromatin within the cell (Dewey *et al.*, 1972). Dettor *et al.* (1972) found that hypertonic treatment of S-phase cells causes both an increase in chromatin condensation and an increase in radiosensitivity, whereas hypertonic treatment fails to radiosensitize mitotic cells in which the chromatin is already condensed. The increase in radiosensitivity with condensation was postulated to reflect an inhibition of lesion repair produced by irradiation, and an increase in the probability of interaction of lesions when the chromatin fibers are close together.

VI. Models for the Formation of Aberrations

The evidence so far presented suggests that chromosome aberrations derive from double-strand breaks or exchanges in the DNA followed by a change in the interaction of the protein and DNA. Furthermore, a chromatin fiber lesion in the heterochromatin seems more likely to be translated into a break or exchange than a lesion in the euchromatin. The actual mechanism involved in the formation of a break or exchange is still in the realm of postulation. In fact, several models have been proposed, each of which is dependent on a different model for the structure of the chromosome.

As mentioned in a previous section, there is some indication that the metaphase chromosome is multistranded. Stubblefield (1973) recently presented evidence that the metaphase chromatid is binemic, each half-chromatid consisting of two parts, a core ribbon and attached loops of nucleoprotein, called the epichromatin. The epichromatin is thought to be a long DNA double helix with associated proteins, with side branches of DNA located at intervals along its length. Furthermore, the epichromatin is thought to be attached to the core by a protein. The core ribbon of parallel chromatin fibers would be attached to the nuclear membrane at both ends of the chromatid. According to the model, heterochromatin is located in the epichromatin loops. At metaphase the two core ribbons are thought to be rolled into a fiber about 500 Å in diameter and, with the attached epichromatin, cohelically spiral along the chromatid length. The interphase state of the chromatin is thought to be a relaxed state of the chromatin fibers of the core ribbon. With this model in mind, Stubblefield proposed that a break in the cellular DNA would not be visible unless the chromonema (chromosome core) is broken. If breaks occur simultaneously in two or more different chromatids, and if the breaks are not repaired, exchanges could result.

Two types of evidence might argue against this model for chromosome damage. First, the kinetics of induction of chromosome breaks in lampbrush chromosomes (Gall, 1963) suggests that the unit of breakage is one DNA helix rather than two as suggested by the bineme model. Second, chromosome breakage induced by a variety of agents appears to be localized in the heterochromatin. Yet, in Stubblefield's model the heterochromatin is localized in the epichromatin, and breakage of the epichromatin would not be expected to lead to a break in the core. One could still argue, however, that heterochromatic regions (including the core) are less likely to repair breaks in the DNA. Also, a break in the core near the heterochromatin is more likely to be involved in an exchange because of the repetitious nature of the DNA in the heterochromatin.

Other investigators view the chromosome as a unineme structure in their models for the production of chromosome breaks. Comings (1974) presents such a model in a recent review. Electron micrographs (Comings and Okada, 1970) have indicated that chromatin is clustered at various points on the nuclear membrane rather than being freely dispersed in the interphase nucleus, and it is proposed that these cluster points may persist to metaphase to lend a stability to the folding pattern. Thus a break within the cluster might be thought to produce a gap in the chromosome (since the structure of the cluster might remain intact), while a break in one of the connecting fibers between clusters could lead to a chromatid deletion. When the chromatin condenses further at prophase, the connecting fibers would be brought close to the cluster and a break would be visualized as a "subchromatid break" or gap which would be turned into a visual chromosome break in the next cycle unless repaired.

One of the ideas of Comings' model is that a gap is really a chromatid fiber break that cannot be expressed because of the clustered nature of the chromatin. Two lines of evidence argue against this notion. First, if a gap were really a chromatid fiber break, one would expect to see an aberration formed in the next cycle. Substitution of BUdR for thymidine in the DNA has been shown to sensitize chromosomes to aberration formation, especially from ultraviolet irradiation. BUdR-induced breaks, however, are not observed until many hours after treatment (Chu, 1965; Hsu and Somers, 1961). In shorter experiments (Zakharov and Egolina, 1972), however, apparent gaps in metaphase chromosomes appear in the locations of BUdR incorporation in the DNA. If the gap in the first mitosis were simply an unresolved break, one would expect to see either a full-fledged chromosome break (without repair of the gap) or an unbroken chromosome (if the gap is required). In fact, at the next metaphase, while the gap still appears where the BUdR is localized, the sister chromatid (without BUdR) appears normal.

This evidence suggests that some of the gaps observed after BUdR treatment are simply areas of underspiralization of the chromosome where BUdR is localized in the chromosome. This underspiralization could possibly be due to the substitution of a bromine for the methyl group in the thymine molecule, leading to an alteration of the DNA–protein interaction (Szybalski, 1962; Gordon *et al.*, 1973), which in turn prevents proper condensation of the chromatin. Similarly, chromosome cross-banding can be induced in metaphase chromosomes by treating cells with a variety of DNA-intercalating agents shortly before fixation (Hsu *et al.*, 1973). The interband regions very much resemble gaps and have led to the suggestion that some gaps simply represent expanded interband regions.

The PCC technique has provided two additional bits of information with regard to the structure of gaps. First, when S-phase cells are fused with mitotic cells, the PCC of the S-phase cells appear to contain many gaps and has often been termed pulverized. These so-called gaps, however, are the sites of DNA synthesis in the S-phase cell (Sperling and Rao, 1974). Rather than being double-strand breaks in the chromatin, they may simply represent regions of the chromatin where the DNA–protein interaction is such that factors from the mitotic cell cannot interact properly with the S chromatin to induce full condensation.

In other experiments treatment of Chinese hamster cells with ultraviolet radiation, or an alkylating agent such as nitrogen mustard or trenimone, resulted in gaps observable in G_2 PCC and in mitotic chromosomes (Hittelman and Rao, 1974b). After ultraviolet treatment the gap frequency observed in G_2 PCC was higher than in the mitotic chromosomes, while the reverse trend was observed after treatment with alkylating agents. These experiments indicate that there are at least two types of gaps visible in mitotic chromosomes—one due to an alteration in the first stage of condensation (probably because of single-stranded regions in the DNA), and another due to an alteration in the second stage of condensation. Perhaps altered DNA–protein interactions after alkylating agents (Grunicke *et al.*, 1973) result in the second type of gap. Thus ultraviolet radiation and x rays produce more of the first type of damage, while alkylating agents produce more of the second type.

These various lines of evidence suggest therefore that, while some gaps may in fact be unresolved chromatid breaks, other gaps may arise from a lack of spiralization in the chromatin due to an altered DNA–protein interaction.

VII. Summary and Conclusions

The spectrum of chromosome damage produced by various natural and synthetic clastogens can now be evaluated by both light and electron microscopy. Various aspects of chromosome damage detectable by both instruments are compared in Table I. In general, ultrastructural analysis of chromosome breaks, gaps, and exchanges confirms the more classic analysis by light microscopy. However, there are notable exceptions. Some aberrations that would have been scored as breaks with light microscopy were found to have chromosome fibers extending across the lesion. Conversely, and perhaps of lesser significance, aberrations that would have been scored as gaps with light microscopy were in fact true breaks. The ultrastructure of gaps indicates that they may be composed of several size classes of chromatin fibrils. Gaps along the chromatid often appeared to be composed of two parallel fiber bundles, suggesting bipartite or subchromatid organization to the chromatid. The nature of side-arm bridges and stickiness is more clearly definable by electron optics. In both types of damage, thin "subchromatid" fibrils were observed between daughter chromatids at metaphase, anaphase, and telophase. Interpretation of the ultrastructure of chromosome damage is greatly limited by our general ignorance of chromosome structure.

In the chromosome the primary target for damage is the DNA, and breaks and exchanges probably involve a double-stranded break in the DNA as a necessary but not completely sufficient intermediate in aberration formation.

Some agents induce aberrations immediately on treatment (e.g., ionizing radiation, bleomycin); other agents (alkylating agents) produce lesions which might induce misreplication during S phase and result in aberrations; still other agents (aminopterin, FUdR, hydroxyurea) induce unbalanced growth and make the S-phase chromatin susceptible to endonuclease action, resulting in aberrations. Thus, while the nature of the initial lesion might be different for different agents, once the DNA is broken and/or exchanged, aberration formation might be common for all agents.

Chromosome aberrations reflect the state of the chromatin at the time of treatment; this might result in localization of chromosome damage in heterochromatic regions, or might determine the type of chromosome aberration produced. Also, aberrations can be hidden within condensed chromosomes and become apparent in the next condensation cycle.

Investigations employing the PCC technique have shown that chromosome breaks and gaps can be repaired within an hour after treatment, while exchanges, once formed, are not repairable. Also, at least two types of gaps exist, one involving an unresolved chromatid break, and the other arising from a lack of spiralization in the chromatin due to an altered DNA–protein interaction.

Models for the mechanism of aberration formation are dependent on chromosome structural models, and models exist for both unineme and bineme chromosome structures. Nevertheless, all models involve breakage of chromatin fibers, resulting in aberrant packaging of chromosomes.

ACKNOWLEDGMENT

We are grateful to several of our friends and colleagues for their helpful advice and counsel during the preparation of this article; to Dr. S. Wolff, Dr. T. C. Hsu and Dr. B. Bowman for stimulating discussion and critical reading of the manuscript; to Dr. R. Humphrey and Dr. M. W. Shaw for their valuable collaboration during many of the initial investigations reviewed in this report; and to Dr. M. McGill for permission to use unpublished micrographs of isolated chromosomes.

The technical assistance of Mr. J. Cartwright, Jr., and Mrs. P. McAfee is gratefully acknowledged. We wish to thank Mrs. B. Ledlie for typing and proofreading the manuscript.

These investigations were supported in part by research grants NIH-NICHD-69-2139 and DHEW CA 14675 to BRB, DRG 1110 from the Damon Runyon Memorial Fund, NSF GB 37636, and NIH-N01-CM-61156 to Dr. P. Rao. WNH holds a U.S. Public Health Service Postdoctoral Fellowship, 5T0 1CA-0523003.

References

Abuelo, J. G., and Moore, D. E. (1969). *J. Cell Biol.* **41**, 73.

Andoh, T., and Ide, T. (1972). *Cancer Res.* **32**, 1230.

Angerer, L. M., and Moudrianakis, E. N. (1973). *J. Mol. Biol.* **63**, 505.

Bajer, A. (1965). *Chromosoma* **17**, 291.

Barnicot, N. A., and Huxley, H. E. (1965). *Quart. J. Microsc. Sci.* **106**, 197.

Beadle, G. W. (1932). *Z. Indukt. Abstamm. Vererbungsl.* **63**, 195.

Bender, M. A., Griggs, H. G., and Walker, P. C. (1973). *Mutat. Res.* **20**, 387.

Berns, M. W., Olson, R. S., and Rounds, D. E. (1969a). *Nature (London)* **221**, 74.

Berns, M. W., Rounds, D. E., and Olson, R. S. (1969b). *Exp. Cell Res.* **56**, 292.

Berns, M. W., Ohnuki, Y., Rounds, D. E., and Olson, R. S. (1970). *Exp. Cell Res.* **60**, 133.

Bhambhani, R., Kuspina, J., and Giblak, R. E. (1973). *Can. J. Genet. Cytol.* **15**, 605.

Bloom, W., and Leider, R. J. (1962). *J. Cell Biol.* **13**, 269.

Bloom, W., and Özarslan, S. (1965). *Proc. Nat. Acad. Sci. U. S.* **53**, 1294.

Brewen, J. G. (1964). *Genetics* **50**, 101.

Brinkley, B. R., and Humphrey, R. M. (1969). *J. Cell Biol.* **42**, 827.

Brinkley, B. R., and Shaw, M. W. (1970). *In* "Genetic Concepts and Neoplasia 23rd Annual Symposium on Fundamental Cancer Research, M. D. Anderson Hospital," pp. 313–345. Williams & Wilkins, Baltimore, Maryland.

Brinkley, B. R., and Stubblefield, E. T. (1966). *Chromosoma* **19**, 28.

Brinkley, B. R., and Stubblefield, E. T. (1970). *In* "Advances in Cell Biology" (D. M. Prescott, L. Goldstein, and E. McConkey, eds.), pp. 119–185. Appleton, New York.

Brinkley, B. R., Murphy, P., and Richardson, L. C. (1967). *J. Cell Biol.* **35**, 279.

Brogger, A. (1971). *Humangenetik* **13**, 1.

Brogger, A., and Johansen, J. (1972). *Chromosoma* **38**, 95.

Burki, H. J., and Okada, S. (1968). *Biophys. J.* **8**, 445.

Chu, E. H. Y. (1965). *Mutat. Res.* **2**, 75.

Cleaver, J. E., Thomas, G. H., and Burki, H. J. (1972). *Science* **177**, 996.

Cohen, M. M. (1963). *Cytogenetics* **2**, 271.

Comings, D. E. (1970). *Can. J. Genet. Cytol.* **12**, 960.

Comings, D. E. (1974). *In* "Chromosomes and Cancer" (J. German, ed.), pp. 95–133. Wiley, New York.

Comings, D. E., and Okada, T. A. (1970). *Exp. Cell Res.* **62**, 293.

Comings, D. E., and Okada, T. A. (1971). *Exp. Cell Res.* **67**, 97.

Comings, D. E., and Okada, T. A. (1972). *In* "Perspectives in Cytogenetics" (S. W. Wright, B. F. Crandall, and L. Boyer, eds.), p. 223. Thomas, Springfield, Illinois.

Corry, P. M., and Cole, A. (1973). *Nature (London), New Biol.* **245**, 100.

Coyle, M. B., and Strauss, B. (1970). *Cancer Res.* **30**, 2314.

Crawford, L. V., and Waring, M. J. (1967). *J. Mol. Biol.* **25**, 23.

Crouse, H. V. (1954). *Science* **119**, 485.

Dettor, C. M., Dewey, W. C., Winans, L. F., and Noel, J. S. (1972). *Radiat. Res.* **52**, 352.

Dewey, W. C., Furman, S. C., and Miller, H. H. (1970). *Radiat. Res.* **43**, 561.

Dewey, W. C., Westra, A., Miller, H. H., and Nagasawa, H. (1971). *Int. J. Radiat. Biol.* **20**, 505.

Dewey, W. C., Noel, J. S., and Dettor, C. M. (1972). *Radiat. Res.* **52**, 373.

Dubinin, N. P., and Soyfer, V. N. (1969). *Mutat. Res.* **8**, 353.

DuPraw, E. J. (1965). *Proc. Nat. Acad. Sci. U. S.* **53**, 161.

Evans, H. J. (1961). *Genetics* **46**, 257.

Evans, H. J. (1962). *Int. Rev. Cytol.* **13,** 221–321.
Evans, H. J., and Bigger, T. R. L. (1961). *Genetics* **46,** 277.
Evans, H. J., and Scott, D. (1964). *Genetics* **49,** 17.
Evans, H. J., and Scott, D. (1969). *Proc. Roy. Soc., Ser. B* **173,** 491.
Gall, J. G. (1963). *Nature (London)* **198,** 36.
Gatti, M., Rizzoni, M., Palitti, F., and Olivieri, G. (1973). *Mutat. Res.* **20,** 87.
Gautschi, J. R., Young, B. R., and Cleaver, J. E. (1973). *Exp. Cell Res.* **76,** 87.
Giménez-Martín, G., López-Sáez, J. F., and González-Fernández, A. (1963). *Cytologia* **28,** 381.
Golomb, H. M., Bahr, G. F., and Borgaonkar, D. S. (1971). *Chromosoma* **34,** 94.
Gordon, J. S., David, J. W., Bell, G., McCarthy, B. J., and Rutter, W. J. (1973). *J. Cell Biol.* **59,** 116a.
Grunicke, H., Bock, K. W., Becher, H., Gang, V., Schnierda, J., and Puschendorf, B. (1973). *Cancer Res.* **33,** 1048.
Hampel, K. E., and Levan, A. (1964). *Hereditas* **51,** 315.
Heddle, J. A. (1969). *Can. J. Genet. Cytol.* **11,** 783.
Hittelman, W. N. (1973). *Mutat. Res.* **18,** 93.
Hittelman, W. N., and Rao, P. N. (1974a). *Mutat. Res.* **23,** 251.
Hittelman, W. N., and Rao, P. N. (1974b). *Mutat. Res.* **23,** 259.
Hittelman, W. N., and Rao, P. N. (1974c). *Cancer Res.* **34,** 3433.
Horikawa, M., Fukuhara, M., Suzuki, F., Nikaido, O., and Sugahara, T. (1972). *Exp. Cell Res.* **70,** 349.
Hsu, L. Y. F., Hirschhorn, K., Goldstein, A., and Barrinski, M. A. (1970). *Ann. Hum. Genet.* **33,** 343.
Hsu, T. C. (1974). *Genetics, Suppl.* **2** (in press).
Hsu, T. C., and Somers, C. E. (1961). *Proc. Nat. Acad. Sci. U. S.* **47,** 396.
Hsu, T. C., Dewey, W. C., and Humphrey, R. M. (1962). *Exp. Cell Res.* **27,** 441.
Hsu, T. C., Arrighi, F. E., Klevecz, R. R., and Brinkley, B. R. (1965). *J. Cell Biol.* **26,** 539.
Hsu, T. C., Brinkley, B. R., and Arrighi, F. E. (1967). *Chromosoma* **23,** 137.
Hsu, T. C., Pathak, S., and Shafer, D. A. (1973). *Exp. Cell Res.* **79,** 484.
Humphrey, R. M., and Brinkley, B. R. (1969). *J. Cell Biol.* **42,** 745.
Humphrey, R. M., Dewey, W. C., and Cork, A. (1973). *Radiat. Res.* **19,** 247.
Johnson, R. T., and Rao, P. N. (1970). *Nature (London)* **226,** 717.
Jokelainen, P. T. (1967). *J. Ultrastruct. Res.* **19,** 19.
Kaplan, H. S. (1966). *Proc. Nat. Acad. Sci. U. S.* **55,** 1442.
Kihlman, B. A. (1970). *Hereditas* **65,** 171.
Kihlman, B. A. (1971). *Advan. Cell Mol. Biol.* **1,** 59.
Kihlman, B. A., and Hartley, B. (1967). *Hereditas* **57,** 289.
Kirby-Smith, J. S., and Craig, D. L. (1957). *Genetics* **42,** 176.
Klevecz, R. R., and Hsu, T. C. (1964). *Proc. Nat. Acad. Sci. U. S.* **52,** 811.
Kuo, M. T., and Haidle, C. W. (1973). *Biochim. Biophys. Acta* **331,** 194.
La Cour, L. F., and Rutishauser, A. (1954). *Chromosoma* **6,** 696.
Lange, C. S., Liberman, D. F., and Sheck, L. E. (1973). *Radiat. Res., Abstr.* Cc-9.
Lawley, P. D. (1966). *Prog. Nucl. Acid Res. Mol. Biol.* **5,** 89–131.
Lett, J. T., Caldwell, I., Dean, C. J., and Alexander, P. (1967). *Nature (London)* **214,** 790.
Lindahl-Kiessling, K., Santesson, B., and Böök, J. A. (1970). *Chromosoma* **31,** 280.
McClintock, B. (1934). *Z. Zellforsch. Mikrosk. Anat.* **21,** 294.
McGill, M., and Brinkley, B. R. (1974). In preparation.
McGill, M., Pathak, S., and Hsu, T. C. (1974). *Chromosoma* **47,** 157.

Manton, I. (1945). *Amer. J. Bot.* **32**, 342.
Mertz, T. (1961). *Science* **133**, 329.
Morad, M., Jonasson, J., and Lindsten, J. (1973). *Hereditas* **74**, 273.
Natarajan, A. T., and Ahnström, G. (1969). *Chromosoma* **28**, 48.
Natarajan. A. T., and Ahnström, G. (1970). *Chromosoma* **30**, 250.
Natarajan, A. T., and Sharma, R. P. (1971). *Chromosoma* **34**, 168.
Natarajan, A. T., Ahnström, G., and Sharma, R. P. (1974), *Mutat. Res.* **22**, 73.
Nebel, B. R. (1932). *Z. Zellforsch. Mikrosk. Anat.* **16**, 251.
Nichols, W. W. (1970). *Annu. Rev. Microbiol.* **24**, 479.
Nowell, P. C. (1964). *Exp. Cell Res.* **33**, 445.
Östergren, G., and Wakonig, T. (1955). *Bot. Notis.* **4**, 357.
Paika, K. D., and Krishan, A. (1973). *Cancer Res.* **33**, 961.
Painter, R. B. (1970). *Photophysiology* **5**, 169.
Paton, G. R., Jacobs, J. P., and Perkins, F. T. (1965). *Nature (London)* **207**, 43.
Peacock, W. J. (1961). *Nature (London)* **191**, 832.
Pederson, T. (1972). *Proc. Nat. Acad. Sci. U. S.* **69**, 2224.
Rattner, J. B., and Berns, M. W. (1973). *J. Cell Biol.* **59**, 281a.
Rattner, J. B., and Berns, M. W. (1974). *J. Cell Biol.* **62**, 526.
Revell, S. H. (1959). *Proc. Roy. Soc., Ser. B* **150**, 563.
Robbins, E., and Gonatas, N. K. (1964). *J. Cell Biol.* **21**, 429.
Roos, U-P. (1973). *Chromosoma* **41**, 195.
Saito, M., and Andoh, T. (1973). *Cancer Res.* **33**, 1696.
Sasaki, M. S. (1973). *Mutat. Res.* **20**, 291.
Scheid, W., and Traut, H. (1970). *Mutat. Res.* **10**, 159.
Scheid, W., and Traut, H. (1971). *Mutat. Res.* **11**, 253.
Scott, D., Fox, M., and Fox, B. W. (1974). *Mutat. Res.* **22**, 707.
Setlow, R. B. (1968). *Progr. Nucl. Acid Res. Mol. Biol.* **8**, 257–295.
Schaeffer, J., Vincent, R. A., and Merz, T. (1971). *Mutat. Res.* **11**, 256.
Shaw, M. W. (1970). *Annu. Rev. Med.* **21**, 409.
Shaw, M. W., and Cohen, M. M. (1965). *Genetics* **51**, 181.
Shaw, M. W., Schwab, L., and Brinkley, B. R. (1972). *In* "Perspectives in Cytogenetics" (S. W. Wright, B. F. Crandall, and L. Boyer, eds.), pp. 251–275. Thomas, Springfield, Illinois.
Sinclair, W. K., and Morton, R. A. (1966). *Radiat. Res.* **29**, 450.
Somers, C. E., and Hsu, T. C. (1962). *Proc. Nat. Acad. Sci. U. S.* **48**, 937.
Spataro, A., and Kessel, D. (1973). *Biochim. Biophys. Acta* **331**, 194.
Sperling, K., and Rao, P. N. (1974). *Chromosoma* **45**, 121.
Stubblefield, E. T. (1973). *Int. Rev. Cytol.* **35**, 1–60.
Stubblefield, E. T., and Wray, W. (1971). *Chromosoma* **32**, 262.
Suzuki, H., Nagai, K., Akutsu, E., Yamaki, H., Tanaka, N., and Umezawa, H. (1969). *Biochim. Biophys. Acta* **174**, 309.
Swanson, C. P. (1947). *Proc. Nat. Acad. Sci. U. S.* **33**, 229.
Swanson, C. P., and Johnston, A. H. (1954). *Amer. Natur.* **88**, 425.
Syzbalski, W. (1962). *In* "Molecular Basis of Neoplasia" 15th Annual Symposium on Fundamental Cancer Research, M. D. Anderson Hospital, p. 14F. Univ. of Texas Press, Austin.
Trosko, J. E., and Wolff, S. (1965). *J. Cell Biol.* **26**, 125.
Unakul, W., and Hsu, T. C. (1973). *Chromosoma* **44**, 285.
Veatch, W., and Okada, S. (1969). *Biophys. J.* **9**, 330.
Vig, B. K., Samuels, L. D., and Kontras, S. B. (1970). *Chromosoma* **29**, 62.

Wahren, B., Lampert, F., and Goetz, O. (1972). *Exp. Cell Res.* **75**, 271.
Waring, M. J. (1965). *J. Mol. Biol.* **13**, 269.
Winans, L. F., Dewey, W. C., and Dettor, C. M. (1972). *Radiat. Res.* **52**, 333.
Wolfe, S. L., and Martin, P. G. (1968). *Exp. Cell Res.* **50**, 140.
Wolff, S. (1961). *In* "Mechanisms in Radiobiology" (M. Errera and A. Forssberg, eds.), Vol. 1, pp. 419–475. Academic Press, New York.
Wolff, S. (1969). *Int. Rev. Cytol.* **25**, 279–296.
Wolff, S., and Luippold, H. E. (1964). *Exp. Cell Res.* **34**, 548.
Wolff, S., and Scott, D. (1969). *Exp. Cell Res.* **55**, 9.
Wray, W., and Stubblefield, E. T. (1970). *Exp. Cell Res.* **59**, 469.
Wray, W., Stubblefield, E. T., and Humphrey, R. M. (1972). *Nature (London) New Biol.* **238**, 237.
Yu, C. K. (1971). *Can. J. Genet. Cytol.* **13**, 690.
Zakharov, A. F., and Egolina, N. A. (1972). *Chromosoma* **38**, 341.
ZurHausen, H. (1973). *Progr. Exp. Tumor Res.* **18**, 240.

Computer Processing of Electron Micrographs: A Nonmathematical Account

P. W. Hawkes

The Cavendish Laboratory, University of Cambridge, Cambridge, England

I. Introduction

During the past few years our understanding of the process of electron image formation at high resolution has gradually advanced, with the result that new techniques, based on computer manipulation, have been developed for improving high-resolution images and for increasing the amount of reliable information that can be extracted from them. The various techniques that have been proposed are usually expressed in mathematical language which many biologists find obscure, however, and the possible benefits to be gained from image processing are not yet widely appreciated. This situation is not helped by the fact that few of the practical attempts at image processing have been applied to specimens that would bring out the real potential of the method; since the subject is still in an exploratory phase, test material tends to be selected for its convenience or its availability rather than for its scientific interest.

For these reasons we have attempted to give an account of the aims and techniques of image processing in almost wholly nonmathematical terms. We have taken pains to ensure that simplification has not produced any distortion of the facts, although some of the more recondite ideas have inevitably had to be explained in terms drastically different from those used in the original papers.

This article falls into two parts. The first part provides a modern account of contrast formation in the electron microscope; in the second part we explain what is meant by computer processing, emphasizing the possible benefits for biology. Methods of three-dimensional reconstruction are not considered here, as they have been extensively discussed (e.g., Huxley and Klug, 1971) and have been reviewed recently in this series (Gordon and Herman, 1974); moreover, they are arguably better discussed separately.

II. Electron Image Formation

The contrast seen on an electron micrograph is produced by two very different mechanisms, one closely analogous to that occurring in an ordinary light microscope, and the other resembling that of a phase-contrast light microscope. It is generally true that relatively coarse detail is produced by the first mechanism, and fine detail by the second.

When electrons fall on a thin, stained biological section, some pass close to the heavy atoms of the strain, uranium, lead, or osmium, and others pass through regions containing only light atoms. Electrons that have passed close to heavy atoms are deflected from their original course more than the remainder so that, if they were traveling more or less straight down the microscope when they struck the specimen, they will subsequently move at an angle to the microscope axis. As a result of this deflection, many of them will strike the objective aperture, which is situated a short distance beyond the specimen in the back focal plane of the objective lens, and make no further contribution to the image. On the fluorescent screen, therefore, we see dark patches where these intercepted electrons should have fallen, indicating that stain was present at the corresponding point on the specimen. Thus, although almost all the electrons that fall on the specimen emerge from the other side, some do not reach the image because they are halted by the objective aperture. The resulting contrast is very similar to that in an ordinary light microscope with a semiopaque specimen, except that there part of the incident light is stopped actually in the specimen. Electron microscope specimens are almost entirely transparent—very few electrons are absorbed within the specimen—but amplitude contrast, as this is called, is produced artificially by the objective aperture.

The second contrast mechanism, which is said to produce phase contrast, may also be understood in terms of the deflections undergone by the electrons in the microscope, but these deflections have an origin very different from those produced by passing close to a heavy atom. Even the light atoms in the specimen will produce some deflection of the electrons in the beam, but if the angles are small the electrons will pass through the objective aperture and reach the image. If the microscope lenses were perfect, these electrons would produce no contrast in an (in-focus) image, but electron lenses are far from perfect. (So large is their spherical aberration that a light microscope with equally bad lenses would not resolve detail smaller than about 50 μm.) It so happens that the effect of this lens aberration is to deflect some electrons from their course more than others, with the result that even electrons that have been only slightly deviated in the specimen strike the image plane in the wrong place, thus producing contrast. This effect may be controlled during observation by altering the focus control of the instrument. Since the image plane—the viewing screen or photographic plate—is fixed this control shifts the region in focus in the vicinity of the specimen; the combination of some deliberate defocus and the unavoidable spherical aberration can be optimized to give the best contrast at the image. This type of contrast is known as phase contrast, because the deflection introduced by the spherical aberration and controlled defocus is characterized mathematically by a phase shift, a term we shall explain later. Qualitatively it can be seen to bear some resemblance to phase-contrast light microscopy, in which the specimen is transparent and produces no visible image under normal conditions. Nevertheless, it may contain substances with different refractive indices, and these are rendered visible in the phase-contrast microscope by inserting a small phase plate. In the electron microscope the combined effect of spherical aberration and defocus performs much the same function as a phase plate.

Unfortunately, an aberration plus some defocusing is a very crude and unreliable substitute for a precisely designed and manufactured phase plate, since the electrons are not deflected consistently. Some produce desirable contrast, giving a true representation of the specimen, but others produce unwanted contrast, which does not represent any genuine feature of the specimen, and there is no way of distinguishing true detail from false merely by examining the image. If a large number of micrographs taken under different conditions exhibit common features, these can normally be regarded as genuine structural details, unlikely to be optical artifacts; in many cases, however, it is difficult or very inefficient to collect enough micrographs to make interpretation reliable. Moreover, it is often precisely those details that are most doubtful that would, if genuine, be of most interest. This, then, is a problem that might be solved by image processing, since the instrumental solutions are not inviting. The latter involve building electron lenses with lower spherical aberration, or using real phase plates in the electron microscope. An immense amount of effort has gone into the problem of designing lenses with low spherical aberration, but the practical difficulties are so formidable that it is unlikely that any commercial instrument with corrected lenses will be put on the market for many years—if ever. Attempts are being made (Thon and Willasch, 1972b) to construct suitable phase plates for electron microscopes, and we cannot yet say whether technological progress will permit these to become routine accessories for very high-quality instruments. Even if such plates were available, it is not clear that they would provide the ideal solution, for a variety of reasons. From the practical point of view, they would be fragile and delicate, difficult to manipulate, prone to contamination, and require very exact positioning in the electron microscope; moreover, each plate would be beneficial for only one setting of defocus, which would have to be known accurately and capable of being set repeatably. Another type of phase-shifting device that has been used successfully is much simpler; in this a fine thread is stretched across the back focal plane of the objective lens and becomes charged by the beam electrons that fall on it. The charge on the thread then repels electrons passing close to it, and the resultant effect bears a distinct resemblance to that of a phase plate (Unwin, 1971, 1972, 1974; Krakow and Siegel, 1972).

Since one of the main objects of image processing is to remedy faults introduced by the shortcomings of the electron microscope, we now describe some other relevant features of electron image formation. So far, the only effect of the specimen on the beam that we have considered is pure deflection; furthermore, we have said nothing about the nature of the electron beam when it falls on the specimen. The process of interaction between incident-beam electrons and specimen atoms is known as scattering. Generally speaking, most of the electrons in the incident beam pass through the specimen without any perceptible change to their motion. They proceed to the image and produce the uniform bright background against which the image is seen in bright-field microscopy; these are the *unscattered* electrons. Some of the remainder are deflected in the specimen, by heavy or light atoms, but emerge from the specimen traveling at virtually the same speed as when they entered; they are said to have been *elastically scattered.* A final group of electrons has been slowed down by some interaction as it passed through the specimen and is said to have been *inelastically scattered*. The electron velocity is a very inconvenient parameter, however, since we always speak of electron beam energies. If the accelerating voltage is 100 kV, the electrons have an energy of 100 keV, and the slowing down of inelastically scattered electrons is always referred to as an energy loss, typically of 5–50 eV. These electrons that have lost energy have an extremely serious effect on image formation, because of a defect of electron lenses that we have not yet mentioned. The focusing properties of electron lenses vary very rapidly with electron velocity (or energy) so that, if the elastically scattered electrons produce a sharp image, the inelastically scattered electrons will produce a thoroughly blurred one superimposed on the sharp picture. (Since the quantity analogous to wavelength in electron optics is determined by the electron energy, this effect is said to be caused by the very high chromatic aberration of electron lenses.) It is for this reason that electron microscope manufacturers attempt to ensure that the energy of the incident beam does not fluctuate by more than a few parts in a million. At 100 kV this means that the energy fluctuation on the incident beam ought not to exceed a few tenths of an electron volt. The only instrumental remedy for the blurred image due to inelastically scattered electrons is an energy filter, which separates electrons that have lost energy from those that have not. We do not discuss this further here, but it is worth pointing out that one of the advantages of the scanning transmission electron microscope (Crewe, 1970; Thomson, 1973; Zeitler, 1975) is the ease with which elastically scattered electrons can be separated from those that have been scattered inelastically. The images formed by each group of electrons can then be observed separately on different fluorescent screens and can be processed in various ways electronically, without the intermediate step of digitization (Section III,A), because each point of the image is produced independently, whereas in ordinary (fixed-beam) transmission electron microscopy, the whole image is produced simultaneously.

We have tacitly assumed in these comments on scattering that each beam electron is deviated at most once during its passage through the specimen. For specimens so thick or containing such a high density of heavy atoms that this is not true, the theory of image formation is very different from that outlined here, and the type of image processing discussed below is not applicable. This requirement, that nearly all the beam electrons are scattered once or not at all, is in practice not a very severe restriction, but it should not be entirely neglected and tables of mean free path are available and should be consulted in doubtful cases. The mean free path is the average distance an electron travels in a given element at a specified energy between collisions with the atoms, and specimens can safely be a little thicker than the mean free path. Throughout the remainder of this article, we assume that single-scattering conditions prevail.

The other parameter we have mentioned, the quality of the incident beam, is extremely important if the types of image processing we describe are to have any real chance of success. The beam is normally produced by heating a filament which emits electrons; these are accelerated to full speed by the gun and then directed along the microscope and onto the specimen by condenser lenses, which incorporate apertures to restrict the size of the beam. However carefully the gun is designed, there is always some velocity spread on the incident beam, which we have to take into account in discussing image formation because even elastically scattered electrons do not all produce sharp images since they have slightly different energies. More important, the interpretation of the image will be least subject to error if the electrons that strike the specimen appear to have come from a point source. Suppose, for example, that we try to arrange that the beam electrons all travel straight down the microscope when they strike the specimen, which they enter perpendicular to its surface. If the electrons have come from a large source, those from the edges will arrive at the specimen at an angle, and the angle through which they are deflected in the specimen will be added to this (variable) angle of incidence. Since the high-resolution contrast mechanism is intimately related to the scattering angle, the angle through which the electrons are deviated, we obtain a blurred image because there is too wide a range of scattering angles for each high-resolution detail. The resulting blur is not just a distortion of the image, but actually suppresses some of the information in the image, which could otherwise have been retrieved by image processing. It has therefore to be guarded against very assiduously, by careful design of the series of condenser lenses and by introducing newer types of filament, in which a flat lancet shape or a very sharp point replaces the conventional hairpin. A few microscopes even offer a different type of electron gun altogether, in which electrons are extracted by applying an electric field instead of by heating the filament. The two effects we have been discussing—the energy spread of the electron beam and the angular spread of this beam when it strikes the specimen—are collectively known as the *coherence* of the incident beam. We try to illuminate the specimen as coherently as possible, that is, with electrons that all travel at virtually the same speed and appear to come from a very small source.

We have been primarily concerned in the foregoing discussion with bright-field microscopy. The reason for this is that under dark-field conditions, in which the unscattered electrons are intercepted before they reach the image plane, the relation between detail seen in the image and structures in the object becomes very complicated at very high resolution unless the specimen is a very weak scatterer (Hanszen and Ade, 1974, 1975). Several procedures using dark-field micrographs are described in Section III, but a bright-field image may be needed as well, however.

In conclusion, we mention a problem of image interpretation, which raises a question that may be soluble more easily by image processing than in any other way. A beam of electrons forms an ordinary electric current, which in an electron microscope is of the order of 10–100 μA. The contrast seen on the fluorescent screen or electron micrograph is thus a record of the current of electrons striking the screen or photographic plate. For coarse detail, which we referred to as amplitude contrast earlier, the current at the image gives a direct indication of the presence of heavy stain in the specimen and hence gives a quite faithful representation of the stained parts of the specimen. For fine detail the relation between the current and hence the contrast in the image and the deviations experienced by the electrons in the specimen combined with the effort of spherical aberration and defocus is much less straightforward. There is, however, a direct and simple reaction between the wave function at the specimen and that at the image. The wave function ψ is related to the electron current in a very simple way. The latter is directly proportional to $\psi\psi^*$, where ψ^* denotes the complex conjugate of ψ. This relation may be pictured as follows. Differences in electron current simply produce variations in brightness on the fluorescent screen. The wave function is characterized by a length, known as its modulus or *amplitude*, and an angle, known as its argument or *phase*. The electron current gives a direct measure of the amplitude of the wave function, but gives no information about its phase. At the specimen, however, the deviations of the beam electrons are represented by variations in the phase of the wave function, and we should therefore like to be able to deduce not only the amplitude variations but also the phase variations across the image plane. We could then make deductions about the atoms of which the specimen was composed, and hence obtain a reliable picture of the structure. Although this particular problem can be solved by electron holography using a biprism, which we mention briefly in Section III,F, the experimental difficulties are formidable and computer image processing may well offer the easier solution.

III. Computer Image Processing

A. Equipment

Before discussing procedures for improving electron micrographs and extracting information that is not immediately recognizable on the print, we describe the apparatus necessary to convert the micrograph into a form suitable for the computer. The photographic plate or film gives a nearly continuous record of incident electron current; this must be divided into a mosaic of small squares (or circles), and the average opacity of the plate must be measured for each square. These squares are typically of the order of 25×25 μm, 50×50 μm, or 100×100 μm in size, and the opacity or "grey level" at each point is divided into 256 steps, numbered 0 to 255, one extreme representing a completely transparent area, and the other, an extremely black zone. This measurement is clearly a delicate task, and specialized microdensitometers are available for it. In some models (the Optronics Photoscan and the Joyce-Loebl Scandig), the micrograph must be recorded on film, which is wrapped round a drum. This drum spins at an accurately controlled speed and, at the end of each revolution, it moves sideways one step (typically 25, 50, or 100 μm). Flat-bed microdensitometers, with which both plates and film can be measured, are also available, and the more refined models offer much finer scanning and a free choice of scanning pattern; these are of course much more expensive than drum scanners and are not seriously slower. Grey level measurements are taken automatically over a region predetermined by the user, and the measurement, a number between 0 and 255, is written directly onto a magnetic tape (like that of a tape recorder). Alternatively, the microdensitometer may be connected directly to a computer, and the measurements stored in the computer memory. In either case the electron micrograph has been coded into a form that can be handled easily by the computer. Modifications are then made to this "digitized" image, which are the subject of the remainder of this section. After all the modifications have been made, some means of looking at the result must be found. The most satisfactory way is to use a device that acts like a microdensitometer in reverse (e.g., the Joyce-Loebl Film-Write or the Optronics Photowrite.) Given a picture coded as numbers on a magnetic tape, or stored in a computer memory, it shines a light on unexposed film with the appropriate intensity, thus "writing" the picture on film. The resulting picture should be visually of at least the same quality as the original micrograph and should of course be in some way better or more readily interpretable, depending on the type of processing that has been attempted.

So far as the computer is concerned, we need only say that, the larger and faster the machine, the more suitable it is. It is, however, extremely convenient if a computer of modest size (and cost) can be made permanently available; in this way, a great deal of time can be saved by preprocessing the digitized micrographs before attempting the more ambitious procedures which require a large computer. For example, it is always wise to measure a slightly larger area of the micrograph than the part actually of interest, and this area can then be trimmed exactly in the small computer; likewise, it is prudent to inspect the digitized form of the picture written on magnetic tape to ensure that it is suitable for further processing, for there are numerous potential sources of trouble against which we must continually take precautions. Provided that the small computer is equipped with a fluorescent screen, or visual-display unit, the digitized picture can be inspected line by line. Moreover, modern minicomputers are so fast and flexible that there are few image processing tasks for which a large computer is indispensable.

This procedure, in which a micrograph is recorded in the microscope on film, which has then to be developed and digitized before it can be introduced into the computer, is very circuitous. It is the only practicable arrangement at the present time but, as each stage is a potential (and usually actual) source of error, and the cost of the equipment is considerable, it would be much more satisfactory if we could somehow put the electron image directly onto magnetic tape in digital form without first recording it on film. This is routinely not possible at the present time, but it seems likely that suitable equipment will become commercially available in the near future. We therefore hope that the microdensitometry stage described above will soon be unnecessary.

B. Spatial Frequency Analysis

In the following sections we describe some of the different procedures being studied for improving electron micrographs or extracting reliable information from doubtful pictures. In order to understand how these procedures work, we first give a qualitative account of the theory on which they are founded.

As mentioned in Section II, there is in general no simple relation between the contrast seen on the image and the distribution of phase at the specimen, which would provide information about the atomic composition of the latter. We shall, however, see that there is a class of interesting specimens for which a simple relation does exist, and much of the present work on developing image processing is applicable only to this class of specimens; these are known as weakly scattering objects or weak-phase and weak-amplitude objects. In what follows we assume that the illumination is perfectly coherent, which in the simplest case means that all the electrons that strike the specimen have the same velocity and are traveling exactly parallel to the axis of the microscope column; we assume too that the nature and thickness of the specimen are such that only very few electrons are deflected more than once in their passage through the specimen, that is, that plural and multiple scattering are negligible. Under these circumstances there is a reasonably simple relation between the wave function at the object and that at the image for all specimens. This relation tells us how the known properties of the microscope—the spherical aberration coefficient and objective focal length (which are given in the manufacturer's brochure) and the defocus—combine with the properties of the specimen, represented by the (unknown) object wave function, to produce the (unknown) image wave function. If we knew the image wave function, we could deduce the object wave function (a computer routine, known as the Gerchberg-Saxton algorithm, can provide the image wave function in a wide range of cases, as we shall see). Alternatively, if we can show that, in some cases at least, there is a simple relation between image *intensity* (which tells us the amplitude of the image wave function), which we do know, and object wave function, then we can deduce the latter; this is the case studied in the weak-scattering approximation.

Although we have stated that the relation between the image and object wave functions is simple, its mathematical form is one that can be made even simpler by applying a transformation that performs a kind of frequency analysis. This is the Fourier transformation and is most easily understood by an example. If white light is shone on a prism, in a spectrometer, say, a pattern of colors will be seen, some colors quite bright, others dim; thus if the original white light had a pinkish tinge, some of the reds will be very bright. Conversely, by combining a number of pure colors, we can synthesize any tone we wish. In the same way any musical chord may be produced by combining individual musical notes in the right way. Moreover, it can be shown that each tone and each chord can be synthesized in only one way. Applying these ideas to the pattern of light and dark regions on the micrograph, we find that we can break any given picture into a series of simple gratings, with different opacity, in such a way that, if these gratings are superposed correctly, the overall effect cannot be distinguished from the original picture; again, the gratings can be chosen in one way only to produce a faithful picture. The spacing of the bars on each grating will be different and, by analogy with the optical and electrical terminology, the number of bars per unit length is known as the spatial frequency. For electron micrographs the spacing of the bars of the gratings corresponds quite closely to the size of the detail of interest, which might fall in the range 2.5–250 Å (0.25–25 nm). The corresponding spatial frequency range is then 0.4–0.004 $Å^{-1}$ (4–0.04 nm^{-1}). [It has been suggested (Hawkes, 1973) that the unit of spatial frequency be named after P. M. Duffieux, to whom we owe the fundamental ideas on this subject. The range in this example would then become 4–0.04 GDf. (Just as 1 Hz = 1 s^{-1}, so 1 Df would equal 1 m^{-1}.)] It is not easy to grasp the reason why the concept of spatial frequency plays such a central role in the theory of image formation without some familiarity with the idea of Fourier transformation or (spatial) frequency analysis. Nevertheless, it is quite reasonable to picture the spatial frequency as the reciprocal of the spacing of a regularly repeated feature of the specimen (for features 5 Å apart, the frequency of occurrence is once per 5 Å or 0.2 $Å^{-1}$). Even if the feature is not regularly repeated, the main spatial frequency components will be around the reciprocal spacing.

If we apply this transformation to the relation between object wave function and image wave function, we obtain the result on which almost all image processing ideas depend. The object wave function can be represented by a spatial frequency spectrum. The electron microscope *distorts* this spectrum by weakening some of the components, and even *cancels* some altogether. The electron microscope is said to act as a filter, and is characterized by a transfer function, that is, by a graph or table, which tells us how much of each spatial frequency reaches the image plane. If we knew the wave function at the image plane, we could calculate its spatial frequency spectrum, and reconstruct that at the specimen by reversing the filtering effect of the electron microscope. Intensities that had been halved by the microscope would be doubled, and so on. We could then transform the true spatial frequency spectrum at the object back into the "real" object.

Such a procedure would have two weaknesses, however. First, we could not reconstitute the frequencies that had been removed altogether. Second, we should have to be able to measure any frequencies that had been severely filtered extremely accurately, because they would have to be multiplied by a large number in the reconstruction; any error in the image wave function would be amplified, and falsify our reconstruction of the object. Such errors are commonly referred to as noise, and we see that we must take special precautions if the ratio of signal to noise becomes low. In addition, as we have already observed, the image wave function cannot be obtained directly.

We have already mentioned the importance of highly coherent illumination. The reason for this lies in the effect of poor coherence on the transfer function of the electron microscope. If the electrons have a spread of velocities, the transfer function will be slightly different for each, so that on the average, the spatial frequencies corresponding to high resolution will be severely weakened, and we shall be unable to distinguish true values from mere errors of measurement. A very similar effect occurs if the electrons strike the specimen at a range of angles so that, unless the coherence is very high, we lose high-resolution detail irretrievably.

C. Weakly Scattering Objects

We have repeatedly stressed that there is in general no simple relation between the contrast of the electron micrograph and the wave function at the object. For specimens that have only a very small effect on the incident beam, however, it can be shown that, to a good approximation, a simple relation does exist. If we perform a Fourier transform, or spatial frequency analysis, of the contrast of the recorded micrograph, the resulting spectrum will be the sum of two spectra arising from the specimen, each modified by a different filter due to the electron microscope. The two spectra arising from the specimen may be pictured as coming from the amplitude of the wave function and from its phase. Each spectrum is distorted, as described in the preceding section, by a filter, which is different for the two cases, and the distorted spectra add to form the image. In practice it may often be quite legitimate to assume that only the phase of the wave function at the object has been affected, so that the image contrast spectrum is produced by only one object spectrum, suitably filtered; this makes the problem very much simpler. The validity of this approximation, for (negatively stained) biological material, has been examined by Grinton and Cowley (1971).

Unfortunately, the effect of the microscope is such that some object spatial frequencies never reach the image at all. Various schemes for remedying this have been proposed, all based on the same principle. If we take two (or more) micrographs of the same specimen area under different conditions, the filtering effect of the microscope will be different in each case, and it should be possible to pool this information to yield a complete corrected image. The simplest way of varying the conditions is to take the two micrographs at different values of defocus. The way in which the two (or more) sets of results (images) are combined depends to some extent on any foreknowledge we may have about the errors of measurement and recording, and very general procedures for exploiting this are available. These procedures yield a pair of filters by which the spatial frequency spectrum of the recorded image is multiplied to give the two spectra associated with the object separately; each of these object spectra can then be transformed back into a picture of the object from which the location and, to a limited extent, the type of atoms comprising it can be identified.

D. General Objects

Before we can remove artifacts or otherwise improve our images, we must devise some means of obtaining the phase of the wave function at the image, even though only the amplitude is measurable. Clearly, since we know one set of quantities and require two sets, we are likely to need a second micrograph of the specimen different in some respect from the first but related to it in a known way. The second micrograph can be chosen in various ways. It may be an ordinary micrograph taken at a different value of defocus, or may be a diffraction pattern taken from the same specimen area as the original image. We now consider the latter alternative.

The contrast of the electron micrograph tells us the amplitude of the wave function at the image, but not its phase; likewise, the electron diffraction pattern tells us the amplitude of the corresponding wave function but not its phase. It is, however, known that these wave functions are related by the Fourier transform. One is the spatial frequency spectrum of the other. A procedure has been devised for exploiting this relationship, which operates as follows. The wave function at the image is guessed by ascribing an arbitrary phase to the (known) amplitude at each point; the Fourier transform of this is then taken, and this gives us amplitude *and* phase at every point of the diffraction pattern. Since the phases were guessed at random, these amplitudes and phases are wrong but since we know the correct amplitudes, the wrong values are replaced by the correct ones. We then transform back to the image and again replace the (now wrong) amplitudes by the measured values. By continuing in this way, it is found that the corrections necessary become progressively smaller and, when they are negligibly small, we have obtained the phase distributions across the image and diffraction pattern that correspond to the measured intensities. It can be shown that this process always converges to an answer, but it is not yet certain whether it always converges to the right answer. This problem is being actively studied at the present time, and ways of avoiding wrong solutions should emerge.

The procedure described above is known as the Gerchberg-Saxton algorithm, after its inventors (Gerchberg and Saxton, 1972, 1973a; Gerchberg, 1972). The alternative procedure, in which two micrographs taken at different values of defocus are used, operates on essentially the same principle (Gerchberg and Saxton, 1973a,b; Misell, 1973).

There is no way of extracting the phase and amplitude of the object wave function from ordinary bright-field micrographs without iterative computer processing, that is, without using a method of successive approximations. It may, however, become possible to do this by electron holography, with or without assistance from the computer. This requires considerable modification of the electron microscope, but the small amount of experimental evidence presently available is quite promising. In the electron microscope part of the incident electron beam travels through the specimen, while part passes next to it and a hologram is formed below the objective lens. This hologram is essentially an interference pattern formed where the two parts of the incident beam overlap. A suitable region of overlap is created by means of an electron biprism, the behavior of which need not concern us here. The hologram is recorded on a plate or film in the usual way and must then be processed to yield phase and amplitude pictures separately. In the experimental demonstration of the method (Wahl, 1974), this processing was done on an optical bench, but it seems likely that the increased flexibility of computer processing offsets the disadvantage of not obtaining a high-quality image instantly. The same is true of a different holographic technique, which has been explored by Tonomura and Watanabe [Tonomura *et al.*, 1968a,b; a paper in Japanese (Tonomura and Watanabe, 1968) gives a better demonstration of the effect, and the relevant figure is reproduced in Hanszen, (1973)] and by Hanszen (1974).

E. Computer Manipulation of Digitized Images

The range of really different tasks that the computer is required to perform is quite small. Furthermore, if computer processing is to become a routine operation, it is obviously desirable that microscopists with little or no computing expertise be able to use it. For this reason several attempts have been made to compose a set of instructions in everyday language, which will cause the computer to perform the various complicated tasks involved in each type of processing. The most successful such language to date is Improc (Saxton, 1974), a very versatile scheme which can be operated with almost no prior knowledge of computers but into which more skilled programmers can insert additional (Fortran) commands at will. Among the most useful words in this language are those that enable us to select common areas from two or more micrographs of the same specimen taken under different conditions: at different values of defocus, for example, or even with different exposures, hence damaged by the beam to a greater or lesser extent. Once approximately common regions have been extracted, they must be rotated and centered on the same point so that they exactly coincide; this is comparatively difficult, but in the Improc language all the difficulties are concealed behind a set of simple commands. "Extract," for example, extracts portions of pictures at arbitrary positions and orientations, and "Orient" determines the rotation and translation needed to match two pictures. As we have seen in previous sections, we are now likely to want the spatial frequency spectrum of the pictures, and this too is obtained by a simple command ("FFT"). At any stage of the process, we may feel that it is a wise precaution to inspect the picture to ensure that the feature we are interested in has not been accidentally trimmed, or that the details of interest have survived the rather drastic process of microdensitometry. A crude representation of the picture can be obtained by typing characters on computer output paper by means of the command "Sketch."

All these rather dull but essential operations precede the real image processing. It is too soon to say which of the various procedures discussed here will prove to be most important, and new techniques will certainly emerge in the next few years. A great advantage of a language such as Improc is the ease with which a new procedure can be incorporated and reduced to a brief and almost self-explanatory command.

A particularly important parameter that must be determined as accurately as possible is the defocus at which a micrograph has been taken. How this is done can be understood by considering the image of a thin, amorphous specimen, of carbon, for example. The image contains no detail of any interest, but the spatial frequency spectrum of this featureless image consists of a set of rings representing directly the phase-contrast transfer function of the microscope. By measuring this and matching the measurements to the form this function is known to follow, the defocus can be deduced (e.g., Frank *et al.*, 1970). A very ingenious procedure in which a tilted specimen is used to sample the transfer function effectively over a range of values of defocus has been developed by Krakow *et al.* (1974).

F. Procedures Requiring Unconventional Electron Microscope Techniques

In the foregoing account we have suggested how information that was invisible or uncertain on the original print could be extracted from bright-field electron micrographs taken in the usual way. It is clear that this is a very complicated process, and we might therefore ask whether or not it can be simplified by some modification of the electron microscope or of the techniques used in producing the image. It seems very reasonable that the types of processing most likely to be fruitful should be a compromise between the difficulties of modifying electron microscopes and the hazards of computer processing. We restrict this account to instrumental modifications designed to be used in conjunction with digital processing; modifications that have been suggested in order to avoid the need for computer processing are not discussed here.

1. *Modifications of the Illumination System*

We explained in Section II that, for successful processing, it is necessary to illuminate the specimen as coherently as possible. It is usually tacitly assumed that it is also desirable to illuminate the object as uniformly as possible, so that the same electron current falls on each point of the specimen area that subsequently contributes to the image. We might, however, enquire whether or not we can make the illumination nonuniform in such a way that interpretation of the image will become easier. For example, the electrons might fall on the specimen mainly along stripes or rings instead of in a uniform stream and, if the specimen exhibits some similar structures, it might be easier to detect it. Such an idea has been suggested by Hoppe (Hoppe, 1969; Hoppe and Strube, 1969), who named the technique ptychography, but in the general case it has proved extremely difficult to operate in practice (Hegerl and Hoppe, 1972). It is nonetheless worth repeating that most computer processing techniques are based on the idea that at least two micrographs of a given specimen area are obtained, some well-understood change being made in the microscope operating conditions between exposures. As we have seen, the defocus may be varied or image and diffraction pattern may be used, and other ways of altering the conditions are described below. Alteration of the illumination offers another possibility, which has been comparatively little studied so far.

A quite different type of illumination, which has been studied in considerable detail, entails irradiation with a hollow conical beam. This is a major departure from the usual type of illumination in which a small circular disc of the specimen is irradiated with electrons; with hollow conical illumination the electrons strike the specimen around a ring, except in the focused case in which the apex of the cone coincides with the specimen, and the incident beam is shaped like the wall of a funnel. The unscattered beam is now cone-shaped, and a bright-field image will be seen if this cone can pass through the objective aperture, a dark-field one if it cannot. The image detail is created by the electrons deflected toward the axis of the microscope by the atoms in the specimen; some contrast is also due to electrons scattered away from the cone and intercepted by the objective aperture. It appears that bright-field conical illumination will be most useful for very low-voltage microscopy. The advantages of the technique are the high electron current density at the specimen, for a given gun current, and the fact that, if the cone is tilted, all spatial frequencies that were present in the specimen reach the image—none is canceled completely. This should considerably simplify reconstruction of weakly scattering objects if the very weak phase contrast produced at normal accelerating voltages can be tolerated.

In dark-field operation the unscattered electrons of the incident cone are intercepted by the objective aperture, but the illumination remains axially symmetric, unlike most dark-field schemes. The advantages of this have been demonstrated by Thon and Willasch (1972a), who obtained images of the individual mercury atoms disposed in triangles in triacetoxy-mercury aurine (TAMA).

2. *Modifications of the Aperture (Bright-Field Case)*

We have already mentioned the possibility of introducing phase plates into the electron microscope; such a plate would be inserted into the back focal plane of the objective, where the objective aperture is normally situated, and would consist of rings of transparent material of varying thickness, carefully controlled. Only a few results have so far been described (Thon and Willasch, 1972b), and it is not at present intended that the plates be used together with computer processing. For the same reason we do not discuss further the type of phase-shifting device in which a fine conducting thread is stretched across the objective aperture.

If, however, the conventional circular aperture centered on the axis of the electron microscope is replaced by a semicircular aperture, with a small nick in the center to allow the main beam through, and if two pictures are taken, in which the half of the aperture that is obscured in the first case is open in the second, we obtain information about the specimen of a different kind from that recorded with a circular aperture. If the specimen is a weakly scattering object, then these two pictures taken with complementary half-apertures can yield both the amplitude and the phase of the specimen wave function. We have already seen that two micrographs taken with different defocus can also provide this information and, just as in that case, we can derive optimum filters which permit us to deduce the most probable values of amplitude and phase when errors of measurement are taken into account.

It is, however, more difficult in practice to use complementary half-apertures than to alter the focus, and the real advantage of the former is apparent only when the object scatters strongly. It is no longer possible to extract the amplitude and phase of the object wave function by suitable combination and filtering of the two images, but it is possible to devise an iterative scheme to do this (Misell *et al.*, 1974a,b). Nevertheless, the method will probably remain quite difficult to apply in practice, because the electrons halted by the aperture close to the diametral edge repel electrons passing close to the edge in the open half of the aperture, hence produce an undesirable deflection or phase shift; the extent of this can, however, be established and can then be allowed for (Downing and Siegel, 1973).

As semicircular apertures would have to be specially made, we might enquire whether or not an ordinary circular aperture could be used off-center. Without going into details, we state that much useful information could probably be collected either with this simpler arrangement or by using a centered aperture and tilted illumination.

3. *Modifications of the Aperture (Dark-Field Case)*

For the reasons given in Section II, we have said little about dark-field microscopy. Although it is not difficult to derive a formula relating the recorded image intensity and the object wave function, this shows that at high resolution many kinds of artifacts can be created and there is no straightforward way of distinguishing between true and false detail in the image. There is no doubt that high-resolution detail in dark-field images should be treated with circumspection. We have seen in the preceding paragraphs that many types of computer image processing require at least two micrographs taken under different conditions, and we might therefore enquire whether or not a bright-field and a dark-field picture of the same specimen area can be used. This proves to be possible in principle, and the processing involved is very simple, although no practical tests have yet been made, mainly because the dark-field image must be made with a central beam stop. Normally, dark-field images are obtained by tilting the electron beam so that it strikes the specimen obliquely and the unscattered electrons hit the objective aperture instead of traveling along the axis of the microscope. For the present application, however, a stop must be placed on the axis of the microscope in the middle of the objective aperture. This can be done crudely by stretching a thread across the aperture, but the charge that collects on this may cause problems.

The method relies on the fact that the intensity distribution seen in a bright-field image is the sum of a uniform bright background (unscattered beam), a variation dictated by both the phase and amplitude of the object wave function and (for strongly scattering objects) a term involving the square of the amplitude only. The dark-field image consists only of the last of these three, so that the central term involving phase and amplitude can be obtained by subtracting the dark-field intensity distribution from the bright-field image. The phase can then be extracted immediately, and the amplitude is known from the dark-field image (Frank, 1973).

Another interesting proposal involving dark-field images may help to remove the undesirable background produced by inelastically scattered electrons. Calculations show that there is a value of defocus that produces the sharpest image of a small detail on a uniform thin substrate if only elastically scattered electrons are considered; at this value of defocus, the other electrons produce a blurred background which is very insensitive to changes in focus. The "elastic image," however, becomes rapidly blurred as the focus is altered. By taking two micrographs, one at the sharp focus and one out of focus, and subtracting them, we should therefore heighten the contrast of the sharp detail by removing some of the diffuse background. Krakow (1974) has attempted this, using photographic rather than computer subtraction, and obtained some evidence of improvement.

G. Radiation Damage Assessment

A serious problem in the electron microscopy of delicate specimens is the extent to which the latter are damaged by the electron beam in the process of image formation. Experiments in which specimens are subjected to the minimum possible exposure have demonstrated beyond all doubt that much fine structure may be destroyed in the period during which the microscopist is focusing on an area of interest, before taking the micrograph. It seems unlikely that techniques for obtaining images of every specimen before it is damaged will emerge, in the near future at any rate. It has therefore been suggested that the computer could be used to help measure the extent of the damage and to follow its progress as the exposure is increased; it might even be possible to extrapolate some way back toward the undamaged object structure.

If a quantitative measure of the damage is to be obtained, an image must obviously be available of the perfect, or nearly undamaged, specimen. Such an image could be obtained only with a very low-electron exposure indeed, for specimens liable to damage, and the contrast of such an image would in all likelihood be so low as to make it undiscernible. The damage measure that has been proposed (Frank, 1974) takes this into account by averaging over the whole illuminated specimen area. The "measure of dissimilarity" μ is defined to be the sum of the squared difference between the object wave functions with and "without" damage at all points on the specimen, divided by the sum of the square of the wave function at all points (Strictly speaking, the square of the modulus of the difference, and so on). As we require, the dissimilarity will be zero if there is no evidence of damage, and it has been arranged that $\mu = 1$ if the damaged specimen bears no resemblance to the undamaged one (maximum dissimilarity). Since we are averaging over the whole specimen area, the dissimilarity μ can be obtained from extremely weak pictures.

Unfortunately, a measure such as μ gives no information about the type of damage, in the sense that it does not distinguish between damage to the coarse structure and that to the fine (high-resolution) detail. At the other extreme we could write μ as a sum over individual components of the Fourier transform (spatial frequency spectrum) of the difference between the wave functions before and after damage and consider these various components. However, the information then available would be far too detailed to be assimilated. The best solution appears to be a compromise between these two extremes, which depends on the fact that coarse detail produces effects close to the center of the spatial frequency spectrum, while fine detail is represented by effects nearer the edge. By averaging the components around narrow concentric circular rings, we obtain a curve showing how the dissimilarity varies with radius. If a series of micrographs of the same specimen is taken with increasing exposure, a set of such curves will be obtained; the region corresponding to the central areas of the spectra will remain fairly constant, while the outer region will alter as the fine detail of the specimen is modified or destroyed by the electron beam.

IV. Concluding Remarks

The present aim of computer processing of electron images may be summarized simply as an attempt to recover some of the information about the specimen that is conveyed by the electron beam into the image plane of the microscope but cannot readily be recognized in the micrograph. This latent information may have been obscured for any of a variety of reasons: the difficulty in obtaining reliable phase information from an intensity record; the fact that some detail is lost in the process of microscope image formation; radiation damage; the difficulty of distinguishing true detail from artifact in a very low-contrast image—this list certainly does not exhaust all the possibilities.

Many ways of interpreting the evidence provided by electron microscopes that would be impossible without computer processing have been proposed, and most of them are described in this article. We may reasonably anticipate that techniques will be developed during the next few years that will permit the microscopist to invoke the help of computer processing routinely. One major question will still remain to be answered, however: What can be deduced about the chemical structure of the specimen from pictures showing the amplitude and phase of the wave function at the specimen, even if we are confident that these pictures are true representations (in some sense) of the undamaged object? Much is already known about the dependence of electron scattering on the atomic weight of the scattering atoms, and it is possible to calculate the image that would be obtained from a given model structure. Proceeding from an image to a detailed statement about the composition of the specimen will always be a very much more difficult task, however. The problem of distinguishing the true specimen from the supporting film has been studied and an image difference technique has been devised and tested by Hoppe's group (Feltynowski *et al.*, 1972). The problems of high-resolution structure research have been discussed in some detail by Hoppe (1970 and general references [14, 15]; Hoppe *et al.*, 1974); the possibility of using computer processing to discriminate between heavy and light atoms at least has been explored by Frank (1972).

Although biologists are likely to benefit the most from computer processing, much of the work on developing techniques is in the hands of physicists. We hope that this nontechnical account of the aims of processing and of its potential usefulness to the biologist will help to whet the interest—and to enlist the cooperation—of the cytologists that computer processing is designed to help.

General References[1]

[1] Beer, M., Frank, J., Hanszen, K.-J., Kellenberger, E., and Williams, R. C. (1974). The possibilities and prospects of obtaining high resolution information (below 30 Å) on biological material using the electron microscope. *Quart. Rev. Biophys.* **7**, 211–238.

[1a] Burge, R. E. (1973). Mechanisms of contrast and image formation of biological specimens in the transmission electron microscope, *J. Microsc. (Oxford)* **98**, 251–285.

[2] Erickson, H. P., and Klug, A. (1970). The Fourier transform of an electron micrograph: Effects of defocusing and aberrations, and implications for the use of underfocus contrast enhancement. *Ber. Bunsenges. Phys. Chem.* **74**, 1129–1137.

[3] Erickson, H. P., and Klug, A. (1971). Measurement and compensation of defocusing and aberrations by Fourier processing of electron micrographs. *Phil. Trans. Roy. Soc. London, Ser. B* **261**, 105–118.

[4] Frank, J. (1973). Computer processing of electron micrographs. *In* "Advanced Techniques in Biological Electron Microscopy" (J. K. Koehler, ed.), pp. 215–274. Springer-Verlag, Berlin and New York.

[5] Frank, J. (1973). Use of anomalous scattering for element discrimination. *In* "Image Processing and Computer-aided Design in Electron Optics" (P. W. Hawkes, ed.), pp. 196–211. Academic Press, New York.

[6] Frank, J., Bussler, P., Langer, R., and Hoppe, W. (1970). Einige Erfahrungen mit rechnerischen Analyse und Synthese von elektronenmikroskopischen Bildern hoher Auflösung. *Ber. Bunsenges. Phys. Chem.* **74**, 1105–1115.

[7] Gerchberg, R. W., and Saxton, W. O. (1973). Wave phase from image and diffraction plane pictures. *In* "Image Processing and Computer-aided Design in Electron Optics" (P. W. Hawkes, ed.), pp. 66–81. Academic Press, New York.

[8] Goodman, J. W. (1968). "Introduction to Fourier Optics." McGraw-Hill, New York.

[9] Hanszen, K.-J. (1971). The optical transfer theory of the electron microscope: fundamental principles and applications. *Advan. Opt. Electron Microsc.* **4**, 1–84.

[10] Hanszen, K.-J. (1973). Contrast-transfer and image processing. *In* "Image Processing and Computer-aided Design in Electron Optics" (P. W. Hawkes, ed.), pp. 16–53. Academic Press, New York.

[11] Hawkes, P. W. (1972). "Electron Optics and Electron Microscopy." Barnes & Noble, New York.

[1] Since most of the publications that have been consulted in preparing this article are of a largely mathematical nature, I have not thought it useful to include them. Instead, I have listed several general references, in which virtually all the fundamental papers on the subject are mentioned. Furthermore, if the work of a particular research group is well-covered in these general references, their papers may not be included at all in the main list, which is restricted to material that would otherwise be inadequately represented and to papers dealing with practical applications of the methods.

[12] Hawkes, P. W., ed. (1973). "Image Processing and Computer-aided Design in Electron Optics." Academic Press, New York.

[13] Hawkes, P. W. (1973). Introduction to electron optical transfer theory. *In* "Image Processing and Computer-aided Design in Electron Optics" (P. W. Hawkes, ed.), pp. 2–14. Academic Press, New York.

[14] Hoppe, W. (1970). Principles of structure analysis at high resolution using conventional electron microscopes and computers. *Ber. Bunsenges. Phys. Chem.* **74,** 1090–1100.

[15] Hoppe, W. (1971). Use of zone correction plates and other techniques for structure determination of aperiodic objects at atomic resolution using a conventional electron microscope. *Phil. Trans. Roy. Soc. London, Ser. B* **261,** 71–94.

[16] Hoppe, W. (1972). Recording, processing and correction of electron microscope images. *Proc. Eur. Cong. Electron Microsc., 5th, 1972* pp. 612–617.

[17] Hoppe, W., Möllenstedt, G., Perutz, M. F., and Ruska, E., eds. (1970). Methoden zur Untersuchung der atomaren Struktur von biogenen Makromolekülen. *Ber. Bunsenges. Phys. Chem.* **74,** 1089–1224.

[18] Hoppe, W., Bussler, P., Feltynowski, A., Hunsmann, N., and Hirt, A. (1973). Some experience with computerized reconstruction methods. *In* "Image Processing and Computer-aided Design in Electron Optics" (P. W. Hawkes, ed.), pp. 92–126. Academic Press, New York.

[19] Huxley, H. E., and Klug, A., eds. (1971). A discussion on new developments in electron microscopy with special emphasis on their application in biology. *Phil. Trans. Roy. Soc. London, Ser. B* **261,** 1–230.

[20] Lenz, F. (1971). Transfer of image information in the electron microscope. *In* "Electron Microscopy in Material Science" (U. Valdrè, ed.), pp. 540–569. Academic Press, New York.

[21] Menzel, E., Mirandé, W., and Weingärtner, I. (1973). "Fourier-Optik und Holographie". Springer-Verlag, Berlin and New York.

[22] Misell, D. L. (1973). Image formation in the electron microscope with particular reference to the defects in electron-optical images. *Advan. Electron. Electron Phys.* **32,** 63–191.

[23] Reimer, L., and Gilde, H. (1973). Scattering theory and image formation in the electron microscope. *In* "Image Processing and Computer-aided Design in Electron Optics (P. W. Hawkes, ed.), pp. 138–167. Academic Press, New York.

[24] Schiske, P. (1973). Image processing using additional statistical information about the object. *In* "Image Processing and Computer-aided Design in Electron Optics" (P. W. Hawkes, ed.), pp. 82–90. Academic Press, New York.

[25] Thon, F. (1971). Phase contrast electron microscopy. *In* "Electron Microscopy in Material Science" (U. Valdrè, ed.), pp. 570–625. Academic Press, New York.

[26] Zeitler, E. (1968). Resolution in electron microscopy. *Advan. Electron. Electron Phys.* **25,** 277–332.

REFERENCES

Crewe, A. V. (1970). *Quart. Rev. Biophys.* **3,** 137.

Downing, K. H., and Siegel, B. M. (1973). *Optik* **38,** 21.

Feltynowski, A., Bussler, P. H., and Hoppe, W. (1972). *Proc. Eur. Congr. Electron Microsc., 5th, 1972* p. 624.

Frank, J. (1972). *Biophys. J.* **12,** 484.

Frank, J. (1973). *Optik* **38**, 582.
Frank, J. (1974). *J. Phys. D* **7**, L75.
Frank, J., Bussler, P., Langer, R., and Hoppe, W. (1970). General reference [6].
Gerchberg, R. W. (1972). *Nature (London)* **240**, 404.
Gerchberg, R. W., and Saxton, W. O. (1972). *Optik* **35**, 237.
Gerchberg, R. W., and Saxton, W. O. (1973a). General reference [7].
Gerchberg, R. W., and Saxton, W. O. (1973b). *J. Phys. D* **6**, L31.
Gordon, R., and Herman, G. T. (1974). *Int. Rev. Cytol.* **38**, 111.
Grinton, G. R., and Cowley, J. M. (1971). *Optik* **34**, 221.
Hanszen, K.-J. (1973). General reference [10].
Hanszen, K.-J. (1974). *Optik* **39**, 520.
Hanszen, K.-J., and Ade, G. (1974). *Proc. Int. Congr. Electron Microsc., 8th 1974* **1**, 196.
Hanszen, K. J. and Ade, G. (1975). *Optik* **42**, 1.
Hawkes, P. W. (1973). *Appl. Opt.* **12**, 2537.
Hegerl, R., and Hoppe, W. (1972). *Proc. Eur. Congr. Electron Microsc., 5th, 1972* p. 628.
Hoppe, W. (1969). *Acta Crystallogr. Sect. A* **25**, 495 and 508.
Hoppe, W. (1970). *Acta Crystallogr., Sect. A* **26**, 414.
Hoppe, W., and Strube, G. (1969). *Acta Crystallogr. Sect. A* **25**, 502.
Hoppe, W., Gassmann, J., Hunsmann N., Schramm, H. J., and Sturm, M. (1974). *Hoppe-Seyler's Z. Physiol. Chem.* **355**, 1483.
Huxley, H. E., and Klug, A. (1971). General reference [19], especially pp. 173–230.
Krakow, W. (1974). *Proc. Electron Microsc. Soc. Amer.* **32**, 304; also see General reference [1].
Krakow, W., and Siegel, B. M. (1972). *Proc. Electron Microsc. Soc. Amer.* **30**, 618.
Krakow, W., Downing, K. H., and Siegel, B. M. (1974). *Optik* **40**, 1.
Misell, D. L. (1973). *J. Phys. D* **6**, L6, 2200, and 2217.
Misell, D. L., Burge, R. E., and Greenaway, A. H. (1974a). *J. Phys. D* **7**, L27.
Misell, D. L., Burge, R. E., and Greenaway, A. H. (1974b). *Nature (London)* **247**, 401.
Saxton, W. O. (1974). *Comput. Graphics Image Process.* **3**, 266.
Thomson, M G. R. (1973). *Optik* **39**, 15.
Thon, F. and Willasch, D. (1972a). *Optik* **39**, 15.
Thon, F., and Willasch, D. (1972b). *Proc. Eur. Congr. Electron Microsc., 5th,* 1972 p. 650.
Tonomura, A., and Watanabe, H. (1968). *Nihon Butsuri Gakkai-shi (Proc. Phys. Soc. Jap.)* **23**, 683.
Tonomura, A., Fukuhara, A., Watanabe, H., and Komoda, T. (1968a). *Jap. J. Appl. Phys.* **7**, 295.
Tonomura, A., Fukuhara, A., Watanabe, H., and Komoda, T. (1968b). *Electron Microsc., Proc. Eur. Reg. Conf., 4th, 1968* Vol. 1, p. 277.
Unwin, P. N. T. (1971). *Phil. Trans. Roy. Soc. London, Ser. B* **261**, 95.
Unwin, P. N. T. (1972). *Proc. Roy. Soc., Ser. A* **329**, 327.
Unwin, P. N. T. (1974). *Z. Naturforsch. A* **29**, 158.
Wahl, H. (1974). *Optik* **39**, 585.
Zeitler, E. (1975). Scanning transmission electron microscopy. *In* "Electron Microscopy and Microbeam Analysis" (B. M. Siegel, ed.). Wiley, New York.

Cyclic Changes in the Fine Structure of the Epithelial Cells of Human Endometrium

MILDRED GORDON

Department of Obstetrics and Gynecology,
Yale University School of Medicine,
New Haven, Connecticut

I. Introduction

The uterus is the portion of the female reproductive tract that capacitates or prepares sperm for fertilization and provides for implantation and growth of the embryo. It is pear-shaped, the corpus uteri corresponding to the rounded body of the organ, while the narrow neck extends into the vagina. The uterus is composed of an internal mucosa, the endometrium, a smooth muscle layer or myometrium, and an incomplete outer serosal cover. The endometrium comprises the luminal epithelial lining and underlying connective tissue stroma. Samples of tissue for study of the endometrium are commonly taken from the corpus.

The uterus is responsive to ovarian steroids which regulate its tissues in a cyclic fashion. Although the entire uterus is affected by ovarian activity, the most notable and reliable changes occur in the epithelial lining of the endometrium (Wynn and Harris, 1967). The epithelial cover of the uterine cavity dips into the stroma, forming uterine glands. The epithelium of the surface and glands are hence identical (Noyes *et al.*, 1950), but the cells on the surface undergo fewer cyclic changes than the glandular cells (Wynn and Harris, 1967; Dallenbach-Hellweg, 1971; Ferenczy *et al.*, 1972; Ferenczy and Richart, 1973). Examination of the glandular epithelium is of interest to clinicians, because the cellular morphology is an indicator of normal or disturbed ovarian function. Prior to ovulation, in the follicular or proliferative stage, the endometrium is dominated by estrogen from the ovarian follicle. Following ovulation, in the secretory or luteal phase, it is under the influence of progesterone secreted by the corpus luteum. On the light microscope level, the important distinction between proliferative and secretory endometrium is the appearance of vacuoles in the latter. They appear first at the basal pole of the cell, but later shift to the cell apex (Noyes *et al.*, 1950). The vacuoles represent accumulated deposits of glycogen, identified histochemically by diastase sensitivity and the periodic acid–Schiff (PAS) reaction (Schmidt-Matthiesen, 1963). Differentiation of secretory endometrium is diagnostic for ovulation.

The uterine epithelium therefore is a dynamic tissue, undergoing predictable cyclic changes as a result of hormonal input. While light microscopy provides an invaluable diagnostic tool for assessing the state of the ovary and the functional condition of the uterus, the refinements of ultrastructure hold the attractive prospect of correlating the effects of hormones on the differentiation of cellular organelles. It furnishes additional details which resolve a more complete picture of the epithelial cells. These provide clinical information, as in infertility (Gore and Gordon, 1974), and permit evaluation of the effects of contraceptive agents on the cellular level.

II. Background

The fundamental cyclic relationship between uterus and ovary was established on firm ground by Schroder in 1921. Following this clarification of the "menstrual cycle," temporal changes in endometrial morphology were thoroughly described by light microscopists (Novak and Everett, 1928; Bartelmez, 1933; Kotz and Parker, 1939; Hertig, 1945; Falconer, 1946; Moricard and DeSenarclens, 1947; Novak, 1947; Noyes *et al.*, 1950; Noyes and Haman, 1953). In a now classic study, Noyes and collaborators (1950) carried out the most comprehensive dating of the endometrium in reference to the ovarian cycle.

The first electron microscope studies of glandular epithelium were made in the later years of the 1950s, primarily on the continent (Lanzavecchia and Morano, 1958; Moricard, 1958; Wetzstein, 1958; Borell *et al.*, 1959; Cartier, 1959; Cartier and Moricard, 1959; DePalo and Stoppelli, 1959). During those years, and for a considerable period thereafter, it was not appreciated that there are inherent errors in the electron microscope technique, which derive from the extremely small sample of tissue under observation. As a result, the epithelium was categorized according to relatively few data. For example, precise measurements of a few organelles, such as mitochondria (Clyman, 1963c; Cavazos *et al.*, 1967), were considered prototypical for a specific stage. Inevitably, classifications of organelles were contradictory and varied from one study to another. The accumulated data emerging in the 1960s revealed that cellular components undergo progressive rather than abrupt modifications throughout the cycle (Wynn and Harris, 1967), and that sampling is complicated by polymorphy of the same cell population at a given stage (Armstrong *et al.*, 1973). Further, tissue from the same donor may not be identical at designated times in successive cycles.

The view of the epithelial cell as revealed by fine-structural examination has profoundly changed the original concepts of this tissue and raised interesting and intriguing questions on the relation of cell structure to function. Evidence on the cellular level for biosynthesis and extrusion of secretory granules during the proliferative phase has altered the traditional view of postovulatory endometrium as the "secretory" endometrium. The production of secretory material culminates at ovulation in uterine secretion of a thin, nutritive fluid (Datnow, 1973) which may be instrumental in the migration of sperm into the uterine cavity and their survival prior to fertilization (Delforge, 1969). The most dramatic disclosures of cellular morphology, however, were obtained on the postovulatory stages. Perhaps the most interesting was the discovery in 1960 (Dubrauszky and Pohlmann, 1960, 1961a), and again in 1963 (Clyman, 1963a), of a new and unique organelle in the nucleus, the nucleolar channel system (NCS), canalicular body, or nucleolar basket. It consists of a complex labyrinth of membrane-bound channels, with a precise arrangement of ribonucleoprotein (RNP) granules framing the periphery. The importance of progesterone for differentiation of this structure was established in organ culture of preovulatory endometrium (Kohorn *et al.*, 1970, 1972).

In the very first studies (Dubrauszky and Pohlmann, 1961b; Gompel, 1962; Nilsson, 1962a; Clyman, 1963a,b), notice was taken of the enormous size of mitochondria in the basal cytoplasm. Their distinctive size and consistent association with the rough endoplasmic reticulum that wraps their outer circumference differentiated them from mitochondria in the same location prior to ovulation, and from mitochondria in other regions of the cell. Their development coincides with glycogen deposition.

Improvement in fixation utilizing glutaraldehyde followed by osmium tetroxide was an important change, because it allowed optimal preservation of glycogen (Moricard and Moricard, 1964; Moricard, 1966) that had been previously solubilized or poorly preserved (Nilsson, 1962b). In addition to clarifying glycogen morphology (Revel *et al.*, 1960), it revealed that glycogen synthesis is initiated during the proliferative phase before it can be detected cytochemically by light microscopy (Gompel, 1962; Wynn and Harris, 1967).

Modifications in classification of the cycle have been suggested (Wynn and Woolley, 1967; Armstrong *et al.*, 1973), namely, that the period close to ovulation, characterized by the secretion of previously formed granules, the beginning of glycogen synthesis, and mitochondrial differentiation, is a distinct phase between proliferative and secretory stages.

Finally, better methods of tissue preparation and embedding procedures, as well as high-resolution microscopy, revealed the presence of finer elements, such as microtubules (Cavazos *et al.*, 1966, 1967), tonofilaments Cavazos *et al.*, 1966; Armstrong *et al.*, 1973), and annulate lamellae (Ancla *et al.*, 1964b; Ancla and de Brux, 1965).

III. Materials and Methods

Endometrial biopsies from the corpus uteri of women of reproductive age with normal 28-day cycles were processed for electron microscopy by placing in glutaraldehyde in Millonig's buffer (pH 7.3) at room temperature. Small blocks of tissue were embedded in Epon 812 or Spurr (Polysciences), sectioned on a Reichert ultramicrotome, stained with uranyl and lead salts, and examined in an Hitachi HU-12 electron microscope. Examination of 1-μm plastic sections permitted selection of the glandular epithelium and a general assessment of the stage of the cycle.

IV. Structure of Epithelial Cells

A few days before menses, in most cases by day 24, the epithelium abruptly contracts, the cells become shorter and more condensed (Plate I, Fig. 1) seemingly in preparation for disruption of the tissue. The reason for the change in the epithelium at this point is unknown (Witt, 1963). The picture during menses is a chaotic one. The epithelium contains cells that are still typical of the last stages of the cycle, and young cells of the new cycle (Delforge, 1969). By day 5, the

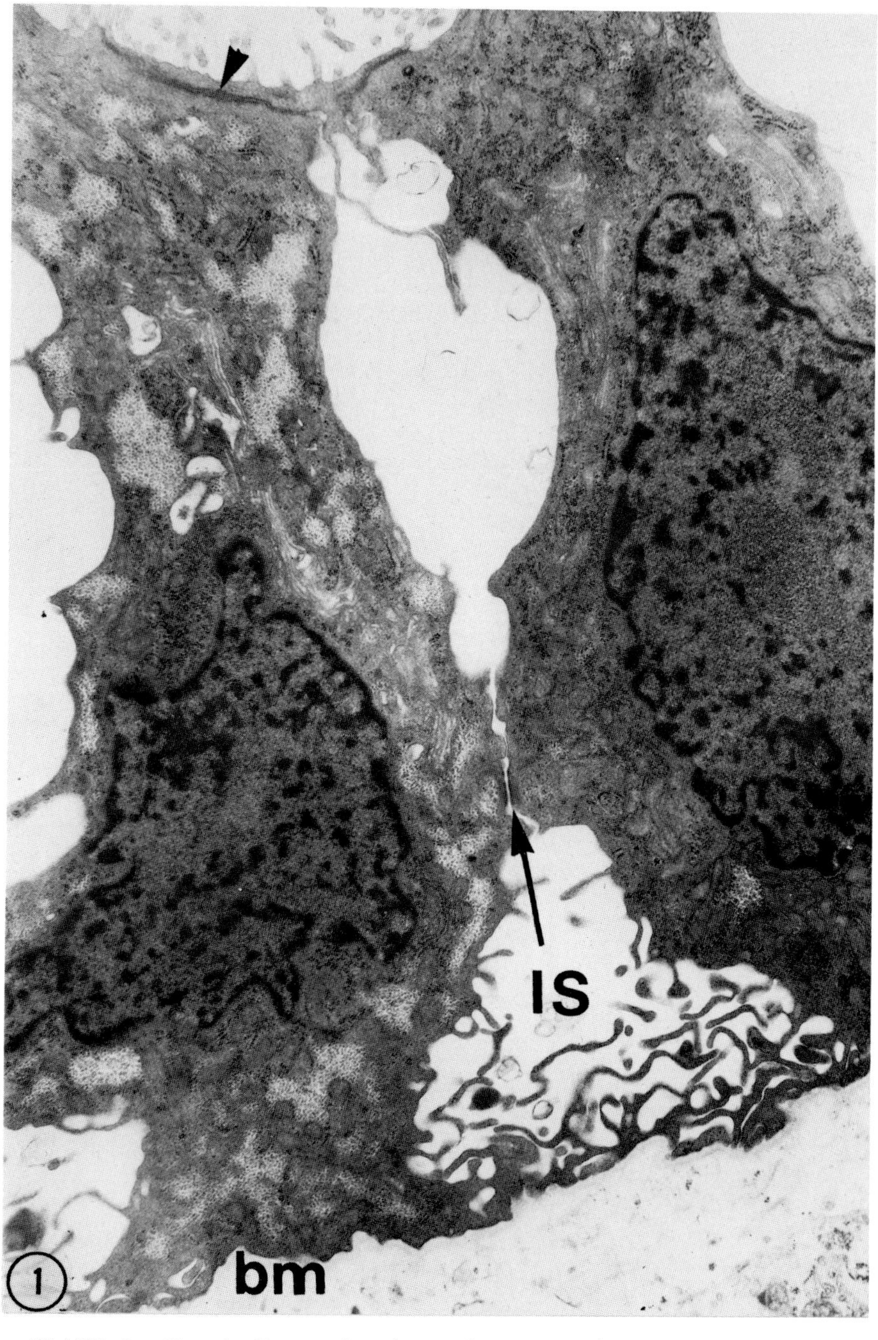

PLATE I. FIG. 1. Twenty-four-day endometrium showing changes prior to menses. Cells are shortened and condensed. Blunt microvilli are present on the luminal surface. The intercellular lacunae are greatly enlarged. Lateral infoldings at the basal portion of the cell are well visualized because of the enlarged intercellular space (IS). Cells are fastened by desmosomes (arrow) and junctional complex (arrowhead) at apex. The basal plasma membrane and basement membrane (bm) are convoluted. ×13,500.

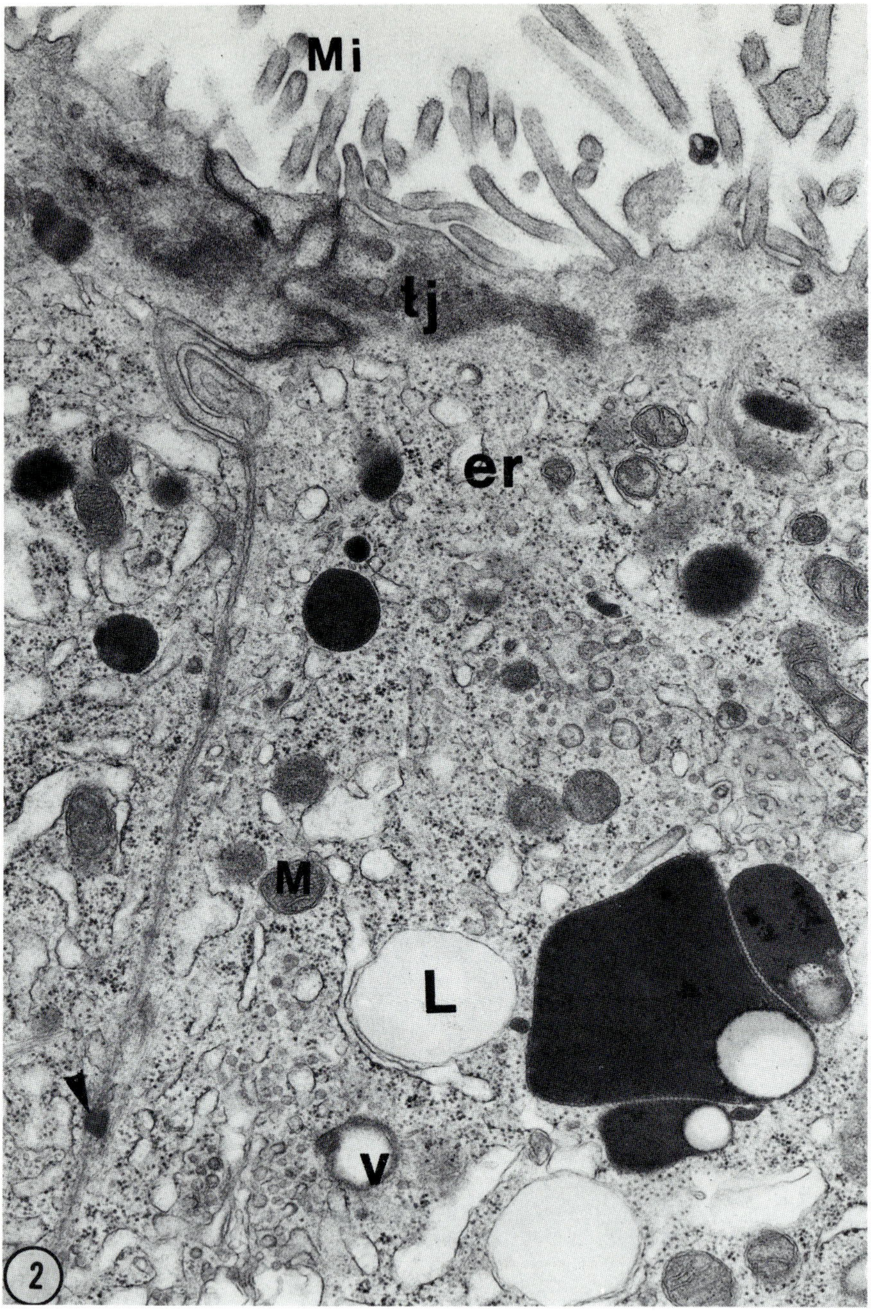

PLATE II. FIG. 2. Apical region of cell in early proliferative (6-day) endometrium. Microvilli (Mi) on luminal surface are blunt. The lateral surface shows a junctional complex (tj) grazed by plane of section, followed by plasmalemmal infolding. A desmosome (arrowhead) is seen further down. The cytoplasm is filled with free ribosomes, vesicular outlines of rough and smooth ER (er), small mitochondria with normal cristae (M), and various inclusion bodies consisting of dense granules, lipid droplets (L), pleomorphic granules, and coated vesicles (v). ×21,600.

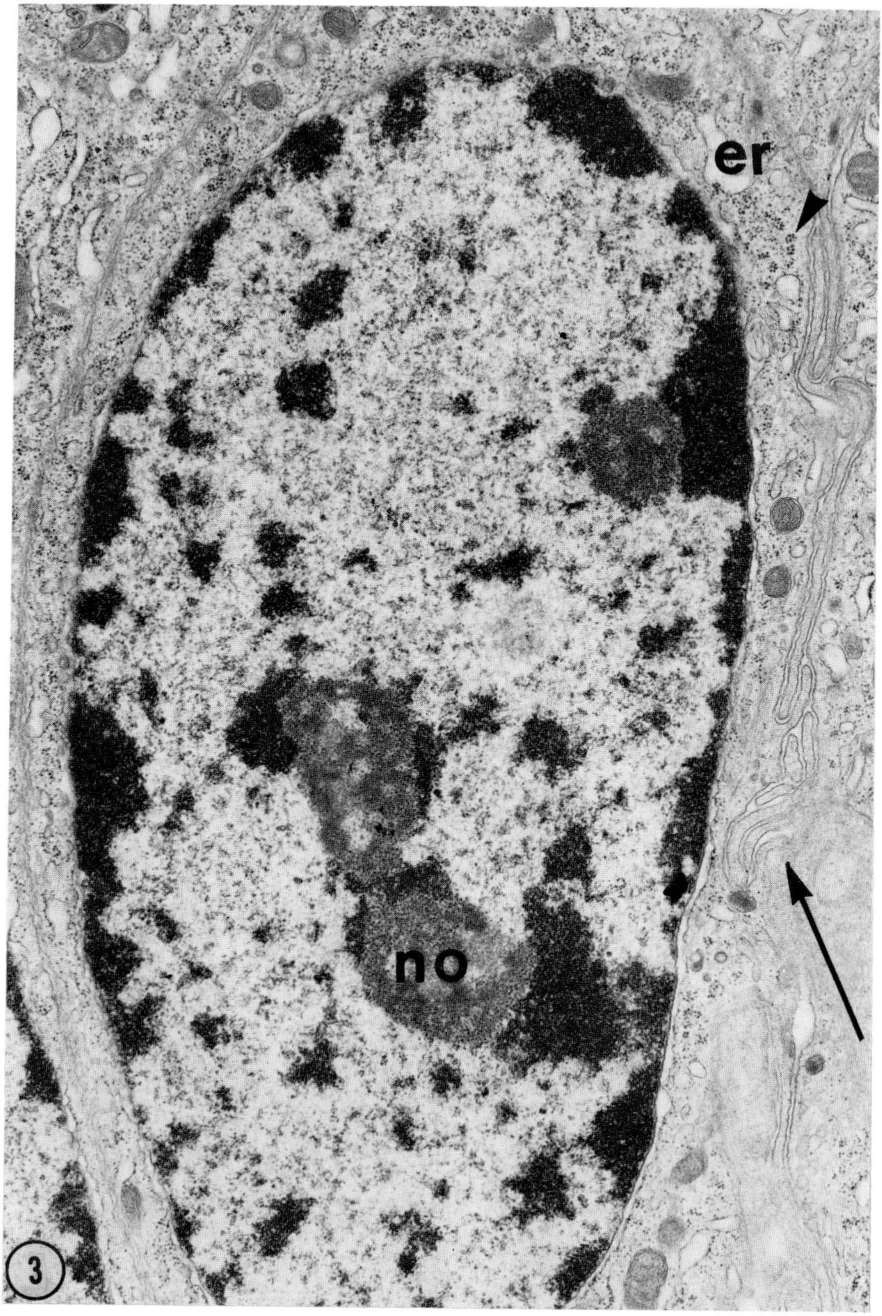

PLATE III. FIG. 3. Nucleus in 6-day endometrium. It is rounded and has smooth contours. The karyoplasm is clear and filled with fine fibrils and granules. Dense chromatin is intermittently distributed on the nuclear envelope, in the karyoplasm, and on the nucleolus. Typical nucleoli (no) are differentiated from the chromatin. The nuclear envelope hugs the surface closely. Cisterns of ER (er) are prominent in the cytoplasm which also contains free ribosomes (arrowhead) and small mitochondria. Lateral interdigitations of the membranes (arrow) are well developed. ×12,600.

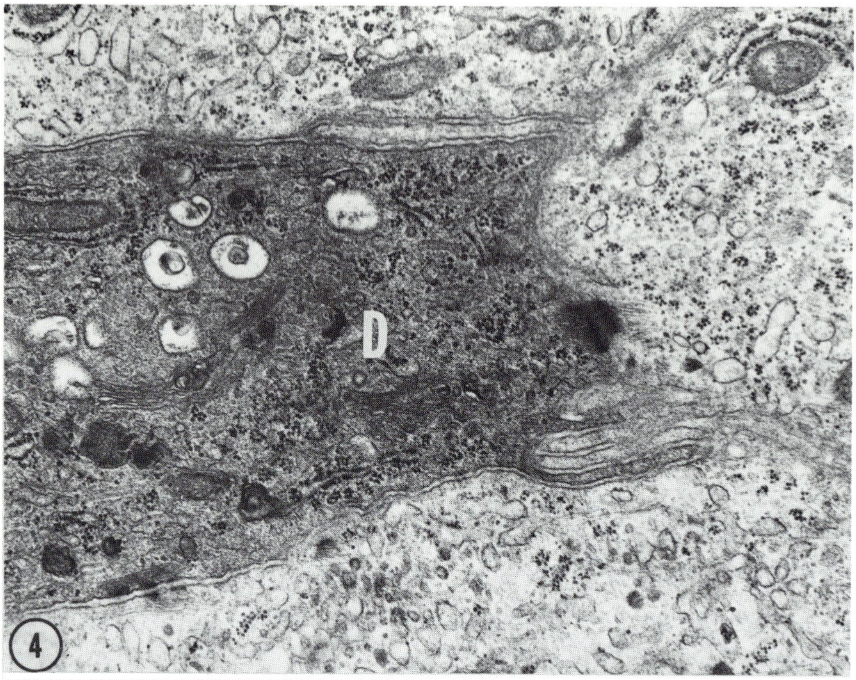

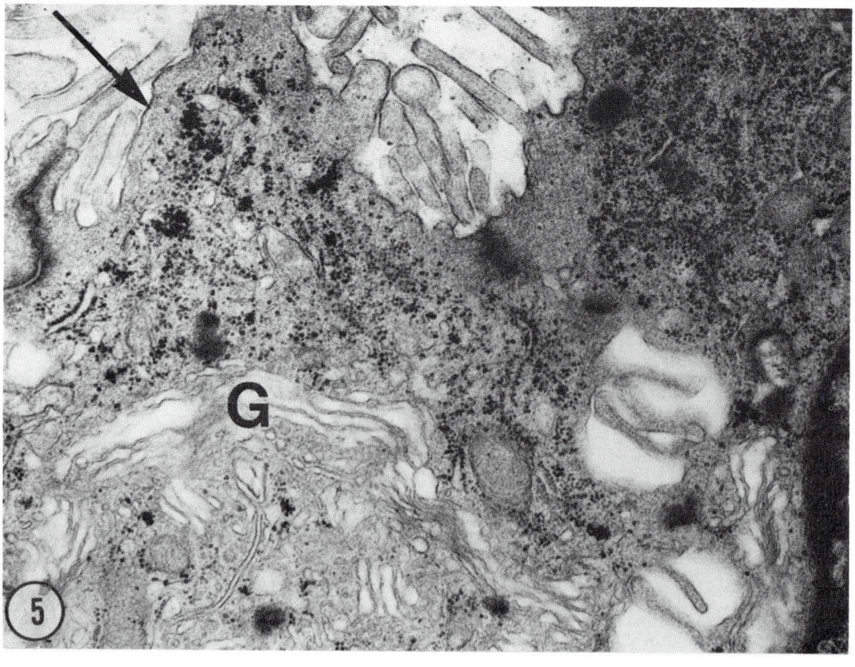

PLATE IV. FIG. 4. Dark cell (D) interposed between light cells in 15-day endometrium. ×19,000. FIG. 5. Twenty-two-day endometrium showing apical cytoplasmic extension (arrow), deposits of glycogen between rough ER, and a hypertrophied Golgi complex (G). ×20,000.

epithelium is reconstituted (Witt, 1963). The stages of the epithelium of interest therefore lie between day 5 or 6 and day 24 to 26.

The cells of the earliest epithelium are low columnar (Wynn and Harris, 1967) or cuboidal (Gompel, 1962). With growth and replication there is a gradual increase in height, noticeable by day 6 (Clyman, 1963c; Cavazos *et al.*, 1967). Migratory white blood cells are occasionally encountered between epithelial cells in the proliferative stages. The majority of the epithelial cells have a rather pale appearance (Plate II, Fig. 2; Plate III, Fig. 3), although there are intermittent "dark" cells. These striking differences in opacity of the background cytoplasm can be found at all stages of the cycle (Plate IV, Fig. 4). Dark cells contain the same organelles as others, and except for density are not remarkably different from less dense cells. Wynn and Harris (1967) thought they represent different stages of secretory epithelium, rather than another cell type. In addition, there are "pale" or "clear" cells in greater number during proliferative stages (Witt, 1963), but present in all phases of the cycle (Hoffmeister and Schulz, 1961). Their functional significance remains undetermined (Colville, 1968). Because they occur in greater concentration in cystic glandular hyperplasia, they were thought to be cells in prophase (Fuchs, 1959). It has also been postulated that they are developing ciliated cells (Hamperl, 1950) or stromal cells displaced in the epithelium (Hamperl, 1950).

A. Endoplasmic Reticulum and Ribosomes

Reports on the endoplasmic reticulum (ER) and ribosomal content of cells in the proliferative phase show great quantitative variation, although the descriptive aspects of their morphology are consistent. Ribosomes are distributed rather uniformly throughout the cytoplasm, which is interlaced with loose, vesicular outlines of rough ER (Plate II, Fig. 2; Plate III, Fig. 3). There is not a noticable amount of smooth ER. At days 5, 6, and 7, the ER is reported as scarce (Wynn and Harris, 1967) or well developed (Themann and Schunke, 1963). There may be a paucity (Cavazos *et al.*, 1967) or a good number of ribosomes (Wynn and Harris, 1967). At this stage of the cycle, these differences may be due to observations on some cells that are replicating and on others that are rapidly growing. Prior to ovulation the ER and ribosomes have reached the same level of differentiation in the majority of the cell population (Plate V, Fig. 7; Plate VI, Fig. 9) when they approximate the degree of development in the more differentiated cells seen earlier. The ER is developed (Borell *et al.*, 1959; Wessel, 1960; Cavazos *et al.*, 1967; Colville, 1968), and ribosomes are numerous (Themann and Schunke, 1963). The stimulus for ER and ribosome synthesis has been attributed to estrogen (Clyman, 1963c). It is possible that the population of ER and ribosomes is stabilized at this time for the remainder of the cycle. Near ovulation, the cytoplasm begins to take on a dense more opaque consistency (compare Figs. 6 and 7, Plate V). In secretory stages, discernment of these elements is complicated by the density of the cytoplasm and the omnipresent deposits of glycogen (Plate IV, Fig. 5; Plate VII, Figs. 10 and 11; Plate VIII, Fig. 12). Following ovulation in early and midsecretory stages, the ER and ribosomal populations are reported to be inconspicuous (Themann and Schunke, 1963; Wynn and Woolley, 1967), or to have greatly dilated cisterns (Colville, 1968) containing irregular fibrillar material (Cavazos *et al.*, 1967). The tissue examined in our laboratory showed contracted, less obvious ER in some but not all specimens, as early as 12 days (Plate V, Fig. 6). This condition is maintained until day 22 (Plate VIII, Fig. 12), when the cisterns dilate and the ER appears as large, clear pockets ramifying through the dense cytoplasm (Plate IX, Figs. 13 and 14; Plate X, Fig. 15). The widening and distension of the membrane system has been attributed to regressive changes (Themann and Schunke, 1963), such as hydration (Sengel and Stoebner, 1970), rather than hyperactivity.

At ovulation there is a dramatic change in the topographical distribution of the ER when it begins to encircle a few mitochondria (Plate VI, Fig. 9) at the base of the cell. This distinctive wrapping of mitochondria with rough ER is a consistent feature of giant mitochondria in the secretory phase (Plate XI, Fig. 17). Initially, the encircled mitochondria are little altered from their previous appearance (compare Plate V, Fig. 7 and Plate VI, Fig. 9). This observation is of some interest, since it suggests that the tropism generated between rough ER and mitochondria is selective in terms of the mitochondrial population and is operative near only a few situated in the basal cytoplasm. The polyribosomes of the ER (Plate XI, Fig. 18) may make a crucial contribution to mitochondrial enlargement. Differentiation of mitochondria begins just prior to ovulation, and basal accumulation of glycogen follows shortly after. Eventually, the giant mitochondria are surrounded by glycogen (Plate VIII, Fig. 12). These specialized mitochondria therefore may play a role in synthesis of the basal

PLATE V. FIG. 6. Thirteen-day endometrium, just prior to or at ovulation. The microvilli (Mi) have elongated. Cytoplasm and mitochondria (M) are opaque. A junctional complex is seen at the right. Deposits of glycogen (arrow) are scattered throughout the cytoplasm. Ribosomes and ER are inconspicuous. ×30,000. FIG. 7. Ten-day endometrium, showing stages preceding Fig. 6. The microvilli (fine arrow) are short. Cytoplasm is clear. A multivesicular body (heavy arrow), ribosomes, and profiles of rough ER (er) are visible. Mitochondria (M) have the same relative background density as the cytoplasm. Pinocytotic vesicles (arrowhead) are beneath the plasmalemma. ×22,000.

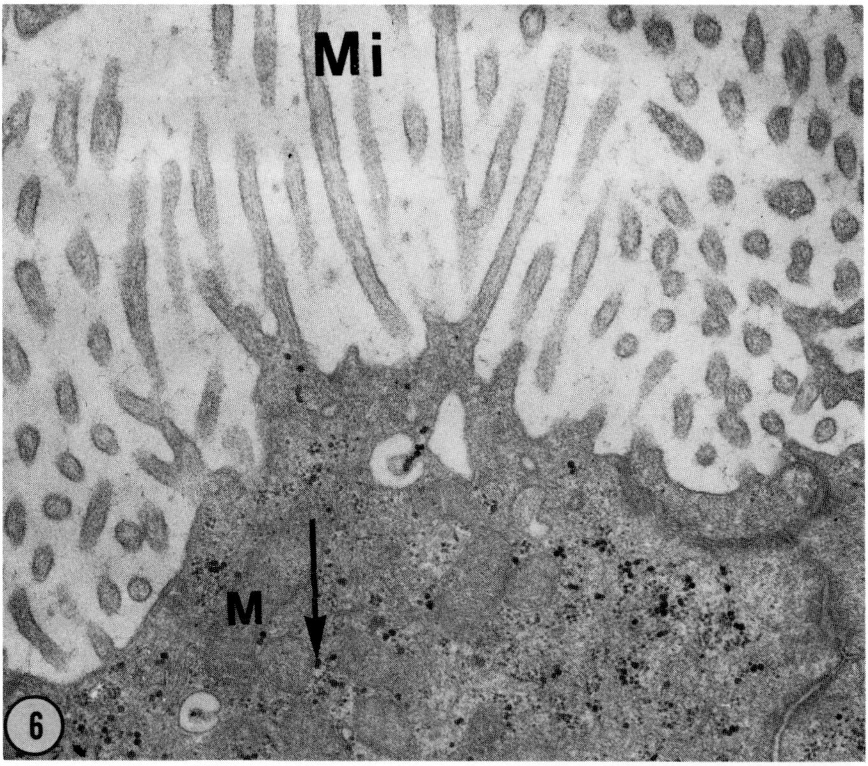
Mi
M
6

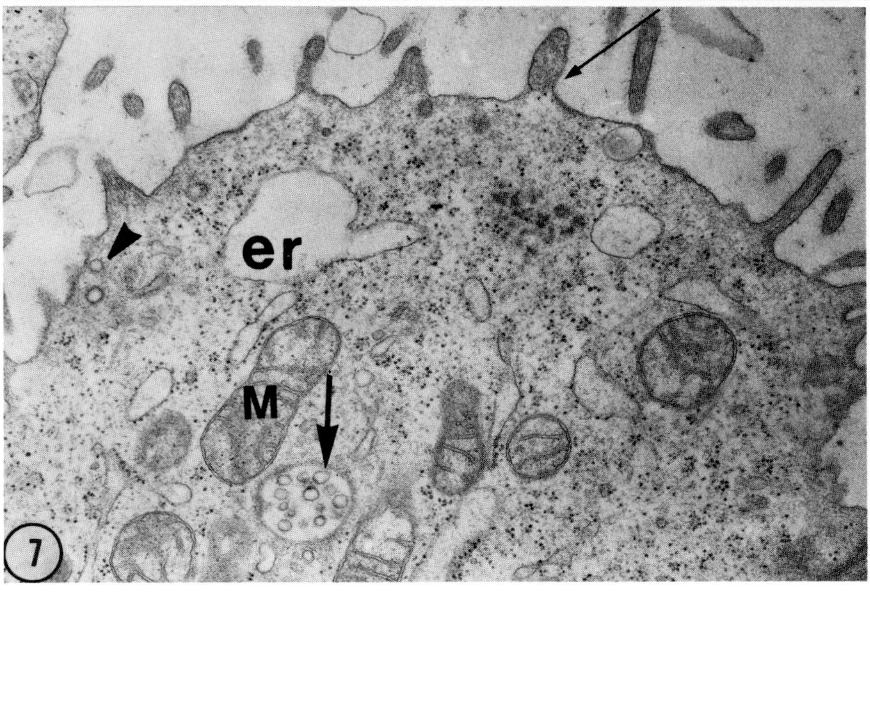
er
M
7

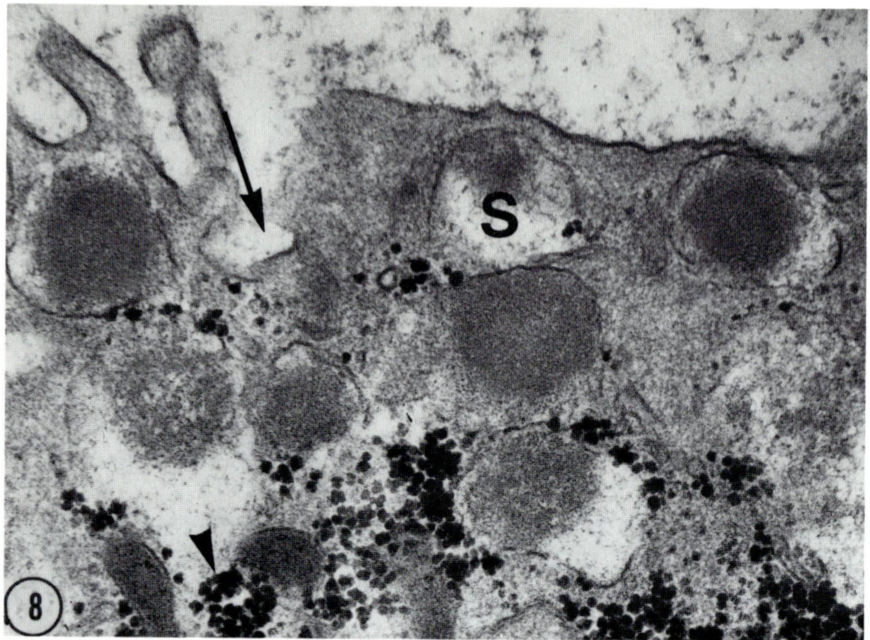

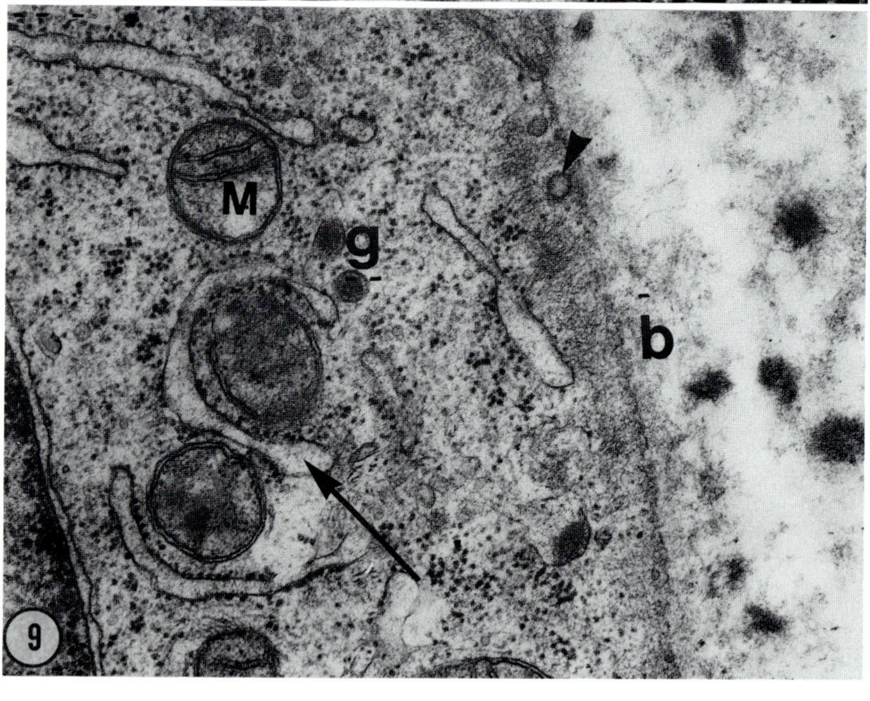

PLATE VI. FIG. 8. Endometrium at ovulation. Secretory vesicles (S) are prominent at the apex. Luminal surface is invaginated, indicating extrusion of granules (arrow). Well-formed glycogen deposits (arrowhead) are scattered throughout. ×45,000. FIG. 9. Basal portion of cell from 13-day endometrium at time of ovulation. The ER (arrow) is in the beginning stages of encircling mitochondria (M). Small dense granules (g) are found. Pinocytotic vesicles (arrowhead) invaginate from the plasma membrane. The basement lamina (b) is frayed and not well organized. ×54,000.

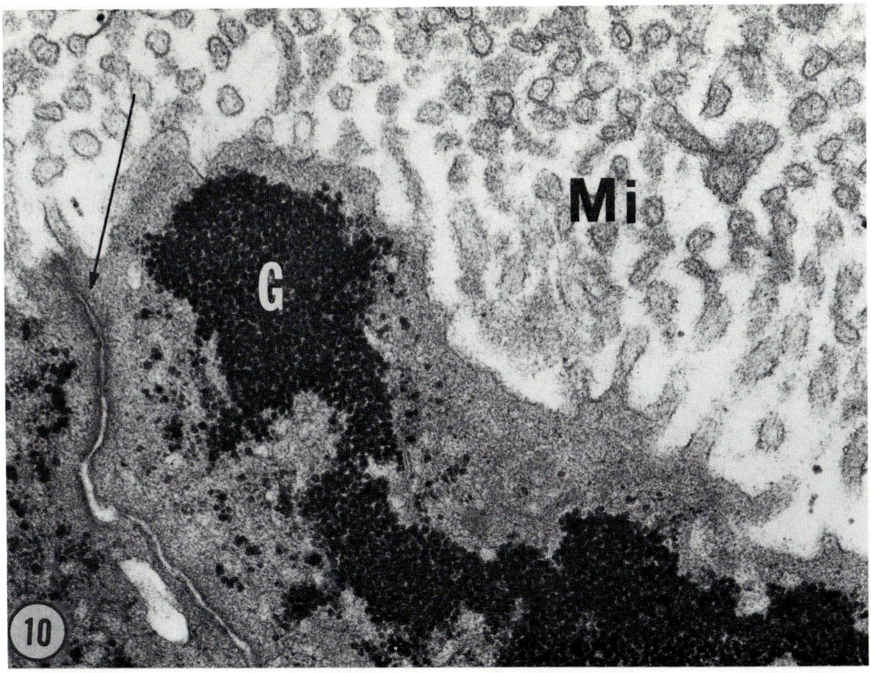

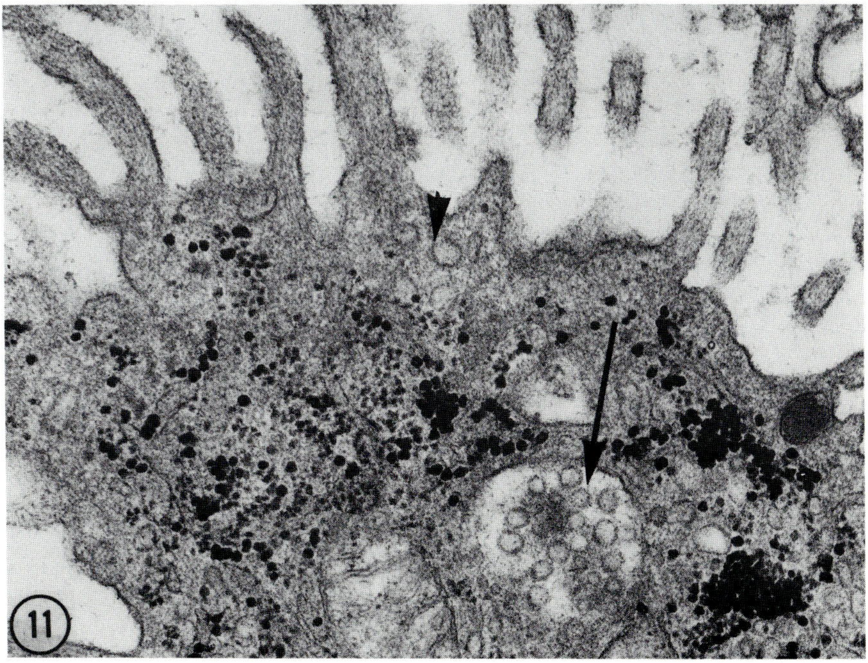

PLATE VII. FIG. 10. Apex of cell from 18-day endometrium. A junctional complex (arrow) is seen between cells. Microvilli (Mi) are extensively developed. Glycogen (G) fills portions of the cytoplasm. ×26,500. FIG. 11. Seventeen-day endometrium showing slightly earlier stages than Fig. 10. Glycogen is not as copious, and microvilli are less elaborated. A multivesicular body (arrow) is present. Apical cytoplasm protrudes into lumen. Pinocytotic vesicles (arrowhead) are numerous under the plasmalemma. ×36,000.

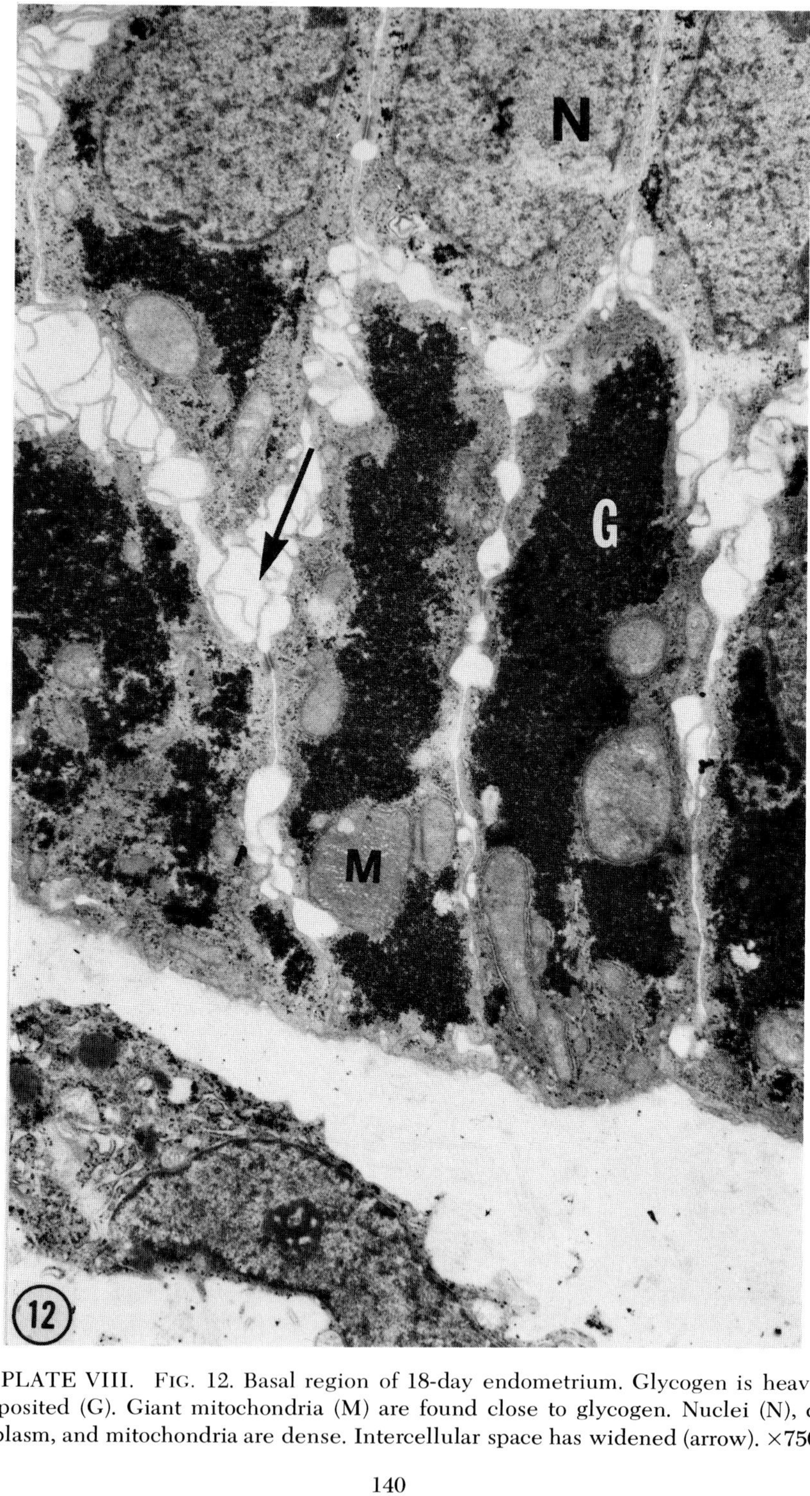

PLATE VIII. FIG. 12. Basal region of 18-day endometrium. Glycogen is heavily deposited (G). Giant mitochondria (M) are found close to glycogen. Nuclei (N), cytoplasm, and mitochondria are dense. Intercellular space has widened (arrow). ×7500.

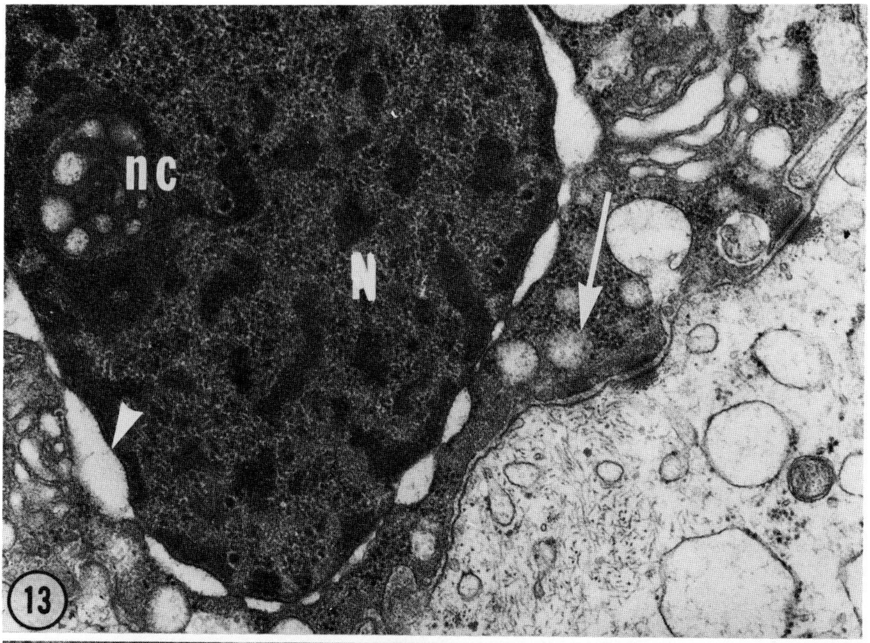

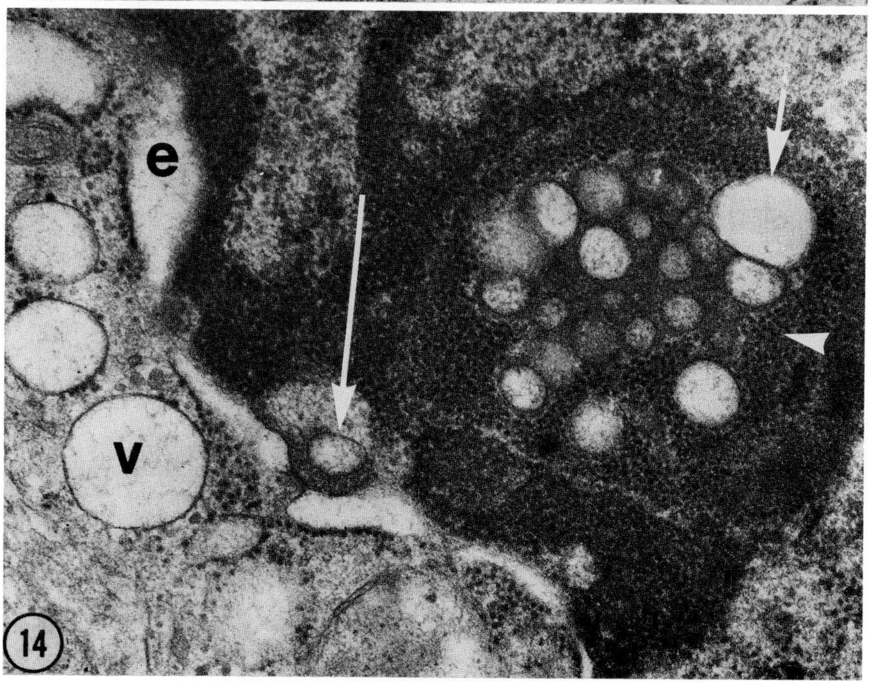

PLATE IX. FIG. 13. Twenty-two-day endometrium demonstrating pycnotic cytoplasm and nucleus (N), dilated cisterns of ER (arrow), and nuclear envelope (arrowhead). Channels of NCS (nc) are similarly dilated. ×20,000. FIG. 14. NCS in 22-day endometrium. The dense granules are no longer in a single row (Fig. 31) but are proliferated around the channels (arrowhead). A vesicle (long arrow) resembles the dilated channels of the nucleolar body (short arrow). It may represent a channel migrating to the nuclear envelope (e). Similar rounded vesicles of ER (v) in the cytoplasm near the nuclear surface. ×42,000.

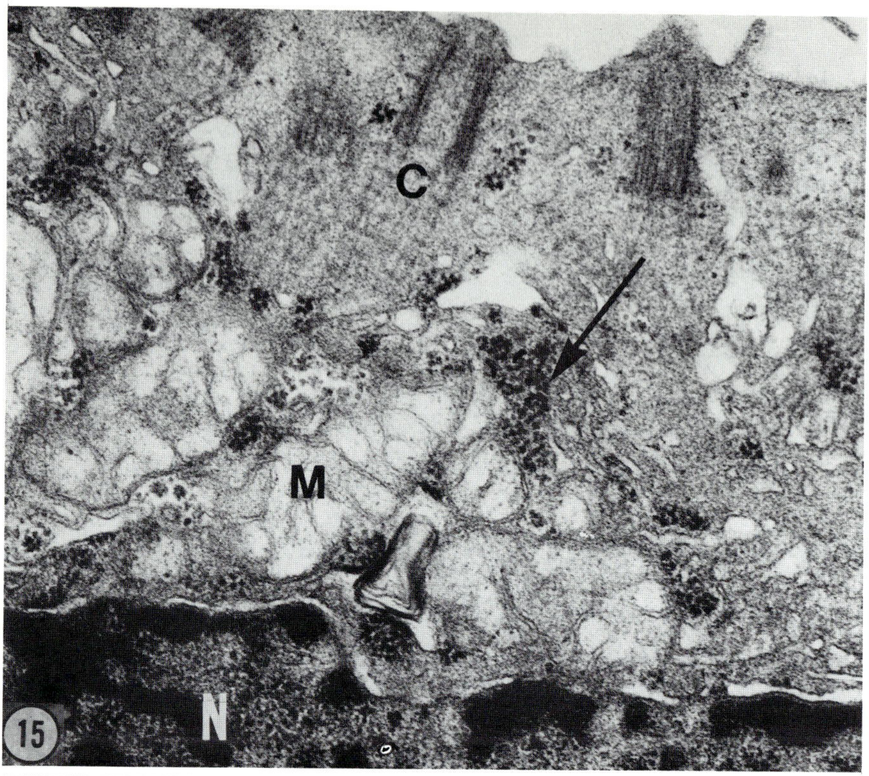

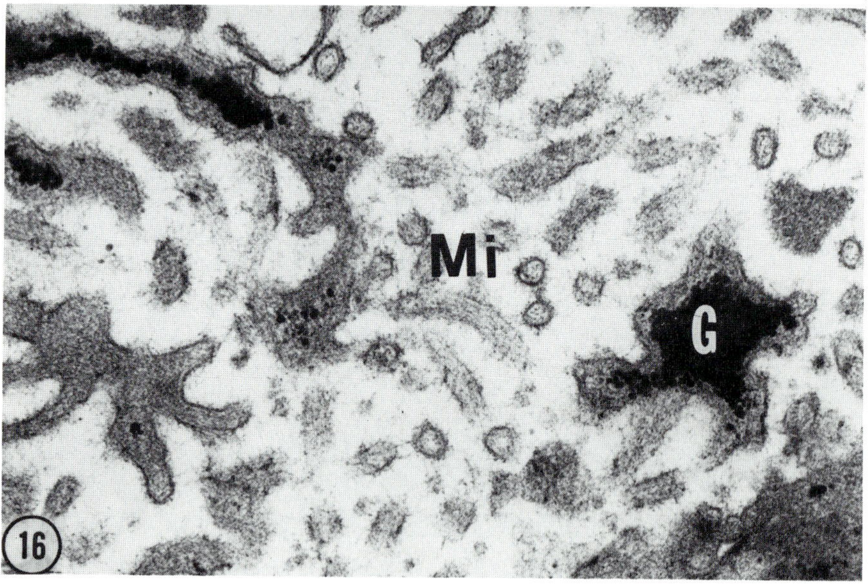

PLATE X. FIG. 15. Ciliated cell from 24-day endometrium. The nucleus (N) is pycnotic. Dense glycogen granules (arrow) are found in random patches. Ballooning mitochondria (M) with clear matrices are abundant. Well-formed ciliary rootlets (C) are in apical cytoplasm. ×48,000. FIG. 16. Cytoplasmic protrustions filled with glycogen (G) and proliferated microvilli (Mi) in 18-day endometrium. ×26,300.

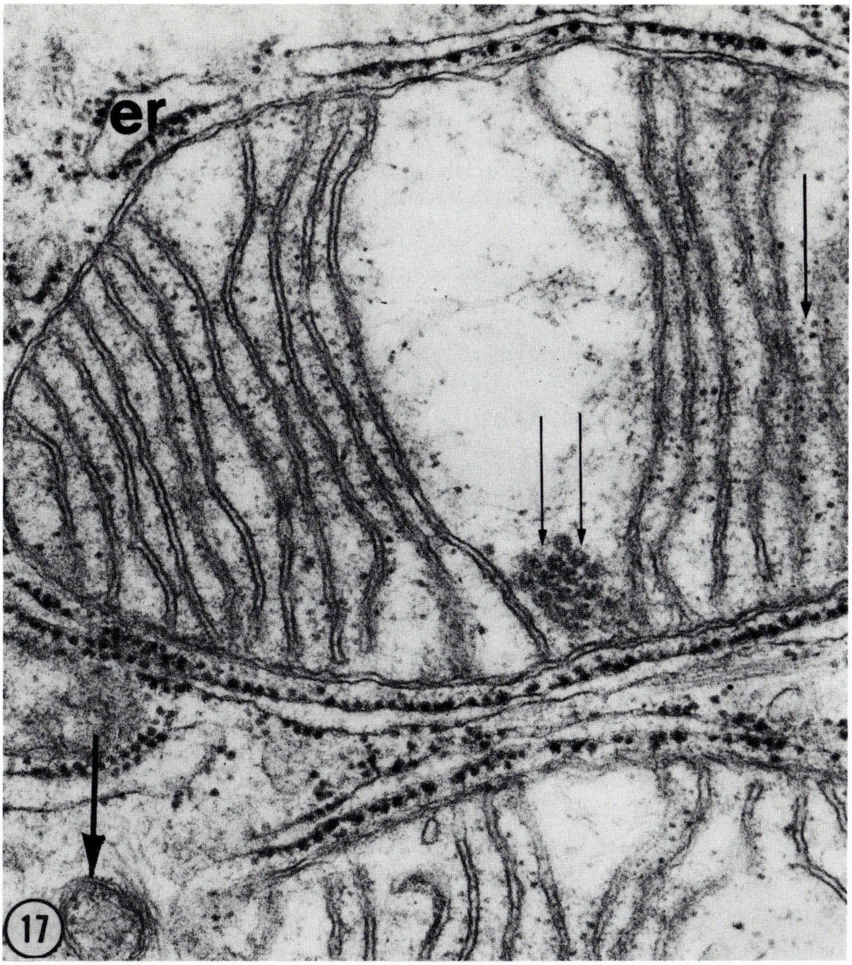

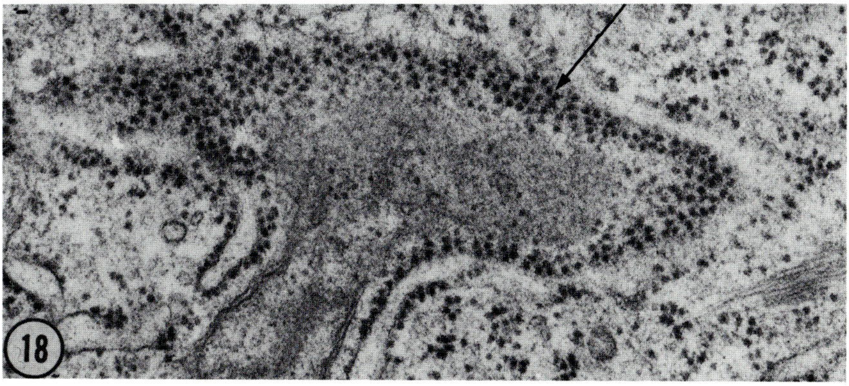

PLATE XI. FIG. 17. Giant mitochondrion encircled by rough ER (er). Parallel cristae are continuous through the diameter. Other cristae are in cross section (double fine arrows). Fine granules are associated with cristae (fine arrow). A normal-sized mitochondrion is seen at heavy arrow. ×56,000. FIG. 18. Polyribosomes (arrow) visible in section through rough ER surrounding giant mitochondrion. ×48,000.

glycogen, perhaps by providing ATP for conversion of uridine 5′-phosphate to UTP, a cofactor in glycogen synthesis.

B. MITOCHONDRIA

In the initial stages of the cycle, mitochondria are of normal size (Lanzavecchia and Morano, 1959), are rounded, elongated, or constricted (Armstrong *et al.*, 1973), and have a modest number of delicate cristae (Clyman, 1963a; Moricard, 1966) (Plate II, Fig. 2; Plate V, Fig. 7; Plate XII, Fig. 21). Concentration of elongated forms at the base of the cell (Wessel, 1960; Cartier and Moricard, 1960; Wynn and Harris, 1967) or at the apex (Themann and Schunke, 1963) does not appear to be the case in all cells (Plate II, Fig. 2). At ovulation those destined to grow into giant or megalomitochondria show a slight increase in size and density (Plate VI, Fig. 9). At the time of ovulation, the mitochondria distributed randomly through the cytoplasm also enlarge to some degree. Their cristae are more highly developed, and the matrix is more opaque (Plate V, Fig. 6). The onset of this change spans the period from day 12 to day 15 or 17, and shows considerable variation from one specimen to another. It parallels the gradual deepening of the ground cytoplasm and the deposition of glycogen particles at the base of the cell. The comment that the mitochondrial matrix has the same density as the cytoplasm (Themann and Schunke, 1963) seems pertinent. Mitochondria reach their maximum development at the midsecretory stage (Plate VIII, Fig. 12; Plate XI, Fig. 17), about the time optimal for nidation (Themann, 1967). Near the end of the cycle, they decrease in size again (Plate I, Fig. 1). They may be small and slender by day 20 (Clyman, 1963c; Wynn and Woolley, 1967; Armstrong *et al.*, 1973), although they can remain enlarged until day 26 (Nilsson, 1962b). In the later stages, when there appears to be enhanced lysosomal activity (Cohen *et al.*, 1964; Bitensky and Cohen, 1965), mitochondria may be found in autophagic granules (Themann and Schunke, 1963; Armstrong *et al.*, 1973). During the few days preceding the menstrual flow,

---

PLATE XII. FIG. 19. High magnification of blunt microvilli in 6-day endometrium. A pronounced extracellular "fuzzy" coat is seen on their external aspect (arrow). The terminal web region (Te) is poorly developed. ×45,000. FIG. 20. Dense body (arrow) in 6-day endometrium showing uniform granular contents and surrounding membrane. ×48,000. FIG. 21. Fragment of annulate lamella consisting of smooth lamellae punctuated by pores (arrows). Mitochondria are small and have a modest number of cristae. ×32,000. FIG. 22. Giant lysosome (GL) in 13-day endometrium containing fragments of rough ER, dense glycogen particles, and amorphous rounded bodies. Membrane-bound inclusions of lower density may be mitochondria. ×28,000.

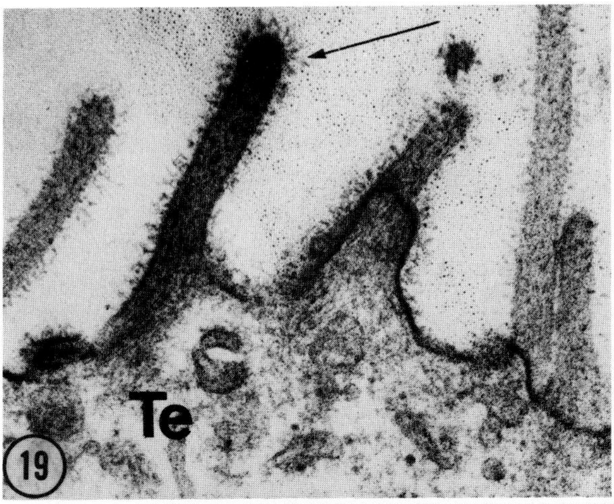
Te
19

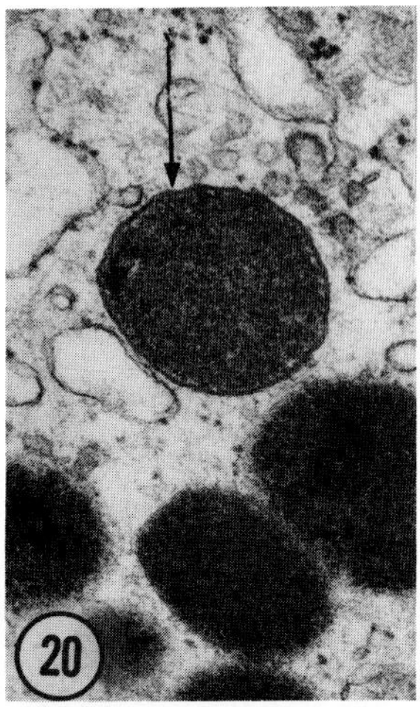
20

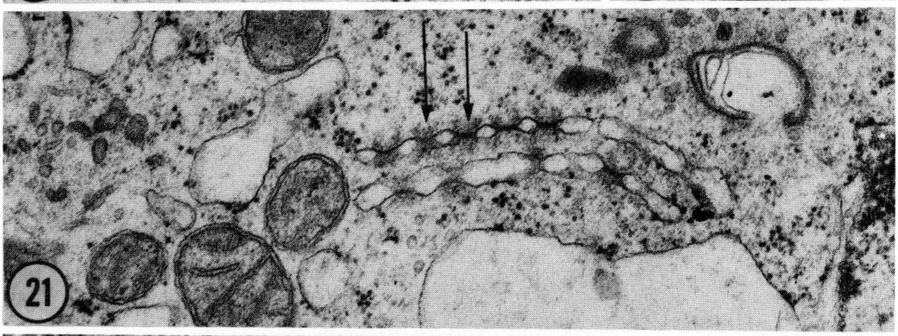
21

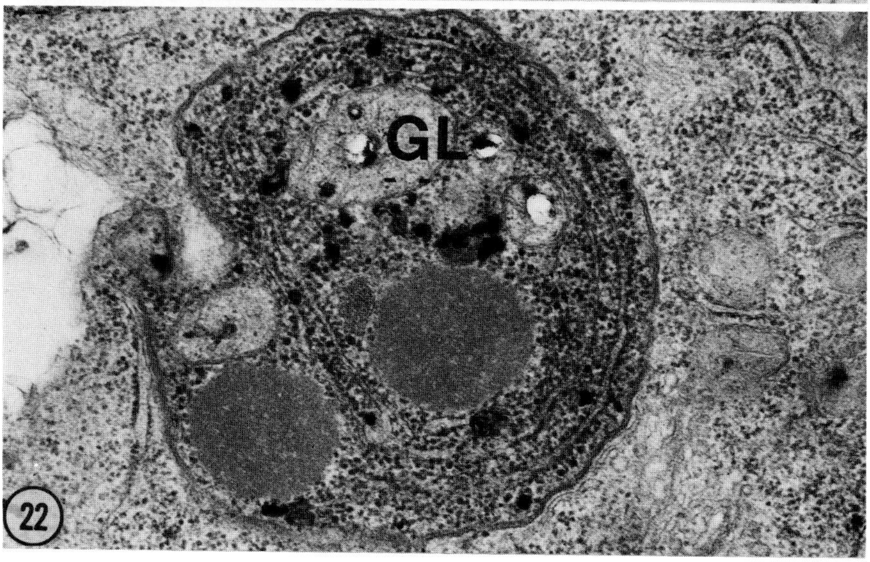
GL
22

coinciding with a decrease in cell height, diminished glycogen, and a widening extracellular space between the lateral cell border (Plate I, Fig. 1), mitochondrial matrices may become clarified and show "ballooning" (Plate X, Fig. 15), perhaps because of fluid intake (Wessel, 1960; Sengel and Stoebner, 1970). The enormous size of megalomitochondria permits clarification of their internal elements. Typically, their cristae run in parallel fashion across the diameter, but they may take a different course and present tubular profiles (Plate XI, Fig. 17) (Armstrong *et al.*, 1973) or digitate formations (Cavazos *et al.*, 1967). Fine electron-dense particles (Gompel, 1967) which may contain RNA are distributed on the cristae (Plate XI, Fig. 17).

C. The Nucleus

Because the tissue approaches a pseudostratified condition during early proliferation, the precise position of the nucleus is not meaningful. It may be located either at the base (Lanzavecchia and Morano, 1958; Gompel, 1962), or centrally (Cavazos *et al.*, 1967). Nuclear morphology, however, is quite consistent, and has been described similarly in previous studies (Lanzavecchia and Morano, 1958; Gompel, 1962; Themann and Schunke, 1963; Wynn and Harris, 1967; Armstrong *et al.*, 1973). The nucleus is somewhat rounded and smooth (Plate III, Fig. 3), without the cytoplasmic digitations so common in later stages. The nucleoplasm is relatively diffuse, containing evenly distributed delicate fibrils and granules. Patches of chromatin are scattered throughout and condensed on the nuclear envelope. The nucleolus consists of a typical pars amorpha and nucleonema (Plate III, Fig. 3) (Estable and Satelo, 1951; Busch and Smetana, 1970; Armstrong *et al.*, 1973), and in well-prepared specimens can be differentiated from the chromatin by a less intense staining reaction (Plate III, Fig. 3). Nucleolus-associated DNA is contained in focal deposits of chromatin on the periphery (Armstrong *et al.*, 1973). A delicate nuclear envelope without prominent dilatations of the cisternal space hugs the contours (Plate III, Fig. 3). Extensive connections with the ER at day 10 may be noticeable (Cavazos *et al.*, 1967). A gradual increase in density of the karyoplasm toward the end of the preovulatory period is the first indication of its intense pycnotic state at the end of the cycle, accompanied by shrinkage and a convoluted periphery (Wessel, 1960; Theman and Schunke, 1963). A comparison of Plate III, Fig. 3 with Plate I, Fig. 1 is illustrative of the modifications that occur from early to late stages of the cycles. The remarkably uneven outline of the nucleus in the later stages is a manifestation of the numerous invaginations of the cytoplasm into the karyoplasm. They have been universally described in nuclei in the secretory phase. The NCS deserves special mention and is discussed separately. After the NCS disappears, approximately near day 22, the nucleus becomes increasingly pycnotic contracted, and irregular (Plate I, Fig. 1).

D. The Golgi Complex and Secretory Bodies

A well-formed Golgi complex of lamellae and vesicles is situated in the supranuclear cytoplasm (Plate IV, Fig. 5). It remains oriented at the upper pole of the cell throughout the cycle although, when it is extensively developed, portions may be found deep in the cytoplasm around the nucleus. Estimates of its activity are most conveniently obtained from the appearance of the organelle. Proliferation and enlargement of the lamellae and a great population of vesicles are indicative of a high functional state. Because of its increased development during cystic glandular hyperplasia (Wessel, 1961) and its response to estrogen administration (Clyman, 1963c; Moricard, 1966), the hormone is considered to affect the state of its activity directly. The Golgi complex is the source of a great diversity of inclusion bodies which are found in large numbers in early proliferative epithelium (Plate II, Fig. 2). They contain a complex of substances, including protein and mucopolysaccharides and lipid droplets (Borell *et al.*, 1959; Gompel, 1962; Nilsson, 1962a; Wynn and Harris, 1967). Typically, the apical cytoplasm is filled with these granulated and lipid bodies, which indicates that they are extruded in great numbers at a specific time, perhaps under hormonal stimulus. A great reduction in their number occurs prior to and during ovulation, when they are extruded by merocrine secretion (Plate VI, Fig. 8) (Themann and Schunke, 1963; Cavazos *et al.*, 1967; Wynn and Harris, 1967). Release of these granules undoubtedly accounts for the copious uterine secretion during this period (Datnow, 1973). The sharp increase in uterine secretion, beginning at day 10, coincides with a high level of circulating estrogen (Strauss, 1962). The role of estrogen in the production of uterine secretion, then, is ultimately in providing a stimulus to the secretory apparatus of the cells.

The timing of these events is complicated by reports that secretory granules may increase in the cell at the late proliferative stage (Nilsson, 1962a; Wynn and Harris, 1967). Further, the Golgi may be compact and of modest proportions, rather than hypertrophied. These observations suggest that secretory activity is not at the same level in all cells, although the additive effect of the total population is expressed as fluid in the uterine cavity. The evidence that cells are filled with secretory granules at ovulation, or shortly after (Schmidt-Matthiesen, 1963; Yaneva *et al.*, 1965; Cavazos *et al.*, 1967), suggests that some cells secrete their granules early, while others maintain the flow through the early secretory phase. This view of secretion, which correlates preovulatory synthesis of granules with the flow of uterine "milk" at ovulation, is complicated by the evidence for a population of secretory granules later in the middle of the postovulatory period (Nilsson, 1962b; Cavazos *et al.*, 1967; Sengel and Stoebner, 1970). Whether they are the same as the ones produced prior to ovulation or contain other substances which modify glandular secretion is unknown. The evidence for apocrine secretion Stoebner, 1970). Whether they are the same as the ones produced prior to ovulation or contain other substances which modify glan- implies that substances are being secreted. Coated vesicles are numerous (Plate II, Fig. 2) in all stages, usually collected around the Golgi. Pinocytosis is common in both the distal and proximal surfaces of the plasma membrane (Plate V, Fig. 7; Plate VI, Fig. 9). The dense granulated bodies noted by Themann and Schunke (1963) in the basal cytoplasm are probably not secretory granules but small lysosomes (Plate VI, Fig. 9).

E. Lysosomes

The presence of true lysosomes in endometrial cells was demonstrated histochemically by the acid phosphatase reaction (Schmidt-Matthiesen, 1963; Cohen *et al.*, 1964; Bitensky and Cohen, 1965; Delforge, 1969). Giant lysosomes, consisting of huge, membrane-bound bodies containing portions of cellular organelles in all stages of dissolution (Plate XII, Fig. 22) have been associated with both proliferative (Nilsson, 1962a; Cavazos and Lucas, 1968; Cavazos and Lucas, 1970) and secretory stages (Wetzstein and Wagner, 1960; Armstrong *et al.*, 1973). In our study they were found at days 6, 10, 12, and 13 and in the secretory phase. That they were seen in early stages in some studies, and later in the cycle in others, demonstrates that they are not ubiquitous throughout the tissue. The fact that they are not associated with one specific phase suggests that their biological role varies. Thus during the proliferative stage they have the obvious role of autodigestion or digestion of materials, including parts of other cells, that penetrate the host cell (Cavazos and Lucas, 1970). In secretory stages they have been implicated in promoting apocrine secretion by liquefaction of apical portions of the cell (Delforge, 1969) and in the autophagia and digestion of glycogen stores (Sengel and Stoebner, 1970; Armstrong *et al.*, 1973).

By monitoring incubation time, Bitensky and Cohen (1965) demonstrated that the histochemical demonstration of lysosomes varied with the stage of the cycle. They correlated the prolonged time required for the acid phosphatase reaction in proliferative endometrium with stabilization of the lysosomal membrane by steroids, presumably estrogen. A decrease in incubation time in early secretory and a marked decrease in later secretory tissue was considered evidence of the fragility of lysosomal membranes which could, under the conditions that prevail at the end of the cycle, provide for autodigestion and cell death.

Multivesicular bodies produced in the Golgi complex (Plate V, Fig. 7; Plate VII, Fig. 11) are observed at the cell apex early in the cycle (Hoffmeister and Schulz, 1961; Gompel, 1962; Themann and Schunke, 1963). They may wander through the cell (Hoffmeister and Schulz, 1961) and into the infranuclear cytoplasm (Themann and Schunke, 1963). Sengel and Stoebner (1970) related these organelles to pinocytosis.

F. Annulate Lamellae

These structures consisting of double smooth membranes, interrupted by pores resembling those in the nuclear envelope, were first described in endometrial cells by Ancla and co-workers (Ancla *et al.*, 1964b; Ancla and de Brux, 1965), who associated them only with hyperestrogenic states. They were found in normal endometrium at random in the cycle (Wynn and Woolley, 1967), but Cavazos *et al.* (1967) saw considerable numbers only in early proliferative stages. Our results concur with those of Cavazos (Cavazos *et al.*, 1967). Many packets of annulate lamellae were disposed in proximity to the upper pole of the nucleus at day 6. An enlargement of a fragment of one of these packages of annulate lamellae is seen in Plate XII, Fig. 21. They are most frequent near the nuclear surface (Ancla and de Brux, 1965; Cavazos *et al.*, 1967), but a direct functional or derivative association with the nucleus has not been established. Annulate lamellae are seen in other cells (for references, see Ancla and de Brux, 1965), such as lower organisms, Sertoli cells of the testis, primordial germ cells, and tumor cells. Their origin and role in the cell is obscure. Wynn and Harris (1967) thought they represented degenerative effects and associated them with myelin figures. According to our observations, they are rather representative of a state of rapid proliferation during the tissue response to estrogen.

G. Luminal Surface

Because of the pressure of cell replication and proliferation of new generations of cells in preovulatory endometrium, the apex may bulge into the lumen, appearing markedly convex. In scanning electron microscopy this results in a surface "cobblestone" appearance (Ferenczy and Richart, 1973). The degree of convexity is modified by the crowding of the cells, and may be extreme, moderate, or absent. As growth proceeds toward midcycle, the epithelium penetrates more deeply into the stroma, and the total surface area becomes more extensive (Witt, 1963), resulting in an alignment of the apical surface with the basal. A secondary, more marked, protrusion, resulting in apical "domes" may begin at the late proliferative stage (Wetzstein and Wagner, 1960) and continue into the midsecretory stage (Cartier, 1959; Schmidt-Matthiesen, 1963; Themann, 1967; Wynn and Woolley, 1967; Sengel and Stoebner, 1970) (Plate IV, Fig. 5; Plate VII, Fig. 11). These may be desquamated later in the cycle (Schmidt-Matthiesen, 1963; Themann and Schunke, 1963; Colville, 1968) and contribute to postovulatory apocrine secretion (Schmidt-Matthiesen 1963; Themann and Schunke, 1963; Themann, 1967; Armstrong *et al.*, 1973).

Throughout the cycle the apical surface contains a population of microvilli or cilia. In early proliferative cells microvilli tend to be rather sparse and slender or stubby (Plate II, Fig. 2; Plate V, Fig. 7) (Lanzavecchio and Morano, 1959; Borell *et al.*, 1959; Gompel, 1962; Nilsson, 1962a). In well-fixed preparations a distinct extracellular coat covers their external aspect (Plate XII, Fig. 19) (Wynn and Harris, 1967). A poorly developed terminal web region is typical (Plate XII, Fig. 19) (Armstrong *et al.*, 1973). With cell growth the microvilli show a parallel development and are appreciably longer just prior to ovulation (compare Figs. 6 and 7, Plate V). Clyman (1963c) saw a notable increase in their number and height at about day 6, but these changes are more usual at midcycle (Plate V, Fig. 6) (Borell *et al.*, 1959; Wetzstein and Wagner, 1960; Themann and Schunke, 1963; Moricard, 1966; Cavazos *et al.*, 1967; Nilsson and Nygren, 1972). The growth of the apical cytoplasm, as well as differentiation of microvilli, are probably under the influence of estrogen stimulation. This is supported by the fact that they are not developed in atrophic endometrium but develop in response to estrogen therapy (Borell *et al.*, 1959; Wetzstein and Wagner, 1960; Nilsson, 1962b; Moricard, 1966). In the midsecretory phase the microvilli become enormously complex (Plate VII, Fig. 10; Plate X, Fig. 16) (Cartier, 1959; Cavazos *et al.*, 1967; Colville, 1968). According to Gompel (1962), the expansion of the microvilli is the only clear evidence of secretion at this stage. Not all investigators see elaboration of microvilli during the luteal phase (Borell *et al.*, 1959; Wetzstein and Wagner, 1960; Wynn and Woolley, 1967). However, our results indicate that at least impressive segments of the postovulatory epithelium possess extensively proliferated microvilli. They are no longer simply elongated (Plate V, Fig. 6), but show a profusion of irregular outlines (Plate VII, Fig. 10) that may be due to branching (Colville, 1968) or to their attachment to protoplasmic projections from the apex (Cavazos *et al.*, 1967). At termination of the cycle, after day 24, the apical surface is more regular, and microvilli are again blunted (Plate I, Fig. 1). This is attributed to apocrine desquamation into the lumen (Schmidt-Matthiesen, 1963; Themann and Schunke, 1963).

Progesterone action on estrogen-primed tissue promotes glycogen synthesis and differentiation of the NCS and giant mitochondria. Elaboration of the microvilli may be due to the same mechanism. Perhaps the diminution in microvilli at the end of the cycle results from a change in hormonal maintenance of the tissue, as well as desquamation.

The increase in length and number of microvilli near ovulation is accompanied by an increase in alkaline phosphatase on the luminal surface (Ober, 1950; McKay *et al.*, 1956; Berger and Mumprecht, 1959; Bontke, 1960; Hughes *et al.*, 1963; Kucera, 1964; Saksena *et al.*, 1965; Baba *et al.*, 1968; Colville, 1968). The growth of microvilli is considered to provide additional surface area for the enzyme (Borell *et al.*, 1959). Alkaline phosphatase is related to changes in cell permeability and transport. Its function is unknown, but it may promote the secretion of glucose into uterine fluid. Stimulation of glucose-6-phosphatase during the follicular phase (Hughes *et al.*, 1963) suggests glucose synthesis. Glucose in uterine secretion at ovulation is thought to be of importance for the viability and maintenance of sperm (Delforge, 1969; Datnow, 1973).

Cilia increase abundantly in proliferative endometrium (Maddi and Papanicolaou, 1961; Schueller, 1961; Wynn and Harris, 1967; Schueller, 1968; Wynn and Woolley, 1967; Sengel and Stoebner, 1970; Schueller, 1973), primarily in glandular cells. Their development during this phase of the cycle, as well as the fact that they proliferate extensively under conditions of abnormal hyperplasia (Novak and Rutledge, 1948; Vesterdal-Jorgensen, 1950; Maddi and Papanicolaou, 1961; Wessel, 1961; Fruin and Tighe, 1967; Fleming *et al.*, 1968; Schueller, 1968), suggest estrogen dependency. Fine-structural identification of cilia does not provide much information on cyclic alterations as regards either their number or distribution. They are relatively scarce in early proliferative tissue, probably because of the high mitotic rate. Quantitative assessment has been more advantageously obtained in wet smears (Fleming *et al.*, 1968) and with the scanning electron microscope (Johanissen and Nilsson, 1972; Ferenczy and Richart, 1973). These studies are not in total accord (Johanisson and Nilsson, 1972), but the evidence is compelling (Fleming *et al.*, 1968; Ferenczy and Richart, 1973) that ciliogenesis is promoted during the follicular phase when they proliferate in the

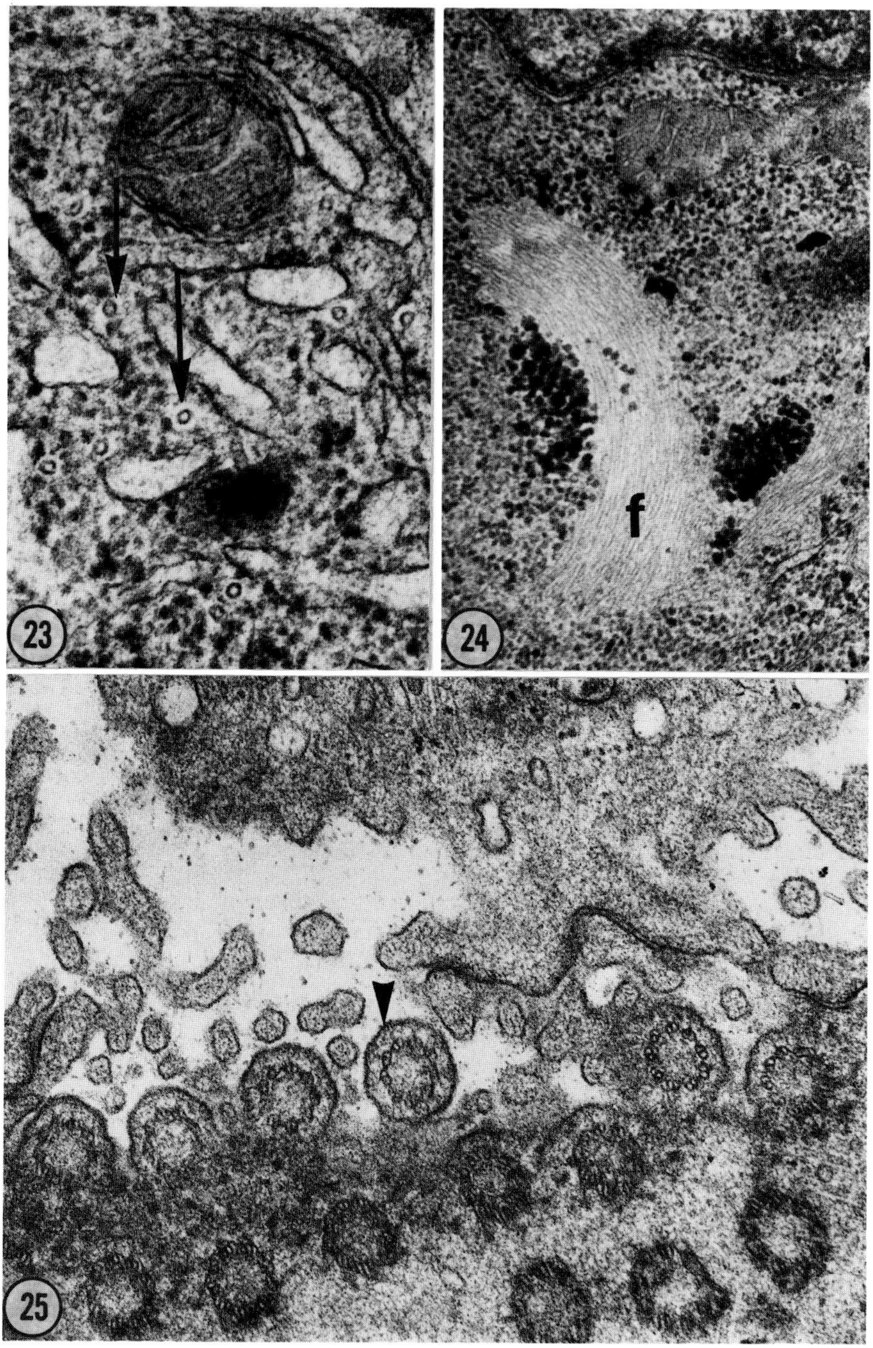

PLATE XIII. FIG. 23. Microtubules, seen in cross-sectional profile (arrows), in 10-day endometrium. ×65,000. FIG. 24. Sheaf of tonofilaments (f) in cytoplasm of 13-day endometrium. Dense granules of glycogen are deposited around the filaments. ×34,000. FIG. 25. Cilia (arrowhead) on the surface of 10-day endometrium. Basal bodies are in the cortical cytoplasm below. ×78,000.

glands and around gland orifices. The peak of ciliogenesis in glands at ovulation is followed by differentiation of cilia on the uterine surface (Fleming *et al.*, 1968; Ferenczy and Richart, 1972). They diminish at the end of the luteal phase, perhaps because of surface desquamation (Hamperl, 1950; Fleming *et al.*, 1968). The elaboration of cilia within the glandular lumen, as well as around orifices, may promote the circulation of secretion, copious at this time, into the uterine lumen and finally through the cervical canal. Cilia on the uterine surface beat toward the vagina and thus could both move uterine fluid toward the cervix, as well as enhance sperm migration by causing a flow of fluid against which the sperm would move upstream (Greep, 1966). Ciliary movement, however, may not be sufficient to affect sperm transport by this mechanism (Schueller, 1973), but it could be important in mixing uterine secretion and thus provide a supportive medium for sperm motility (Schueller, 1973) and metabolism (Datnow, 1973). It should be mentioned that atypical cilia are not unusual in these cells (Hando *et al.*, 1968; Sengel and Stoebner, 1970; Schueller, 1973), although most of the organelles have a normal axonemal complex and are well anchored by ciliary rootlets (Plate X, Fig. 15; Plate XIII, Fig. 25).

H. Basal Surface

The basal cell membrane runs a relatively straight course in the proliferative phase (Gompel, 1962; Cavazos *et al.*, 1967). In some preparations a lucent region intervenes between the plasma membrane and an amorphous dense layer which follows its contours. This extracellular layer is PAS-positive (Gompel, 1962) and corresponds to the basement membrane of light microscopists. It is rather uniform and has a diameter of about 300–400 Å. In other specimens the extracellular dense material may be disorganized and uneven (Plate VI, Fig. 9) (Colville, 1968). In early studies (Cartier and Moricard, 1960; Dubrauszky and Pohlmann, 1961b) no cyclic changes were noted in the basal lamina. However, it has been reported to swell at the midsecretory stage, when it may reach 700–800 Å (Wessel, 1960; Themann and Schunke, 1963; Wynn and Woolley, 1967). Striking changes are seen in the late secretory phase about 4 days prior to menses, when the basal cell membrane and its accompanying extracellular lamina become increasingly more irregular (Armstrong *et al.*, 1973). At this stage dilation of the intercellular lacunae is marked, and the cells are greatly contracted. These changes throw the basal surface into small folds or tortuous undulations (Gompel, 1962) (Plate I, Fig. 1). The undulatory basement membrane and cell surface reported in less advanced stages of the secretory phase (Cavazos *et al.*, 1967) undoubtedly are preliminary to the transformations seen later. In the midsecretory phase the convolutions of the plasma membrane were thought to be related to mitochondria (Wynn and Woolley, 1967).

I. Lateral Surface

Tight junctions are a distinguishing landmark of the lateral membranes at the apex of the cell, which persist throughout the cycle (Plate I, Fig. 1; Plate II, Fig. 2; Plate V, Fig. 6; Plate VII, Fig. 10). A grazing section through a junctional region shows its extensive development (Plate II, Fig. 2). In the final stages, when cells become increasingly separated (Wessel, 1960; Nilsson, 1962b; Themann and Schunke, 1963), the junctional complexes at the luminal surface continue to maintain cell contact (Plate I, Fig. 1). This causes the cells to diverge progressively in a basal-to-apical direction (Armstrong *et al.*, 1973). The lateral membranes, which are poorly (Vecchietti and Morano, 1959) or moderately interdigitated in early epithelium (Plate III, Fig. 3) (Gompel, 1962; Nilsson, 1962b), show a greater complexity toward midcycle (Cavazos *et al.*, 1967; Wynn and Harris, 1967). They become more pronounced in the secretory phase (Lanzavecchia and Morano, 1958; Hoffmeister and Schulz, 1961; Gompel, 1962; Themann and Schunke, 1963; Armstrong *et al.*, 1973). The changing outline of the lateral plasma membrane is undoubtedly induced by modifications in cell growth and changes in the overall height of the cell (Gompel, 1962). Therefore the most elaborate interdigitations are seen at the end of the cycle, when the cells have shortened perceptibly (Plate I, Fig 1).

Desmosomes are scattered intermittently down the lateral surfaces. Themann and Schunke (1963) made the interesting suggestion that desmosomes are found at the same level on each side of the cell synchronously throughout the epithelial tissue, and that this forms a rigid ringlike girdling or intercellular skeleton which gives mechanical support to the tissue. Such a site may be present in Plate I, Fig. 1 approximately midway between the basal and apical limits. They have also proposed that the many foldings are not simply a static representation of tissue architecture, but provide a greater surface to facilitate transport of intercellular fluid. The tendency for the lateral membranes to interdigitate more extensively as they approach the basal portion of the cell (Hoffmeister and Schulz, 1961; Armstrong *et al.*, 1973) is more easily illustrated when the intercellular space is pronounced (Plate I, Fig. 1). The progressive widening of the intercellular space during the latter half of the cycle is a prelude to the final regressive changes at menses. By day 24 (Plate I, Fig. 1) (Wynn and Woolley, 1967), and in the succeeding days up to day 28 (Wessel, 1960), lateral lacunae become extremely dilated. The mechanism of these changes, which are associated with cell shortening (Gompel, 1962) and loss of apical organelles, is not known (Witt, 1963).

J. The Nucleolar Channel System

The NCS is a highly organized structure found in the nucleus of the epithelial cell during the luteal phase of the cycle. During the years following its discovery (Dubrauszky and Pohlmann, 1960, 1961a,b; Clyman, 1963a,b,c), it was described in increasingly greater detail, and the complexity of its structure gradually revealed (Ancla *et al.*, 1964a, b; Moricard and Moricard, 1964; Ancla and de Brux, 1965; Terzakas, 1965; Cavazos *et al.*, 1967; Wynn and Woolley, 1967; Sengel and Stoebner, 1970; Armstrong *et al.*, 1973). The fact that it was overlooked in several early studies of normal secretory endometrium (Cartier, 1959; Lanzavecchio and Morano, 1958; Cartier and Moricard, 1960; Gompel, 1962; Nilsson, 1962b) indicates that it is not found at all times in the secretory phase. According to Wynn and Woolley (1967), it is most frequent on days 17 to 20, diminishes by day 23, and is absent on day 24. It has been seen as early as the sixteenth day (Clyman, 1963a). The screening of a large sampling of tissue from many donors in our laboratory suggests that the optimal period for its differentiation is approximately 3 days after ovulation. It persists for at least 3–4 days afterward, and begins to degenerate by day 22 (Plate IX, Figs. 13 and 14). It is therefore present for a very short, select period in the entire cycle. In clinical diagnosis of infertility (Gore and Gordon, 1974), if it is not seen at first, tissue is examined in successive cycles. Clyman (1963c) considered that its manifestation in epithelial cells is a direct result of ovulation as monitored by a rise in basal body temperature and culdoscopic examination for ruptured follicles and a rise in pregnanediol secretion. This view is supported by the fact that it is not seen in the luteal phase of women who have been on prolonged birth control treatment (Kohorn *et al.*, 1972), in cultures containing birth control pharmaceuticals (Kohorn *et al.*, 1972), and in women receiving norethindrone acetate with ethinyl estradiol (Clyman, 1963c). Two patients with prolonged unexplained infertility were observed to develop NCSs during the cycle they become pregnant (M. Gordon, unpublished observations). However, it is more specifically related to the action of progesterone rather than ovulation, since it was observed in menopausal women on norethindrone therapy (Ancla and de Brux, 1965), in ovarian agenesis after administration of estradiol and progesterone (Ancla and de Brux, 1965), and in organ cultures of preovulatory endometrium containing progesterone (Kohorn *et al.*, 1970, 1972). Its deterioration in women receiving estrogen during the postovulatory phase (Gordon *et al.*, 1973), and its absence in cases of primary unexplained infertility, suggest that either an insufficient luteal phase or an end organ insensitivity to progesterone (Gore and Gordon, 1974) may be a factor in preventing its differentiation *in vivo*. Normally, development of the NCS coincides with maximal enzymic activity of the endometrium (Stuermer and Stein, 1952; Hagerman and Villee 1953; Ancla *et al.*, 1964b) when conditions are optimal for nidation. At the same period that the NCS is differentiated, the cytoplasm contains giant mitochondria and massive glycogen deposits (Plate VII, Fig. 10; Plate VIII, Fig. 12). However, the NCS has a shorter life-span than these cytoplasmic elements. Further, giant mitochondria and glycogen may be present without formation of the NCS (Kohorn *et al.*, 1972; Gore and Gordon, 1974). It was not seen in early pregnancy (Clyman, 1963a).

The NCS is morphologically related to two structures, the nucleolus (Plate XIV, Figs. 26 and 28) and cytoplasmic invaginations into the nucleoplasm (Plate XIV, Fig. 27) (Dubrausky and Pohlmann, 1961a,b; Clyman, 1963a,c; Ancla *et al.*, 1964a,b; Moricard and Moricard, 1964; Ancla and de Brux, 1965; Terzakis, 1965; Wynn and Woolley, 1967; Armstrong *et al.*, 1973). Clyman (1963a,b) interpreted these observations to mean that the channel system developed in the nucleolus and subsequently migrated to the nuclear envelope.

The NCS consists of a labyrinth of membrane-bound channels (Plate XV, Fig. 30b) (Terzakis, 1965; Armstrong *et al.*, 1973) embedded in or surrounded by a dense amorphous matrix (Plate XVI, Fig. 31). The canals or tubules course in a parallel fashion, but can bend, sometimes in an S-shaped curve (Plate XV, Fig. 29). In some instances there is but a single channel or a few channels, widely separated, in which case they may not be noticed (Ancla and de Brux, 1965). Because of the flection of the tubules, both a honeycomb and a longitudinal tubular array may appear simultaneously (Plate XIV, Figs. 27 and 28; Plate XVI, Fig. 33). More than one NCS are often present in the same nucleus (Terzakis, 1965) (Plate XIV, Fig. 28). Although the diameter of the canals has been estimated (Clyman, 1963a,b; Ancla and de Brux, 1965; Terzakis, 1965), as well as the total dimensions of the formed body (Ancla and de Brux, 1965), they vary enormously (Plate XIV, Figs. 26 and 27) and precise measurements may be misleading. The canaliculi are surrounded by a dense material whose delimiting zone, next to the channels, has decreased density (Plate XVI, Fig. 31a and b) (Ancla and de Brux, 1965). Electron-dense, 150-Å particles of great contrast, more typical of cy-

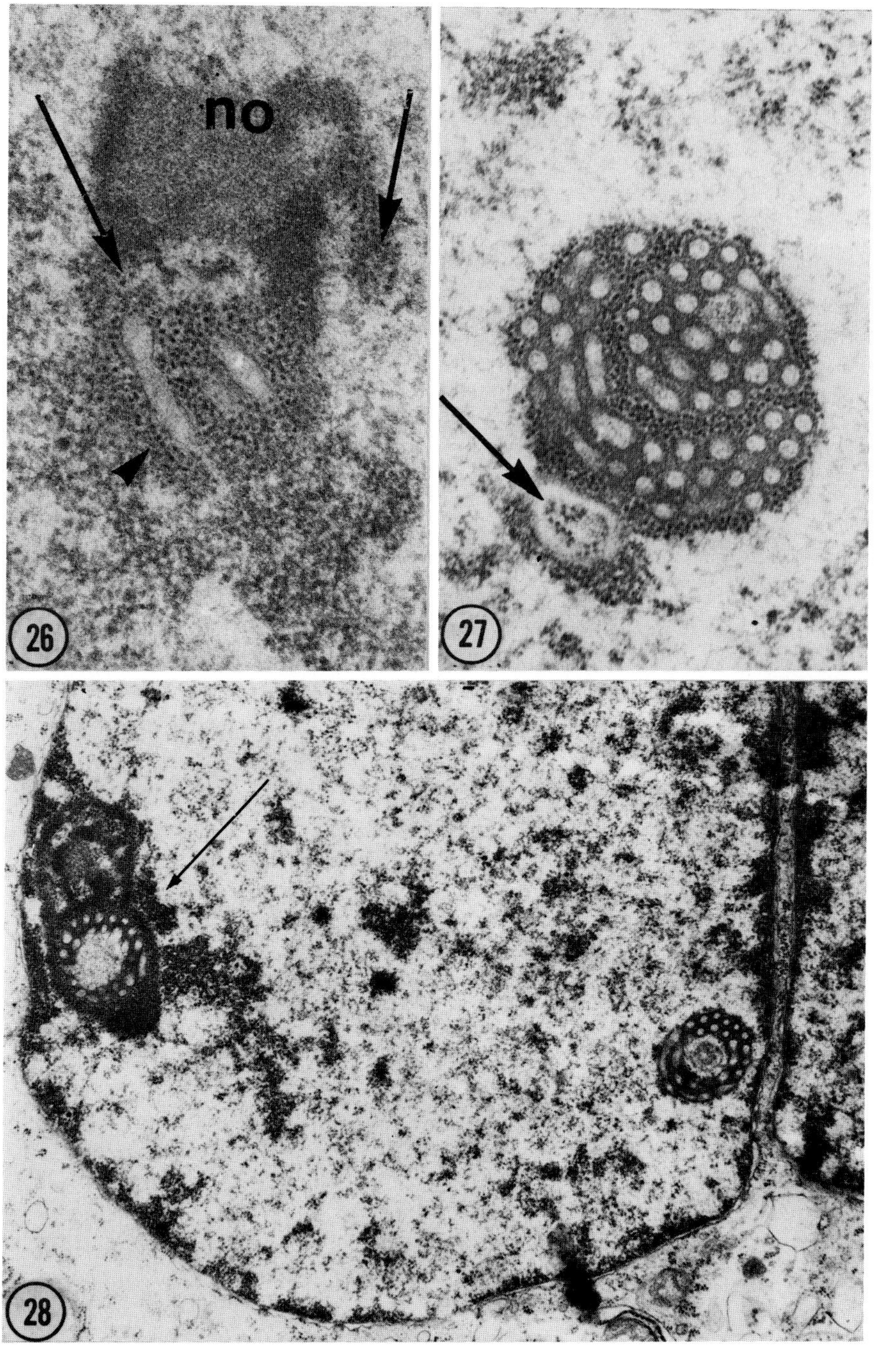

PLATE XIV. FIG. 26. NCS forming in association with nucleolus (no). Distinct, dense RNP granules are developing in nucleolus (arrows) and migrating around the NCS (arrowhead). ×34,500. FIG. 27. NCS on surface of cytoplasmic invagination into nucleus (arrow). A halo of karyoplasm surrounds the channel system. ×39,000. FIG. 28. Two NCSs in one nucleus from 21-day endometrium. One of the systems (arrow) is seen in association with a nucleolus. ×13,300.

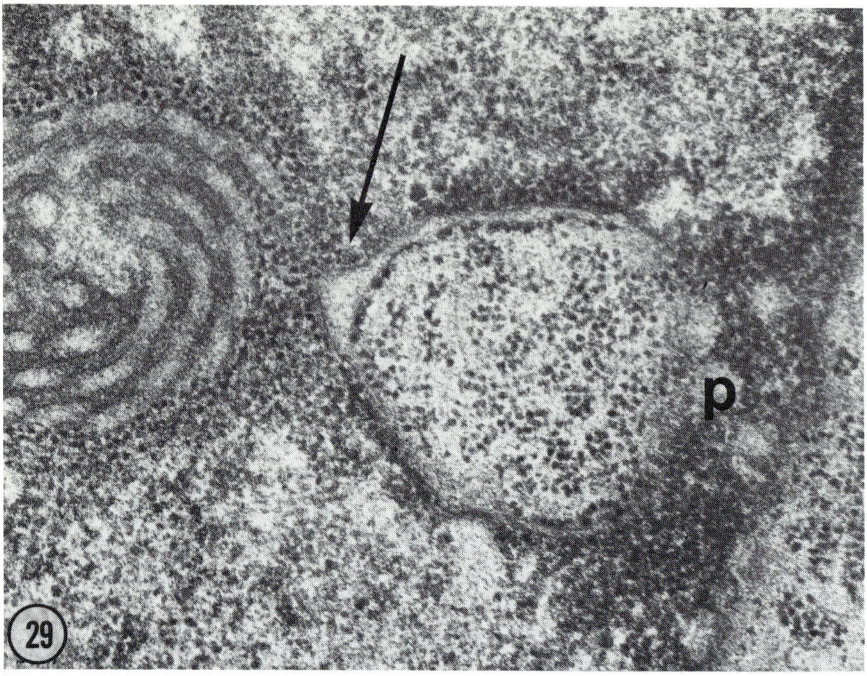

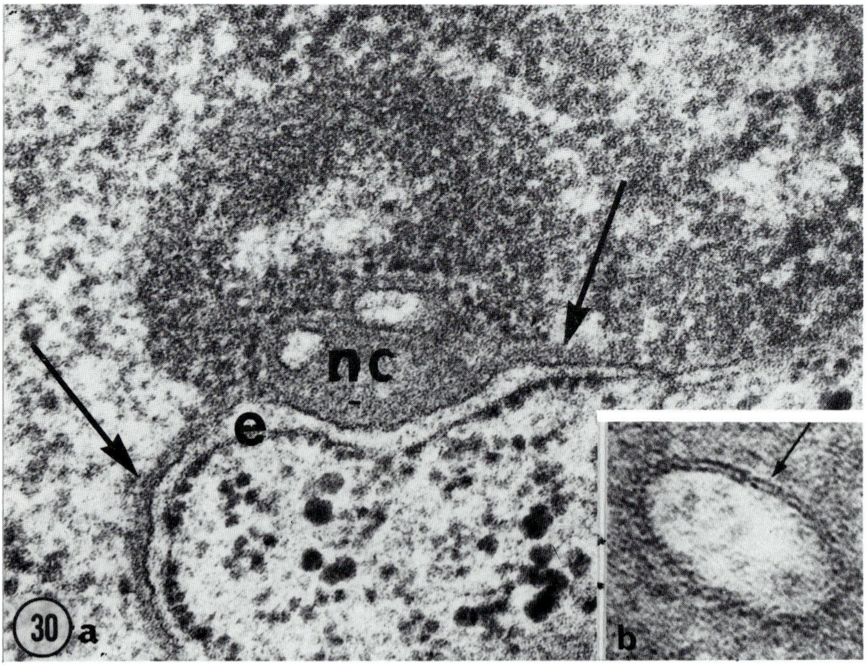

PLATE XV. FIG. 29. NCS developing from internal membrane of nuclear envelope (arrow). Nuclear pores (p) in cross section are numerous at this site. ×57,000. FIG. 30. (a) Developing NCS. A dense layer (arrows) deposited on the internal aspect

toplasmic ribosomes, are found at the periphery. These granules are evenly spaced in a single row and may surround the limits of the system of canals (Plate XVI, Fig. 31a) or girdle single tubules (Plate XVI, Fig. 31b). Their origin in the nucleolus is suggested in Plate XIV, Fig. 26 in which they are seen migrating from the nucleolus to the periphery of the NCS. In some views typical nucleoli impinge on the nucleolar channel body (Plate XIV, Figs. 26 and 28) and, although this association may be absent (Plate XIV, Figs. 27 and 28) this is probably due to errors of sampling. The canals are filled with fine fibrils (Plate XVI, Figs. 31a, 32, and 33 which Terzakis (1965) thought formed helical configurations. The canals feed into a central cavity (Plate XVI, Fig. 31a) filled with the same fibrillar elements.

The relationship of the NCS to the cytoplasm is obscure. It was not observed previously that the core is confluent with the cytoplasm through nuclear pores (Plate XVI, Figs. 32 and 33), which suggests that the NCS may consistently abut on the nuclear envelope (Plate XIV, Fig. 27). Serial sections are needed to establish this relationship firmly, as well as its association with normal nucleoli (Terzakis, 1965). Several NCS were followed in serial section in this study and were always found on the nuclear surface. This association was seen in only one or two sections, which indicates that the locus of contact is severely limited. In the secretory phase, invaginations of the nuclear envelope deep into the nucleus are extensive, and it is entirely possible that they are associated with the development of the NCS.

The channels were previously viewed as open to the cistern of the nuclear envelope (Terzakis, 1965; Sengel and Stoebner, 1970; Armstrong *et al.*, 1973), and their formation from the inner membrane of the envelope was inferred (Armstrong *et al.*, 1973). This interpretation is supported by sections which demonstrate a separation of the internal membrane of the envelope from the outer as it invaginates into the karyoplasm (Plate XV, Fig. 29). Further evidence is presented in Plate XV, Fig. 30a, in which the material surrounding the canals is seen to be derived from a dense amorphous lamina deposited on the internal aspect of the envelope. This suggests that the basic architecture of the NCS could be achieved by invagination of the inner nuclear membrane, carrying an even, dense coat on its internal surface. As it thrusts into the nucleus, it then turns back on

---

of the nuclear envelope (e) is continuous with dense matrix surrounding channels (nc). ×80,000. (b) High magnification of nucleolar channel demonstrating unit membrane (arrow). ×150,000.

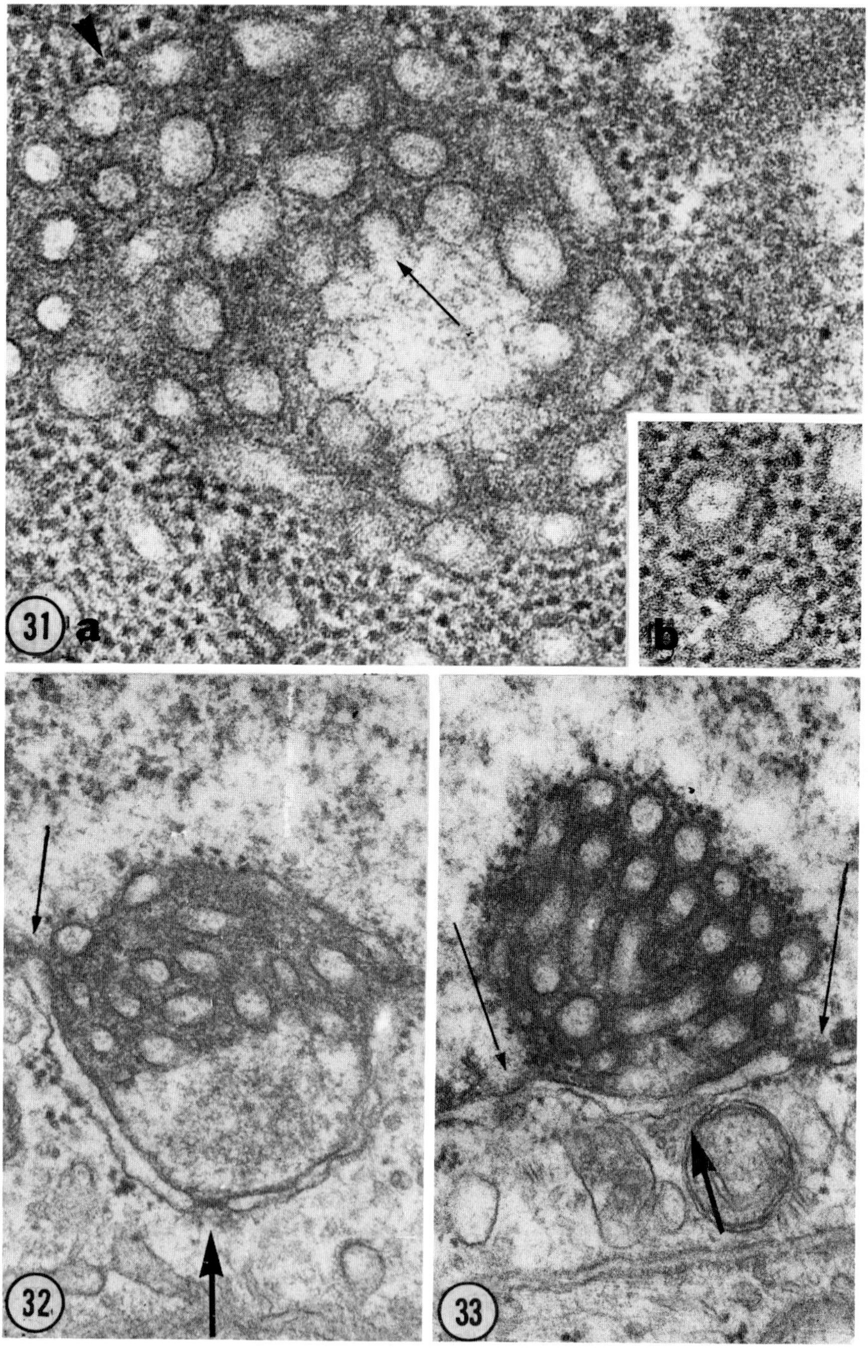

31
a
b
32
33

itself many times before rejoining the nuclear surface, forming a curious configuration of alternating lucent channels and dense layers. This postulate implies production of a dense lamina on the nuclear surface of the envelope, which develops in association with the membrane system of the NCS. The channel system is always found at a locus where the nuclear envelope contains numerous nuclear pores (Plate XV, Fig. 29). Further suggestive evidence for nucleocytoplasmic communication at this site, is the curious "halo" around the NCS (Plate XIV, Fig. 27; Plate XVI, Figs. 32 and 33), implying that substances from the nucleus are rapidly passing to the NCS or to the nuclear pores adjacent to the body (Plate XVI, Fig. 33).

One of the most puzzling aspects of the life cycle of the NCS is that terminal stages were not found. This led to the conclusion that they must disappear abruptly (Kohorn *et al.*, 1972). Because they can protrude from the nucleus, they were throught to pass into the cytoplasm and disintegrate, but they were never found outside the nucleus (Clyman, 1963a,b; Ancla *et al.*, 1964b; Ancla and de Brux, 1965).

Evidence for its final stages may be seen in Plate IX, Figs. 13 and 14, taken from a 22-day endometrium. The precise, single-file arrangements of dense particles around the channels is disorganized. Instead they have accumulated in large masses around the periphery and between channels. The channels are greatly dilated. The resemblance between the cisterns of the nuclear envelope, ER, and canaliculi is striking. Comparison of the vesiculated channels with the envelope and ER suggest that they may be undergoing identical changes. This further indicates that the channel system is a modification of the membrane system of the cell with which it may be continuous. The NCS may disappear as it gradually returns to the nuclear surface and become integrated with the rest of the nuclear envelope. This process would be a reversal of its formative stages.

PLATE XVI. FIG. 31. (a) Section of NCS where channels are opening into central cavity (arrow). The channels as well as the central cavum are filled with a fibrillar substance. Channels are embedded in dense material which has a margin of reduced density containing high-contrast particles (arrowhead). ×76,000. (b) Single channels completely surrounded by dense granules. ×80,000. FIG. 32. NCS abutting on nuclear envelope. A nuclear pore (thick arrow) connects the central core with the cytoplasm. Another nuclear pore is seen bounding the channel system (fine arrow) and the limits of a region of less dense karyoplasm around the channel system. ×43,000. FIG. 33. NCS continuous with cytoplasm through nuclear pore in envelope (thick arrow). The channel system is surrounded by less dense karyoplasm and bounded on either side by nuclear pores (fine arrows). ×45,000.

A typical nucleolus is composed of granules and fibrils (for review, see Busch and Smetana, 1970) which are RNAase-sensitive (Marinozzi and Bernhard, 1963; Marinozzi, 1964). The granules and fibrils of the NCS are sensitive to protease and RNAase digestion (M. Gordon, unpublished observations), indicating that they are composed of RNA–protein complexes. The nucleolus is the site of production of most of the rRNA in the cell (Muramatsu *et al.*, 1964; Perry, 1964), as well as some species of mRNA (Busch and Smetana, 1970). Transport of nucleolus-synthesized substances to the cytoplasm has always presented a logistic problem (Busch and Smetana, 1970), since the nucleolus is normally deep within the nucleus. The approximation of the nucleolus to the nuclear envelope in fine-structural studies has been viewed as a fixation artifact (Busch and Smetana, 1970). In the case of the NCS, however, its intimate association with the envelope cannot be questioned. Since it is also confluent with the nucleolus, it may function as a nucleolus-cytoplasm transport system (Ancla and de Brux, 1965; Armstrong *et al.*, 1973). The significance of the precise arrangement of high-contrast particles around the periphery of the channel system is puzzling. In other cells nucleolar ribosomelike particles have been seen (Smetana *et al.*, 1968) migrating from the nucleolus to the nuclear envelope (Lane, in Busch and Smetana, 1970, p. 5). If it is borne in mind that cellular organelles are in a dynamic rather than a static state, the membranes of the NCS may be in constant flux with the rest of the nuclear envelope, hence with the ER. The granules of the channel system, which undoubtedly represent a highly organized form of RNP, could be funneled to the cytoplasm via the membranes of the NCS and nuclear envelope. As mentioned, when its role in cellular metabolism is over, it returns to the nuclear envelope. The loss of the granules and disintegration of the channels when estrogen is administered (Gordon *et al.*, 1973) suggest an abnormal precipitous interruption of the final stages.

Another, perhaps additional, function of the granules is the synthesis of moieties which form all or part of the fibrillar elements in the channels. The fibrils may migrate to the central core (Plate XVI, Fig. 31a), where they may contribute to substances in the cytoplasm through nuclear pores (Plate XVI, Figs. 32 and 33). Intracanalicular fibrils have not been seen on the cytoplasmic site of the pores. Alternatively, the nuclear pores at the central cavum provide for a progression of proteins from cytoplasm to nucleus (Means and O'Malley, 1972), possibly to activate genetic mechanisms for cellular differentiation (Armstrong *et al.*, 1973). In estrogen-primed tissue, progesterone accentuates the development of cytoplasmic elements, such as giant mitochondria and glycogen. This is documented in studies of normal endometrium (Cartier and Moricard, 1960; Moricard and Moricard, 1964; Ancla and de Brux, 1965; Moricard, 1966; Wynn and Woolley, 1967; Sengel and Stoebner, 1970; Armstrong *et al.*, 1973), as well as in clinical observations (Gordon *et al.*, 1973; Gore and Gordon, 1974). A similar mechanism for formation of the NCS was demonstrated *in vitro* (Kohorn *et al.*, 1970). Subsequently, the 17-acyl group on the progesterone molecule was revealed as a critical factor in promoting its differentiation (Kohorn *et al.*, 1972). The maturation of other nuclear tubular structures which appear in mouse and rat trophoblast II also necessitates a balance between circulating estrogen and progesterone (Carlson and Ollerich, 1969). Amplification of nucleolar synthesis by estrogen (Hamilton *et al.*, 1968; Means and O'Malley, 1972), androgen, and other hormones (Tata, 1968) has been established. In the endometrium the data that show stimulus for the formation of the NCS to be dependent on progesterone suggest this hormone may have a direct effect on nucleolar activation.

In summary, the NCS may not be a nucleolar structure, but a unique differentiation of the membrane system of the cell which penetrates deep within the nucleus to approximate the site of the nucleolus. The extraordinary organization of this structure implies that it must play a significant role in uterine function. Although, as pointed out by Terzakis (1965), the appearance of the NCS coincides with a decline in total RNA synthesis (Stein and Stuermer, 1951), it is manifest when enzymic activity is maximal (Ancla and de Brux, 1965) and the production of message for the synthesis of specific proteins cannot be ruled out. The NCS could function as: (1) a feedback mechanism between cytoplasm and nucleus for the stimulation of nucleolar synthesis, (2) a site for the rapid transport of RNA to the cytoplasm, and (3) an independent site for the production of specific RNA and/or proteins. Its absence in infertile epithelium (Gore and Gordon, 1974), its deterioration by estrogenic compounds such as diethylstilbesterol (Gordon *et al.*, 1973) which are known to interefere with implantation (Morris and Van Wagenen, 1966), and its timing in the cycle all suggest a role in implantation of the blastocyst.

V. Glycogen Synthesis

The deposition of glycogen in the endometrium is unique, because it is not influenced by carbohydrate intake or exercise (Gregoire *et al.*, 1973). The amount of glycogen synthesized during the secretory portion of the cycle is an impressive phenomenon which has no counterpart in other cells. Fine structure has supported biochemical evidence (Stuermer and Stein, 1962; Hagerman and Villee, 1953; Hughes *et al.*, 1963) that synthesis is begun during the follicular stage (Gompel, 1962; Nilsson, 1962a; Themann and Schunke, 1963; Wynn and Harris, 1967; Colville, 1968; Armstrong *et al.*, 1973; Schueller, 1973), before it can be visualized cytochemically in the light microscope (Gompel, 1962; Wynn and Harris, 1967). Estrogen probably is the controlling factor through its stimulation of glycogen synthetase (Hughes *et al.*, 1963). Estrogen alone, as tested in ovariectomized or menopausal women is relatively ineffectual for glycogen storage, but the addition of progesterone to estrogen-primed tissue results in an immediate accumulation of glycogen (Hughes *et al.*, 1963, 1969; Moricard and Moricard, 1964; Luginbuhl, 1968; Kohorn and Tchao, 1969; Kohorn *et al.*, 1970, 1972).

Before ovulation glycogen synthesis is extremely uneven and varies from one specimen to another. It has been noted as early as the seventh (Sakuma, 1970; Armstrong *et al.*, 1973) or ninth day (Themann and Schunke, 1963). It may be copious in some tissue at earlier times than in others (Colville, 1968). In our study it was seen in one specimen in good amount on day 12, but was sparse in another at day 15. Generally, it can be found on day 11 or 12 (Plate V, Fig. 6). Basal accumulation after ovulation is common but not universal (Cavazos *et al.*, 1967) by day 17. The variability in timing suggests that the physiological state of the tissue is not identical in all cycles, or that the glandular epithelium does not respond in a synchronous fashion throughout. Massive glycogen deposition is established by day 18 or 19 (Plate VIII, Fig. 12) (Cartier, 1959; Wessel, 1960; Moricard and Moricard, 1964; Moricard, 1966; Wynn and Woolley, 1967; Armstrong *et al.*, 1973). Following its collection at the basal pole, glycogen is then found in gradually increasing concentrations at the apex (Plate VII, Fig. 10). Localization at the apex is coordinated with basal diminution. Traditionally, this has been interpreted to mean that glycogen migrates through the cytoplasm from base to apex. The intracellular displacement of glycogen has been attributed to microtubules (Cavazos *et al.*, 1967) or cytoplasmic flow (Armstrong *et al.*, 1973). Both mechanisms raise complications which are not easily resolved. Microtubules and tonofilaments, although present (Plate XIII, Figs. 23 and 24), are not necessarily involved in glycogen transport (Armstrong *et al.*, 1973). The association with glycogen observed by Cavazos *et al.* (1967) may be fortuitous, since glycogen is scattered throughout the cytoplasm. Cavazos *et al.* (1967) were of the opinion that the undulating character of the microtubules, which differentiates them from those that run a straight course, implies functional specialization. Our observations concur that the microtubules undulate, but this does not offer sufficient ground to give them functions other than those common to microtubules. The concept of movement of glycogen from the base to the apex by either microtubules or cytoplasmic streaming implies that there is a net flow of cytoplasm to the apical region, and some trapping mechanism for the capture of glycogen at the new site. It is significant that the sites where glycogen is located are notably free of other organelles. Displacement of organelles and homogeneous filling of the ground cytoplasm is suggestive of *in situ* synthesis rather than collection from other loci. The hypothesis presented here, that synthesis of glycogen is biphasic and glycogen at the base is degraded while synthesis at the apex is stimulated, may be a means of reconciling these data. Such a mechanism is not more cumbersome than migration of highly polymerized glycogen from one pole of the cell to the other, and is related to metabolic pathways for both the breakdown and synthesis of glycogen that occur in this tissue.

The control of glycogen synthesis by ovarian steroids is complex. Both anabolic and catabolic pathways are active at the same time in proliferative endometrium (Delforge, 1969), since there is a rise in glucose-6-phosphatase activity (Hughes *et al.*, 1963) as well as glycogen synthetase. The relationship between glycogen degradation and glycogen buildup is even more complicated in the secretory phase. Many years ago, Hagerman and Villee (1953) pointed out that glycogen is probably consumed to the greatest extent at the time of its maximum production. The stimulation of phosphorylase during the midsecretory stage (Koide and Nakayama, 1966; Hughes *et al.*, 1969) provides evidence for glycogen breakdown during a period of enormous glycogen production. The intimate morphological relationship between giant mitochondria and basal glycogen deposits could promote reciprocal metabolic relationships. Mitochondrial ATP would promote glycogen synthesis, by production of UTP from uridine 5′-phosphate. Glycogen could be degraded to substrates for the citric acid cycle, either by glycolysis or the pentose phosphate shunt. It is of historical interest that early biochemical studies showed an association between massive glycogen synthesis and a rise in the enzymes of the citric acid cycle (Marcuse, 1957, Fuhrmann, 1960). These data were collected just prior to the discovery under the electron microscope of giant mitochondria and their association with glycogen. Glycogen could also produce a pool of substrates for the secretion of glycoproteins and mucopolysaccharides (Francini, 1958; Cartier and Moricard, 1960; Schmidt-Matthiesen, 1963; Themann and Schunke, 1963; Moricard and Moricard, 1964; Cavazos *et al.*, 1967; Sengel and Stoebner, 1970). The further possibility that glycogen storage is related to events occurring near nidation cannot be ruled out (Zondek and Hestrin, 1947; Koide and Nakayama, 1966; Delforge, 1969).

VI. Uterine Secretion

Traditionally, the middle of the postovulatory phase, between ovulation and menses, is considered the time of maximal secretion (Noyes *et al.*, 1950; Witt, 1963). Domination of the cells by glycogen has led to the premise that glycogen is a major component of uterine secretion at this time. While the glands become engorged during the luteal phase, there is little evidence that the secretory product enters the lumen of the uterus in appreciable amounts. Estimates of uterine secretion by monitoring the flow through the cervical canal choose the time near ovulation as the period of maximal uterine secretion. Flow through the cervical canal increases a few days before ovulation and continues for a few days afterward (Strauss, 1962; Datnow, 1973). Fine-structural examination of glandular epithelium reveals that the fluid is probably derived from a large population of granules and vesicles of varying density and size which proliferate in the early stages (Gompel, 1962; Nilsson, 1962a) (Plate II, Fig. 2) and reach their height before ovulation (Wessel, 1960). At ovulation there is a marked decline in secretory bodies in the apical cytoplasm (Cavazos *et al.*, 1967) (Plate V, Fig. 6), and evidence for merocrine secretion (Plate VI, Fig. 8) (Wessel, 1960; Themann and Schunke, 1963; Cavazos *et al.*, 1967; Wynn and Harris, 1967). These data indicate that the proliferative phase is the period of heightened endogenous secretion, perhaps the secretory peak of the entire cycle (Sengel and Stoebner, 1970). Wessel (1960) proposed that secretory granules extruded after ovulation are produced earlier and retained in the cells.

At midcycle fluid flows readily from the glandular lumen into the cavity of the uterus. It is composed of glycoproteins, mucopolysaccharides, and lipids. By autoradiography, using $Na_2{}^{35}SO_4$ as label, Moricard (1966) showed conclusively that sulfated mucopolysaccharides are a secretory product of the epithelium. Glucose, produced from glycogen, may also be a component of secretory fluid. This may aid sperm migration by flushing the cervical canal and providing a medium of low viscosity for motility (Datnow, 1973). It may also favor viability of sperm by providing nutrients and an optimal physiological medium (Moricard, 1966; Delforge, 1969; Datnow, 1973). Since sperm are profoundly altered by the uterine environment (Gordon *et al.*, 1974), where they become capacitated (Austin, 1964), the secretion of the glands must be of some importance. Uterine secretion at ovulation has been related to the effects of estrogen (Strauss, 1962), which predominates in the proliferative phase.

In conflict with this view are data showing that glands continue development during the luteal phase, becoming more and more engorged (Witt, 1963). It is therefore not surprising to find secretory bodies at the apical pole of the cell (Borell *et al.*, 1959; Cartier, 1959; Gompel, 1962; Nilsson, 1962b; Cavazos *et al.*, 1967; Sengel and Stoebner, 1970; Armstrong *et al.*, 1973). The released granules are PAS-positive and diastase-resistant (Borell *et al.*, 1959; Schmidt-Matthiesen, 1963), and are undoubtedly mucopolysaccharides (Boutselis *et al.*, 1963; Yaneva *et al.*, 1965). The development of luminal cytoplasmic processes is further evidence of secretory activity (Borell *et al.*, 1959; Colville, 1958). It therefore appears reasonable that merocrine secretion continues into the middle of the secretory phase (Nilsson, 1962b; Cavazos *et al.*, 1967; Wynn and Woolley, 1967).

In a reappraisal of the secretory function of the endometrium, Datnow (1973) has questioned that secretion is released from the glands into the uterine cavity in mid- and late secretory stages. Rather, he suggests that distention of the glands serves to narrow the uterine cavity and thus provide for blastocyst "grasp." Progesterone acts on the mucilagenous substances in the glandular cava, rendering them less soluble (Schmidt-Matthiesen, 1963; Sengel and Stoebner, 1970; Datnow, 1973). This creates a firm, gelated substratum which maintains an occluded uterine lumen. This picture of secretion rules out glycogen as a component of the secretory product (Datnow, 1973).

Fine-structural studies of secretory endometrium have added many details but have not provided clarification of the question of glycogen secretion. In light microscope tissue sections of early and midsecretory endometrium, glycogen is enclosed in cytoplasmic blebs which protrude into the lumen (Noyes *et al.*, 1950; Bontke, 1960; Schmidt-Matthiesen, 1963; Witt, 1963). These are identified in fine structure with the added evidence for manifold polypoid extensions into the lumen (Cartier, 1959; Wessel, 1960; Gompel, 1962; Themann and Schunke, 1963; Cavazos *et al.*, 1967; Colville, 1968;

Armstrong *et al.*, 1973). On ultrastructural examination the large apical segment is seen to be free of organelles and homogeneously filled with glycogen. Therefore glycogen deposition at the apex appears to be a true intracellular specialization.

The final event that would provide glycogen to the glandular lumen (Sakuma, 1970) is desquamation of the apical surface. There is not complete agreement, however, that this occurs. Gompel (1964) and Sengel and Stoebner (1970), although they note the apical deposition of glycogen, see no evidence of glycogen extrusion. However, fragmentation of the luminal surface has been reported in a great number of studies (Themann and Schunke, 1963; Cavazos *et al.*, 1967; Wynn and Woolley, 1967; Colville, 1968; Armstrong *et al.*, 1973) and apocrine sloughing of the apex was considered by Delforge (1969) to be promoted by the endogenous activity of enzymes derived from lysosomes. Dehiscing of the apical portion of the cell is considered to be a factor in the loss of glycogen and the return to an appreciably shorter cell dimension a few days before menses (Themann and Schunke, 1963; Witt, 1963). However, the marked decrease in cell height is correlated with shrinkage and pycnosis of the cytoplasm, which suggests that other factors are involved in this transformation. In support of the idea that glycogen does not enter the glandular lumen, it should be pointed out that desquamation, the morphological index of apocrine secretion, could be caused by artifactual disruption of the luminal surface during tissue preparation. Several investigators, for example, have evidence that the apical limits of the cell are unduly fragile at the end of the secretory phase (Themann and Schunke, 1963; Cavazos *et al.*, 1967; Colville, 1968).

The marked decline in glycogen stores may be the result of glycogen utilization for endogenous needs, possibly by supplying substrates for mucopolysaccharide and glycoprotein secretion (Francini, 1958; Cartier and Moricard, 1960; Schmidt-Matthiesen, 1963; Themann and Schunke, 1963; Moricard and Moricard, 1964; Cavazos *et al.*, 1967; Sengel and Stoebner, 1970). The localization of glycogen at the apex and in apical proliferations that follows basal synthesis suggests its release from the cell, since such a change in location would not be a prerequisite to serve endogenous needs. The presence of α-amylase in the fallopian tube (Green, 1957) has been presented as a mechanism for the breakdown of glycogen in the uterine cavity. Finally, as in proliferative endometrium, glycogen could be degraded to small carbohydrate moieties and be secreted into the uterine fluid.

VII. Hormone Action

In the proliferative phase estrogen stimulates the genome (McKerns, 1967) for the synthesis of protein, resulting in growth and replication of the epithelium. Estrogen has an effect on specialized proteins as well, such as production of membrane alkaline phosphatase and perhaps stimulation of preexisting or *de novo* synthesis of enzymes related to glucose metabolism (McKerns, 1967; Wynn and Woolley, 1967). It also elaborates the secretory mechanism of the cell and differentiates the organelles of the luminal surface.

The effect of progesterone on cells already treated to high levels of estrogen results in formation of giant mitochondria and of the NCS, proliferation of microvilli, and enhanced synthesis of glycogen and phosphorylase activity. Although the total activity of RNA synthesis is greater during the follicular phase (Segal and Scher, 1967), progesterone may stimulate production of unique species of proteins that would facilitate implantation. This speculation is based on the circumstantial evidence that infertile endometria (Gore and Gordon, 1974) or tissue deficient in progesterone stimulation (Kohorn *et al.*, 1970, 1972) do not differentiate the NCS.

ACKNOWLEDGMENTS

The author is indebted to colleagues in the Department of Obstetrics and Gynecology, Yale University School of Medicine, for their generous help in obtaining biopsies of endometrium.

This research was supported by a grant from the Ford Foundation.

REFERENCES

Ancla, M., and de Brux, J. (1965). *Obstet. Gynecol* **26**, 23.

Ancla, M., de Brux, J., and Belaisch J. (1964a). *Rev. Fr. Etud. Clin. Biol.* **9**, 490.

Ancla, M., de Brux, J., Belaisch, J., and Musset, R. (1964b). *Gynecol. Obstet.* **63**, 365.

Armstrong, E. M., More, I. A., McSeveney, D., and Chatfield, W. R. (1973). *J. Obstet. Gynaecol. Brit. Commonw.* **80**, 446.

Austin, C. R. (1964). *Nature (London)* **170**, 326.

Baba, N., Vidyarthi, S. C., and Vonhaam, E. (1968). *Fed. Proc., Fed. Amer. Soc. Exp. Biol.* **27**, 541.

Bartelmez, G. W. (1933). *Contrib. Embryol.* **24**, 143.

Berger, J., and Mumprecht, E. (1959). *Schweiz. Med. Wochenschr.* **89**, 433.

Bitensky, L., and Cohen, S. (1965). *J. Obstet. Gynaecol. Brit. Commonw.* **72**, 769.

Bontke, E. (1960). *In* "Les fonctions de nidation uterine et leurs troubles," p. 269. Masson, Paris.

Borell, U., Nilsson, O., and Westmann, A. (1959). *Acta Obstet. Gynecol. Scand.* **38**, 364.

Boutselis, J. G., DeNeef, J. C., Ullery, J. C., and George, O. T. (1963). *Obstet. Gynecol.* **21**, 423.

Busch, H., and Smetana, K. (1970). "The Nucleolus." Academic Press, New York.

Carlson, E. C., and Ollerich, D. A. (1969). *J. Ultrastruct. Res.* **28**, 150.

Cartier, R. (1959). *Bull. Fed. Soc. Gynecol. Obstet. Lang. Fr.* **11**, 463.

Cartier, R., and Moricard, R. (1959). *Gynecol. Obstet* **58**, 477.

Cartier, R., and Moricard, R. (1960). *Bull. Soc. Roy. Belge Gynecol. Obstet.* **30**, 325.

Cavazos, F., and Lucas, F. V. (1968). *Amer. J. Pathol.* **52**, 31A.

Cavazos, F., and Lucus, F. V. (1970). *Amer. J. Obstet. Gynecol.* **106**, 434.

Cavazos, F., Green, J. A., Hall, D. G., and Lucas, F. V. (1966). *Amer. J. Obstet. Gynecol.* **94**, 824.

Cavazos, F., Green, J. A., Hall, D. G., and Lucas, F. V. (1967). *Amer. J. Obstet. Gynecol.* **99**, 833.

Clyman, M. J. (1963a). *Amer. J. Obstet. Gynecol.* **86**, 430.

Clyman, M. J. (1963b). *In* "Progress in Gynecology" (J. V. Meigs and S. Sturgis, eds.), p. 36. Grune & Stratton, New York.

Clyman, M. J. (1963c). *Fert. Steril.* **14**, 352.

Cohen, S., Bitensky, L., Chayen, J., Cunningham, G. J., and Russell, J. K. (1964). *Lancet* **2**, 56.

Colville, E. A. (1968). *J. Obstet. Gynaecol. Brit. Commonw.* **75**, 342.

Dallenbach-Hellweg, G. (1971). "Histopathology of the Endometrium," p. 22. Springer-Verlag, Berlin and New York.

Datnow, A. D. (1973). *J. Obstet. Gynaecol. Brit. Commonw.* **80**, 865.

Delforge, J. P. (1969). *Bull. Soc. Roy. Belge Gynecol. Obstet.* **39**, 97.

DePalo, A., and Stoppelli, I. (1959). *Monit. Obstet.-Ginecol. Endocrinol. Metab.* [N.S.] **1**, 1.

Dubrauszky, V., and Pohlmann, G. (1960). *Naturwissenschaften* **47**, 523.

Dubrauszky, V., and Pohlmann, G. (1961a). *Electron. Microsc., Proc. Eur. Reg. Conf. 2nd, 1960* Vol. II, p. 862.

Dubrauszky, V., and Pohlmann, G. (1961b). *Arch. Gynaekol.* **196**, 180.

Estable, C., and Satelo, J. R. (1951). *Inst. Invest. Cienc. Biol.* (*Montevideo*) **1**, 105

Falconer, B. (1946). *Acta Obstet. Gynecol. Scan.* **26**, 453.

Ferenczy, A., and Richart, R. M. (1973). *J. Clin. Endocrinol. Metab.* **36**, 999.

Ferenczy, A., Richart, R. M., Agate, F. J., Jr., Purkerson, M. L., and Dempsey, E. W. (1972). *Fert. Steril.* **23**, 515.

Fleming, S., Tweeddale, D. N., and Roddock, J. W., Jr. (1968). *Amer. J. Obstet. Gynecol.* **102**, 186.

Francini, M. A. (1958). *Ann. Ostet. Ginecol.* **80**, 333.

Fruin, A. H., and Tighe, J. R. (1967). *J. Obstet Gynaecol. Brit. Commonw.* **74**, 93.

Fuchs, M. (1959). *Acta Anat.* **39**, 244.

Fuhrmann, K. (1960). *Zentralbl. Gynaekol.* **82**, 1874.

Gompel, C. (1962). *Amer. J. Obstet. Gynecol.* **84**, 1000.

Gompel, C. (1964). *J. Microsc.* (*Paris*) **3**, 427.

Gompel, C. (1967). *Proc. World Congr. Fert. Steril., 5th, 1967* p. 41.

Gordon, M., Kohorn, E. I., Gore, B. Z., and Rice, S. I. (1973). *J. Reprod. Fert.* **34**, 375.

Gordon, M., Dandekar, P. V., and Bartoszewicz, W. (1974). *J. Reprod. Fert.* **36**, 211.

Gore, B. Z., and Gordon, M. (1974). *Fert. Steril.* **25**, 103.

Green, C. L. (1957). *Amer. J. Obstet. Gynecol.* **73**, 402.

Greep, R. O. (1966). "Histology." McGraw-Hill, New York.

Gregoire, A. T., Ledger, W. D., and Morgan, M. J. (1973). *Fert. Steril.* **24,** 198.

Hagerman, P. D., and Villee, C. A. (1953). *Endocrinology* **53,** 667.

Hamilton, T. H., Widnell, C. C., and Tata, J. R. (1968). *J. Biol. Chem.* **243,** 408.

Hamperl, H. (1950). *Virchows Arch. Pathol. Anat. Physiol.* **319,** 265.

Hando, T., Okada, D. M., and Zamboni, L. (1968). *J. Cell Biol.* **39,** 475.

Hertig, A. T. (1945). "Proceedings of the Conference on Diagnosis in Sterility." Thomas, Springfield, Illinois.

Hoffmeister, H., and Schulz, H. (1961). *Beitr. Pathol. Anat. Allg. Pathol.* **124,** 415.

Hughes, E. C., Jacobs, R. D., Rubelis, A., and Husney, R. M. (1963). *Amer. J. Obstet. Gynecol.* **85,** 594.

Hughes, E. C., Demers, L. M., Csermeley, T., and Jones, D. B. (1969). *Amer. J. Obstet. Gynecol.* **105,** 707.

Johanisson, E., and Nilsson, L. (1972). *Fert. Steril.* **23,** 613.

Kohorn, E. I., and Tchao, R. (1969). *J. Endocrinol.* **45,** 401.

Kohorn, E. I., Rice, S. I., and Gordon, M. (1970). *Nature (London)* **228,** 671.

Kohorn, E. I., Rice, S. I., Hemperly, S., and Gordon, M. (1972). *J. Clin. Endocrinol. Metab.* **34,** 257.

Koide, Y., and Nakayama, T. (1966). *J. Jap. Obstet. Gynecol.* **13,** 123.

Komae, Y. (1968). *J. Jap. Obstet. Gynecol.* **20,** 741.

Kotz, J., and Parker, E. (1939). *Endocrinology* **24,** 447.

Kucera, F. (1964). *Zentralbl. Gynaekol.* **86,** 1332.

Lanzavecchia, G., and Morano, E. (1958). *Minerva Ginecol.* **10,** 882.

Luginbuhl, W. H. (1968). *Amer. J. Obstet. Gynecol.* **102,** 192.

McKay, D. G., Bardawil, W. A., Velardo, J. T., and Hertig, A. T. (1956). *Obstet. Gynecol.* **8,** 22.

McKerns, K. W. (1967). *In* "Cellular Biology of the Uterus" (R. M. Wynn, ed.), Appleton, New York.

Maddi, F. V., and Papanicolaou, G. N. (1961). *Amer. J. Obstet. Gynecol.* **82,** 99.

Marcuse, P. M. (1957). *Amer. J. Clin. Pathol* **28,** 539.

Marinozzi, V. (1964). *J. Ultrastruct. Res.* **10,** 433.

Marinozzi, V., and Bernhard, W. (1963). *Exp. Cell Res.* **32,** 595.

Means, A. R., and O'Malley, B. W. (1972). *Metab. Clin. Exp.* **21,** 357.

Moricard, R. (1958). *Bull. Soc. Roy. Belge Gynecol. Obstet.* **28,** 407.

Moricard, R. (1966). *Arch. Gynaekol* **203,** 85.

Moricard, R., and DeSenarclens, F. (1947). *C. R. Soc. Biol.* **141,** 57.

Moricard, R., and Moricard, F. (1964). *Gynecol. Obstet.* **63,** 203.

Morris, J. Mc., and Van Wagenen, G. (1966). *Amer. J. Obstet. Gynecol.* **96,** 804.

Muramatsu, M., Hodnett, J. L., and Busch, H. (1964). *Biochim. Biophys. Acta* **91,** 592.

Nilsson, O. (1962a). *J. Ultrastruct. Res.* **6,** 413.

Nilsson, O. (1962b). *J. Ultrastruct. Res.* **6,** 422.

Nilsson, O., and Nygren, K. G. (1972). *Uppsala J. Med. Sci.* **77,** 3.

Novak, E. (1947). "Gynecologic and Obstetric Pathology." Saunders, Philadelphia, Pennsylvania.

Novak, E., and Everett, H. S. (1928). *Amer. J. Obstet. Gynecol.* **16,** 499.

Novak, E., and Rutledge, F. (1948). *Amer. J. Obstet. Gynecol.* **55,** 46.

Noyes, R. W., and Haman, J. O. (1953). *Fert. Steril.* **4,** 504.

Noyes, R. W., Hertig, A. T., and Rock, J. (1950). *Fert. Steril.* **1,** 3.

Ober, K. G. (1950). *Klin. Wochenschr.* **28,** 9.

Perry, R. P. (1964). *Nat. Cancer Inst., Monogr.* **14,** 12.

Revel, J. P., Napolitano, L., and Fawcett, D. W. (1960). *J. Biophys. Biochem. Cytol.* **8,** 575.
Saksena K. C., Arora, M. M., Gupta, J. C., and Rangam, C. M. (1965). *Indian J. Med. Sci.* **19,** 121.
Sakuma, S. (1970). *Beitr. Pathol. Anat. Allg. Pathol.* **140,** 454.
Schroder, R. (1921). *Monatsschr. Geburtsh. Gynaekol.* **56,** 183.
Schueller, E. F. (1961). *Arch. Gynaekol* **196,** 49.
Schueller, E. F. (1968). *Obstet. Gynecol.* **31,** 215.
Schueller, E. F. (1973). *Obstet. Gynecol.* **41,** 188.
Schmidt-Matthiesen, H. (1963). "The Normal Human Endometrium." McGraw-Hill, New York.
Segal, J., and Scher, W. (1967). *In* "Cellular Biology of the Uterus" (R. M. Wynn, ed.), Appleton, New York.
Sengel, A., and Stoebner, P. (1970). *Z. Zellforsch. Mikrosk. Anat.* **109,** 245.
Smetana, K., Unuma, T., and Busch, H. (1968). *Exp. Cell Res.* **51,** 105.
Stein, R. J., and Stuermer, V. M. (1951). *Amer. J. Obstet. Gynecol.* **61,** 414.
Strauss, G. (1962). *Arch. Gynaekol.* **197,** 524.
Stuermer, V. M., and Stein, R. J. (1952). *Amer. J. Obstet. Gynecol* **63,** 359.
Tata, J. R. (1968). *Nature (London)* **219,** 331.
Terzakis, J. A. (1965). *J. Cell Biol.* **27,** 293.
Themann, H. (1967). *Proc. World Congr. Fert. Steril., 5th, 1967* p.42.
Themann, H., and Schunke, W. (1963). *In* "The Normal Human Endometrium" (H. Schmidt-Matthieson, ed.), p. 99. McGraw-Hill, New York.
Vecchietti, G., and Morano, E. (1959). *Int. J. Fert.* **4,** 109.
Vesterdal-Jorgensen, J. (1950). "Modern Trends in Obstetrics and Gynecology." Butterworth, London.
Wessel, W. (1960). *Z. Zellforsch. Mikrosk. Anat.* **51,** 633.
Wessel, W. (1961). *Virchows Arch. Pathol. Anat. Physiol.* **334,** 181.
Wetzstein, R. (1958). *Anat. Anz.* **105,** Suppl., 218.
Wetzstein, R., and Wagner, H. (1960). *Anat. Anz.* **108,** 362.
Witt, H. J. (1963). *In* "The Normal Human Endometrium" (H. Schmidt-Matthieson, ed.), p. 24. McGraw-Hill, New York.
Wynn, R. M., and Harris, J. A. (1967). *Fert. Steril.* **18,** 632.
Wynn, R. M., and Woolley, B. A. (1967). *Fert. Steril.* **18,** 721.
Yaneva, H., Lumbroso, P., and Netter, A. (1965). *Minerva Ginecol.* **17,** suppl., 78.
Zondek, B., and Hestrin, S. (1947). *Amer. J. Obstet. Gynecol.* **54,** 173.

The Ultrastructure of the Organ of Corti[1]

ROBERT S. KIMURA

Department of Otolaryngology, Massachusetts Eye and Ear Infirmary, Boston, Massachusetts, and Department of Otolaryngology, Harvard Medical School, Boston, Massachusetts

I. Introduction

The foundation of present-day knowledge of the structure of the inner ear was established by Corti (1851), Hensen (1863), Boettcher (1869), Retzius (1881, 1884), Held (1926), and Kolmer (1927). Further contributions to advance our knowledge of these structures did not come until the advent of electron microscopy which was first introduced by Engström and his associates (Engström and Wersäll, 1953a,b,c; Engström *et al.*, 1955). Their investigations were soon followed by those of Smith (1955), Smith and Dempsey (1957), Spoendlin (1957), Iurato and Bairati (1959), and Friedmann (1959). These individuals are considered the pioneers of fine morphology in

[1] This work was supported by U. S. Public Health Grant 5 R01 NS03932-13.

otology. Their publications following these initial contributions are numerous, and other investigators such as Flock *et al.* (1962), Hilding and Wersäll (1962), Kimura and Wersäll (1962), Duvall and Wersäll (1964), Kimura *et al.* (1964), Kikuchi and Hilding (1965), Rodriguez-Echandia (1967), Bredberg *et al.* (1970). Lim and Melnick (1971), Angelborg and Engström (1974), and Densert (1974) have added to our knowledge of the ultrastructure of the organ of Corti. Their transmission electron microscope studies have been further augmented by studies of surface structures with the scanning electron microscope (Lim, 1969, 1972; Engström *et al.*, 1970; Bredberg *et al.*, 1972).

Although a wealth of information is available about the sense organs of lower animals, this article is limited to the mammalian organ of Corti in order to describe all the essential details, with special emphasis on mechanoreceptors and neural elements. Extensive descriptions of the fine structure of the organ of Corti were provided earlier by Engström and Wersäll (1958), Iurato (1961), Spoendlin (1966), Engström (1967), Smith (1968), and Wersäll (1973). This article is based on these descriptions; however, a large amount of information is provided from our own materials which have accumulated over the years.

There is a wide gap between neurophysiological and anatomical information on this sense organ. To date, no clear distinction between neural responses from either outer or inner hair cells can be made. Obviously, the lack of morphological information appears to contribute greatly to the confusion in interpretation of electrical activities generated by the organ of Corti. It is hoped that this report will add a new dimension to our existing knowledge of fine morphology, and provide a firm foundation for future study.

The organ of Corti is a papillary structure resting on the basilar membrane, which is composed of sensorineural and supportive elements (Fig. 1). It is specially designed to convert mechanical vibrations into electrical events which are transmitted to the central nervous system as coded messages (Kiang, 1965). There are two types of mechanoreceptors, outer and inner hair cells. The inner hair cells are supported by inner pillar cells, inner phalangeal cells, and border cells. The outer hair cells are supported by outer and inner pillar cells and Deiters' cells. More peripheral to these structures are inner sulcus cells, Hensen's cells, Claudius cells, and Boettcher cells. The bases of all these cells rest on the basilar membrane; at the top the stereocilia of the hair cells attach to the tectorial membrane. The general organization of the organ of Corti suggests that it is most

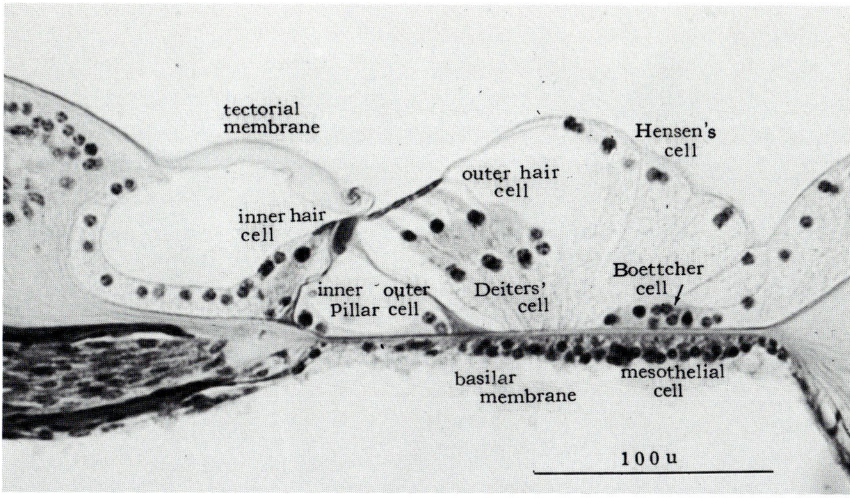

FIG. 1. Light micrograph showing the organ of Corti at the basal turn in the guinea pig.

uniquely designed to transmit small mechanical movement effectively into the hair cells, and at the same time to withstand strong mechanical vibrations.

II. The Tectorial Membrane

A. Structure

The tectorial membrane is an acellular layer essentially composed of fibrils, filaments, and a homogeneous substance (Fig. 2A). It extends from the phalanx of the interdental cells of the limbus spiralis to the outer margin of the reticular lamina, and possibly beyond to the Hensen's and Claudius cells (Fig. 2B). The main part of the membrane is occupied by fibrils radially directed and inclined at about 30° toward the apical turn of the cochlea (Iurato, 1960; Lim, 1972). These fibrils are about 100–130 Å in diameter and unbranched; their length is unknown (Iurato, 1960). These fibers do not show periodicity. At the outer margin of the tectorial membrane, the fibrils run longitudinally at right angles to the transverse ones. They do not make direct contact with the interdental cells nor with the stereocilia of the hair cells. The fibrils appear to originate and terminate in a dense homogeneous substance which is granular or filamentous, depending on the particular method used in preparing the

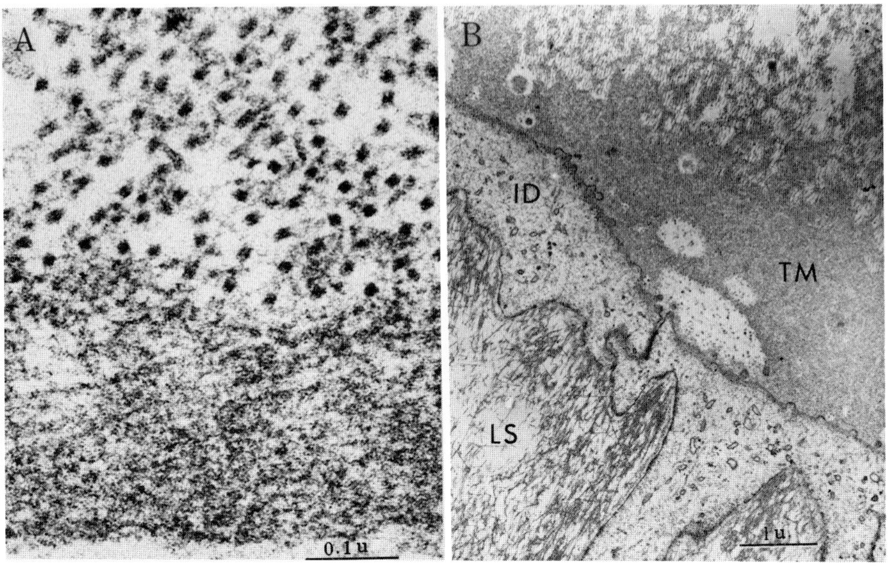

FIG. 2. The tectorial membrane. Squirrel monkey. (A) The homogeneous substance faces the hair cells. The fibrils above this layer are similar to those of the basilar membrane. (B) Attachment of the tectorial membrane (TM) to the interdental cell (ID). LS, Fibrils of the limbus spiralis.

specimen. The homogeneous substance is diffusely spread throughout, but is more condensed at the periphery than in the center, particularly at the interdental cells, Hensen's stripe, stereocilia tops, and border network (*Randfasernetz*) of the outer margin.

B. Attachments

The attachment of the tectorial membrane to the organ of Corti is not clear, primarily because of an artifact which causes the tectorial membrane to retract toward the limbus spiralis where the contact surface is wider. Thus it is common to see the tectorial membrane lifted high above the organ of Corti. However, when the membrane is examined under the electron microscope, either the imprints of outer hair cell stereocilia or torn stereocilia can be seen in the homogeneous layer at the lower margin (Kimura, 1966). While attachment of the cilia of outer hair cells is usually seen, it is rare to find the cilia of inner hair cells touching the tectorial membrane. Iurato (1967) and Tanaka *et al.* (1973) demonstrated the contact, while Engström *et al.* (1962), Kimura (1966), Lindeman *et al.* (1971), and Lim (1972) did not observe it. The imprints of the cilia are not seen, even under the scanning electron microscope (Lindeman *et al.*, 1971). If they attach to the membrane, the attachment is not very strong, and firm attachment may not be essential as the inner hair cells are situated on the less vibratory part of the basilar membrane.

Hensen's stripe, according to Lim (1972), attaches to the border cell or the inner phalangeal cell or both. Tanaka *et al.* (1973) believe that the membrane also attaches to the third row of inner sulcus cells from the inner hair cells. The attachment of the outer margin of the tectorial membrane is likewise still in dispute. Without chemical fixation and by the light microscopic study, de Vries (1949), Hilding (1952), and Tonndorf *et al.* (1962) indicated its attachment to the Hensen's cells. Von Békésy (1960) believed the membrane was attached to the reticular lamina. Such attachments were observed in the organ of Corti when it was in the final phase of cellular differentiation; however, no attachment was seen in adult specimens (Kimura, 1966). These observations were confirmed by a scanning electron microscope study of the newborn kitten and adult cat by Lindeman *et al.* (1971). However, Lim (1972) and Tanaka *et al.* (1973) believe that attachments to the outer margin of the reticular lamina and/or Hensen's cells exist in adult specimens.

C. Nature

The nature of the tectorial membrane is a subject of controversy. Histochemical studies of the membrane by Wislocki and Ladman (1955), Plotz and Perlman (1955), and Igarashi and Alford (1969), show the presence of mucopolysaccharide which is sulfated, according to autoradiographic studies by Bélanger (1953) and Friberg and Ringertz (1956). Iurato (1967), however, from extensive studies utilizing polarized light, x-ray diffraction, and chemical analytical techniques, concludes that the major component of the tectorial membrane is protein of the k, e, m, and f groups, which is neither collagen nor elastic, and that it does not contain acid mucopolysaccharide.

III. Hair Cells

There are two types of hair cells, the outer and the inner, each of which differs in number, size, shape, and also in distribution of cell organelles. In the human the outer hair cells number 12,000 and are organized in three rows at the basal turn and increase to four to five rows at the second and apical turns. The inner hair cells are about 3500 in number and form a single row in the longitudinal direction

(Retzius, 1884; Bredberg, 1968). The length of the organ of Corti is 33.5 mm (Kolmer, 1927). The number of cochlear nerve fibers is about 31,400 (Rasmussen, 1940).

Both the outer and inner hair cells are inclined toward Corti's tunnel. Although Tonndorf *et al.* (1962) indicated that the inner hair cells are also tilted toward the apical turn, this observation was not confirmed by Engström *et al.* (1965) nor by Spoendlin (1966). Our observations support the latter. However, Spoendlin reported an inclination of the outer hair cells toward the basal end of the cochlea. Our measurements in the squirrel monkey show that the long axis of the cell is tilted toward the tunnel about 60°, and also tilted about 70° toward the basal end with respect to the basilar membrane.

The outer hair cells have a test-tube shape, and the inner hair cells a flask shape. The apical surface of the outer hair cell is heart-shaped, and that of the inner hair cell is oblong, with the long axis directed along the cochlear duct; the cell surface of the inner hair cell is larger. These hair cells are shorter at the basal turn and gradually increase in length toward the apical turn. The outer hair cells also increase in height from the tunnel radially outward. According to Bredberg (1968), the density of hair cells increases from the basal to the apical turn.

A. Outer Hair Cells

The outer hair cells (Fig. 3) are highly specialized cells with apical surfaces modified to accommodate stereocilia which are anchored to a thick cuticular plate. Their basal ends show specific modifications for the attachment of different types of nerve endings. Mitochondria are localized at both poles, where energy requirements are high. The apical portion of the cell, the cuticular plate, establishes a tight junction (zonula occludens) with the reticular lamina. The middle portion is free and is surrounded by the fluid of Corti's tunnel. The basal portion with attached nerve endings is supported by the cup-shaped part of the Deiters' cells. The cell junctions between the hair cell and Deiters' cup show no specialized attachment zone; the gap is wide, about 130 Å.

FIG. 3. Outer hair cells of the bat. Note the large nuclei in the bottom portion and localization of mitochondria in the subcuticular and infranuclear zones. The hair cells rest on the cup-shaped part of the Deiters' cells, where the afferent (A) and efferent (E) endings are located. PH, Phalangeal processes of the Deiters' cells (D); H, Hensen's cell; R, reticular lamina.

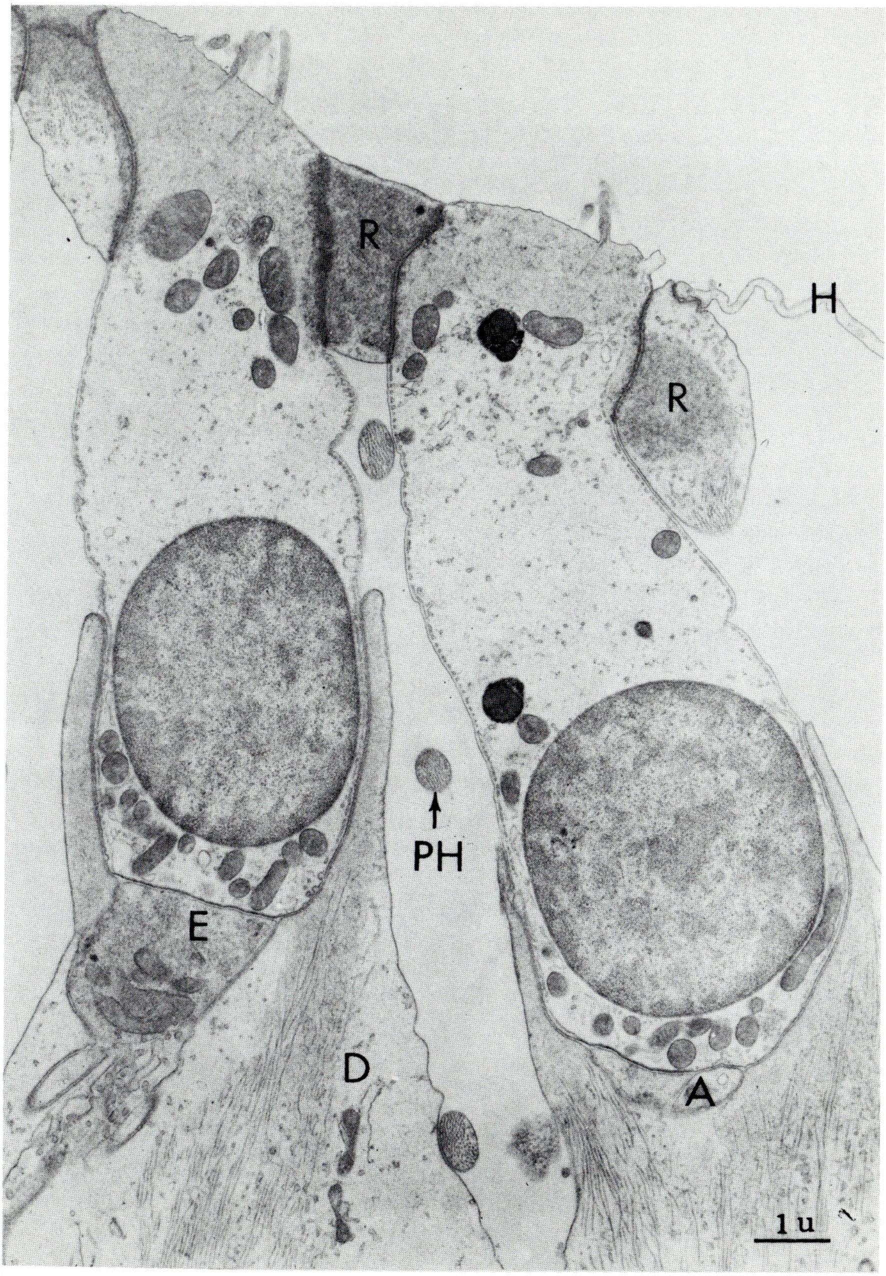
R
H
R
PH
E
D
A
1u

1. *Stereocilia*

The stereocilia of the outer hair cells are anchored to the cuticular plate about 30° from the long axis of the cell body. This arrangement is quite unique to the hearing organ, and is unlike that in other epithelia such as the vestibular sensory cells, respiratory epithelia, and epithelial lining of the Eustachian tube in which cilia or microvilli are aligned with the main axis of the cell body. This angular insertion of the stereocilia is considered a specific functional adaptation, presumably to provide an effective means to transmit mechanical energy into the cell by stretching or bending. Other functional adaptations are the length and shape. When the height of the stereocilia is measured from the cuticular plate to the attachment of the tectorial membrane, the high-frequency area (basal turn) shows short cilia (1.6 μm) which gradually increase in length toward the apical turn (5.7 μm) (Kimura, 1966). The number of stereocilia on a hair cell of the squirrel monkey is about 135 at the basal turn, 120 at the second turn, and 80 at the third turn, thus decreasing gradually from the basal to the apical turn. They are shaped like baseball bats, large (3200 Å) at the tectorial membrane attachment and narrow (1300 Å) at the cuticular plate.

The cilia penetrate the tectorial membrane about 0.1–0.4 μm. In cross section the cilia form a "W", the base of which faces the otic capsule (Fig. 4A). The angle of the "W" is wide at the basal turn, about 120°, and small at the apical turn, about 60° (Angelborg and Engström, 1973). The number of rows of cilia in small animals is generally three; in primates, including the human, it may reach five or more rows. Only the tall peripheral row of the "W" and a few tall cilia at the base of the "W" attach to the tectorial membrane (Fig. 4B). A homogeneous substance similar to that of the tectorial membrane is found between the tall and short cilia; it suggests possible attachment of the short cilia to the membrane. However, when a detached tectorial membrane is examined, only a single row of tall cilia or their imprints are observed (Kimura, 1966; Lindeman *et al.*, 1971; Lim, 1972). If the short cilia do not touch the membrane, they may receive mechanical vibrations which increase as the intensity increases through the tall cilia. The attachment of the tall peripheral cilia to the tectorial membrane is presumed to be firm (Kimura, 1966; Lim, 1972); this conclusion is based essentially on the evidence of torn cilia remaining attached to the lifted tectorial membrane. Experimental evidence provided by von Békésy (1960) appears to support this contention. When he moved the Hensen's cells downward beyond the outer edge of the tectorial membrane, the tectorial mem-

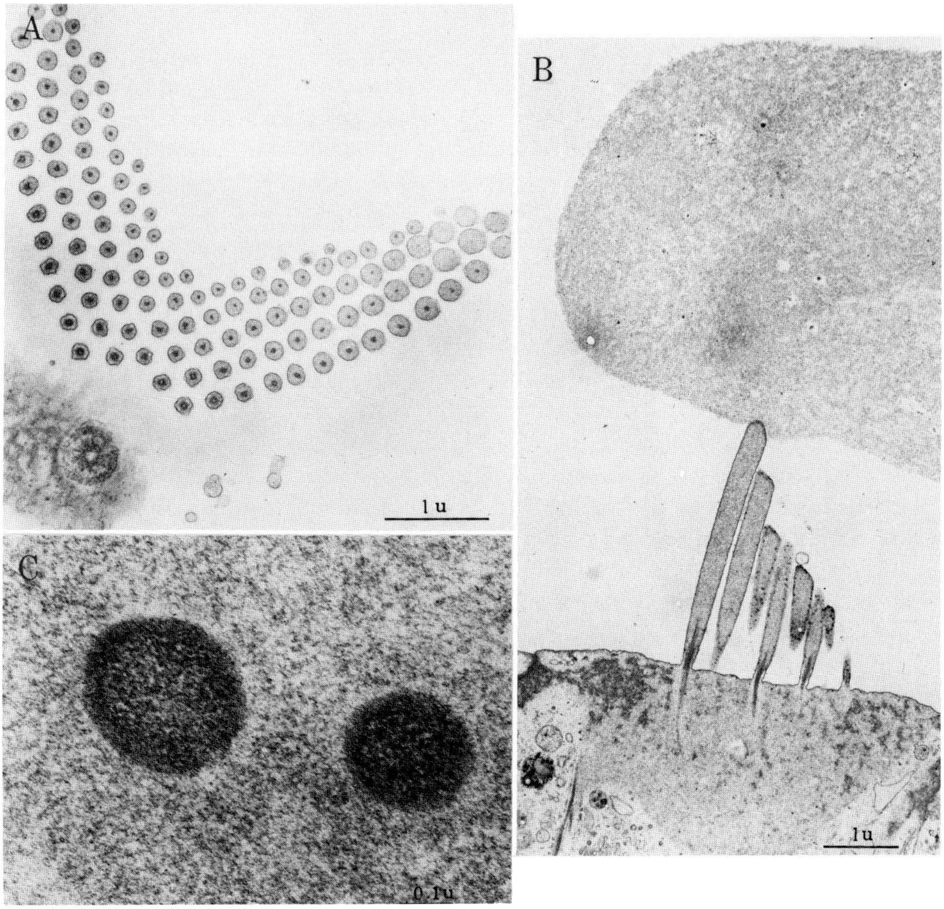

FIG. 4. Stereocilia of the outer hair cells and their attachment to the tectorial membrane. (A) Cross section of stereocilia from a single hair cell, showing the "W" form. Rhesus monkey. (From Kimura, 1966.) (B) Attachment of the peripheral tall stereocilia to the tectorial membrane. Squirrel monkey. (From Kimura, 1966.) (C) Cross section of stereocilia tops within the homogeneous layer of the tectorial membrane. Note a light zone around the highly electron-dense cilia tops. Squirrel monkey.

brane followed this movement. He further stated that the attachment is "suddenly" loosened by pushing down the modiolar side of the membrane. If the cilia were merely touching the membrane without a firm attachment, the membrane would not follow the downward movement of the Hensen's cells or the reticular lamina, and it also would not be released suddenly. The strong attachment of the outer row of cilia is probably essential, since these hair cells are located on the part of the basilar membrane that receives strong mechanical vibrations. The attachment of the tall cilia would have a damping effect on the organ of Corti, so that it would not oscillate up and down continuously. This effect may be more pronounced at the basal turn, where more stereocilia per hair cell are noted than at the apical turn.

The ultrastructure of the stereocilia is rather simple. The main body is filled with a homogeneous substance, similar to that of the cuticular plate, and numerous longitudinally arranged filaments (Kimura, 1966; Spoendlin, 1966; Bredberg *et al.*, 1972). The cilia are covered by a continuous unit membrane which is delineated from the tectorial membrane. The ciliary top is often round, but it can be flat or slanted. The top portion is more electron-dense (Fig. 4C), and sometimes vesicles or blebs are seen in this area as well as on the sides. There is some indication that the uppermost portion may be detached or pinched off and shed into the tectorial membrane; it is not uncommon to see a small acellular circular body form a short distance from the top of the cilia. The top of the cilia may be damaged by strong mechanical vibrations, or lost from natural causes, and later the attachment is restored after recovery. The top of the cilia is surrounded by a light zone of about 190 Å, which is halolike in cross section (Kimura, 1966; Angelborg and Engström, 1973). The light zone is traversed by filaments radiating in all directions, which enter into a slightly denser homogeneous layer at the margin of the light zone. This light zone may be an artifact, or it may represent a natural state. The width of the light zone is just about the same as the one described at the side of the cilia, where the dense homogeneous layer lies between the cilia.

The stereocilia contain central cores which begin as a dense aggregate within the homogeneous substance at about midlength and extend toward the cuticular plate. These dense aggregates assume a cylindrical shape as the cilia taper down and are inserted into the cuticular substance, either maintaining the circular form or as a solid dense core. The short cilia terminate within the cuticular plate, but the long cilia penetrate the entire thickness of the plate and extend a short distance as roots into the cytoplasm (Fig. 5A). Some roots are noted to reach the vicinity of the subsurface cisterns (Fig. 5B). Aggregates of small granules are located along the roots in the cytoplasm. The presence of the core, particularly of the cylindrical shape, provides stiffness to the cilia (Fig. 5C).

Engström *et al.* (1962) proposed that the cilia do not bend, but cause the entire cuticular plate to move by a leverlike action, which is contrary to the opinion held by Davis (1959) that the bending of the cilia is the final mechanical event. Although the cilia may be stiff,

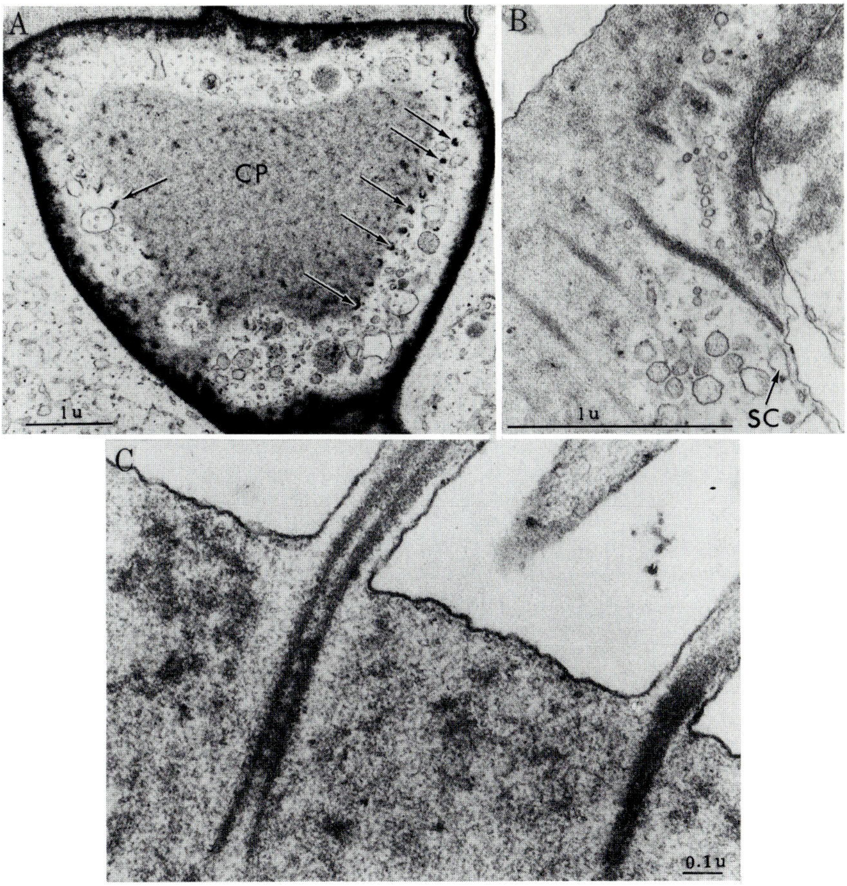

FIG. 5. Roots of the stereocilia of outer hair cells. Squirrel monkey. (A) Cross section of the hair cell top showing the tall ciliary roots (arrows) penetrating the cuticular plate (CP) into the cytoplasm. (B) The ciliary root comes very close to the peripheral subsurface cisterns (SC) and the plasma membrane. (C) The roots, showing both cylindrical and solid forms. Note the light zone around the core near the cilia attachment.

they can be bent at the narrow neck without bending the main part and without causing an obvious tilt in the cuticular plate (Fig. 6A). The general shape of the cilia appears to be designed to convert a large mechanical movement at the top to a smaller movement at the neck where the unit membrane and the central core are subject to stretch and stress. It is also conceivable that, whenever the cilia move, the rootlets move or are subject to shearing or lifted, since the

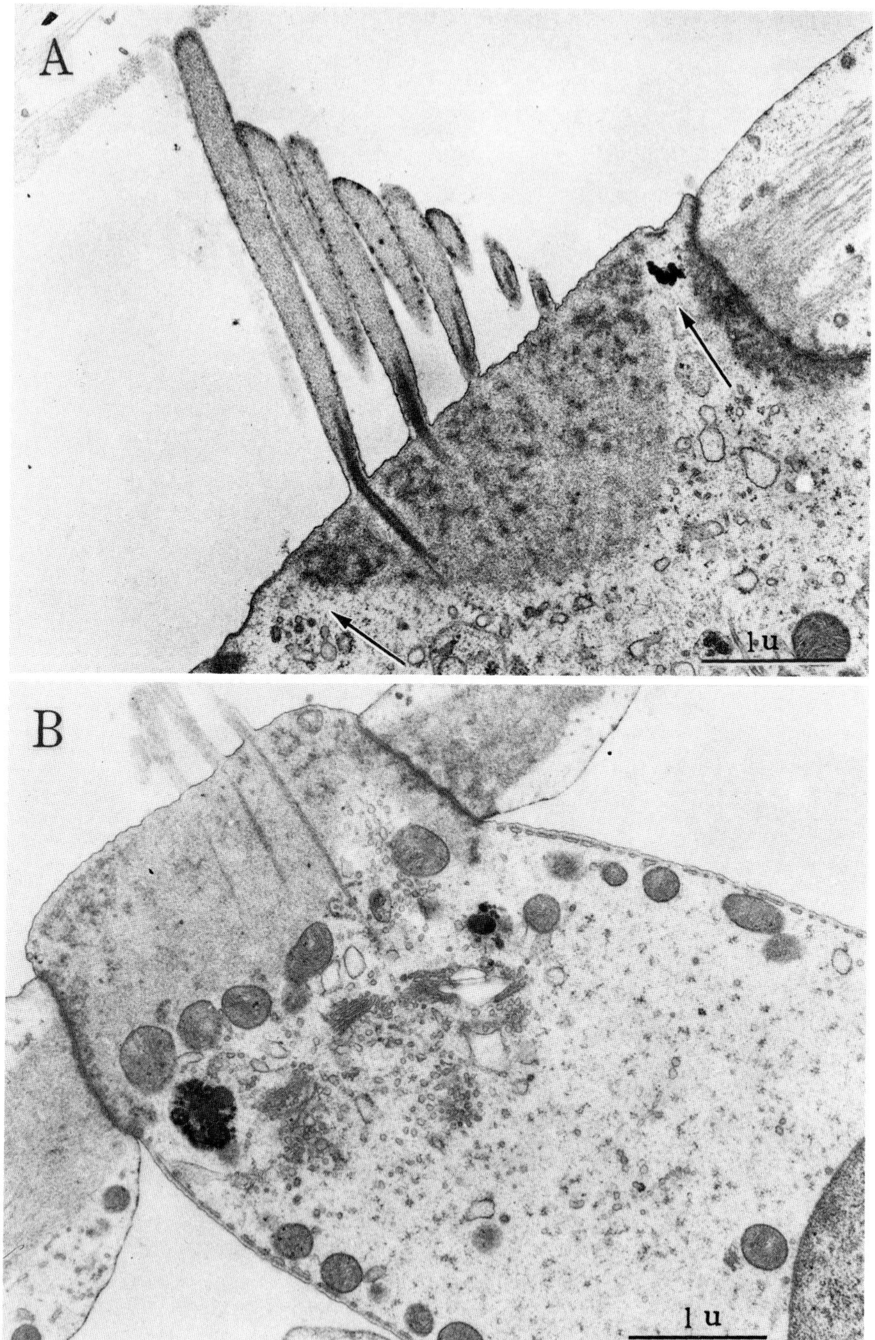
A
1u
B
1 u

attachment of roots surrounded by a light zone appears to be looser in comparison to that of the top portion. The mechanism of hair cell excitation may involve a change in the molecular organization of the unit membrane, or the cells may possess an excitable structure in their cytoplasm. Readers are referred to an article by Flock (1971) for his excellent analysis of sensory transduction.

Cochlear hair cells, unlike vestibular sensory cells, do not contain kinocilia adjacent to stereocilia in adult specimens (Iurato, 1961; Engström *et al.*, 1962; Flock *et al.*, 1962). However, kinocilia were found at the base of the "W" form, where the cuticular substance is absent in cochlear hair cells of fetuses and newborn animals (Kikuchi and Hilding, 1965; Kimura, 1966; Wersäll and Flock, 1967; Lindeman *et al.*, 1971). The remnant of the kinocilium, the basal body (Fig. 4A), is found in adult specimens and is regarded as the excitable structure of the hair cell in response to the movement of the cuticular plate (Engström *et al.*, 1962). This theory was questioned by Spoendlin (1966), who stated that the basal body cannot be found in hair cells of the adult cat. The position of the kinocilium or the basal body is considered to be of great importance in determining the directional sensitivity of hair cells. Movement of the cilia toward the kinocilia and taller cilia produces excitation, displacement in the opposite direction, and inhibition (Lowenstein and Wersäll, 1959; Flock, 1965, 1971). Although this concept applies well to the canal organs, it is not clear in the cochlea. Hair cell excitation resulting from displacement of the stereocilia toward the basal body in the outer hair cells fits in well with the maximum production of cochlear microphonics as reported by von Békésy (1960). In the inner hair cells, where the basal body is located in a similar position, maximum potentials are produced by displacement in the longitudinal direction, which is at right angles to that of the outer hair cells.

The uncertainty about the attachment of the tectorial membrane to the cilia is reflected in the theory of cochlear potentials being generated as the basilar membrane oscillates and the cilia bend. Tonndorf (1970) reports that the downward movement of the basilar membrane tilts the cilia toward the otic capsule (basal body) side. Lim (1972), however, believes that the bending of the cilia in the same

---

FIG. 6. Apical portion of outer hair cell. Squirrel monkey. (A) The stereocilia appear to be stiff, unlike microvilli. Bending takes place at the narrowest part, near the attachment. Note the cuticle-free zone at the periphery (arrows) of the cell. (B) The subcuticular zone showing mitochondria, lipofuscin granules, numerous vesicles, and a Golgi network. The stereocilia are inserted at an inclined angle to the main axis of the hair cell.

direction occurs as the basilar membrane is moved upward (toward the scala vestibuli). Tonndorf's concept is based on the idea that the tectorial membrane is attached only to the limbus spiralis. Lim's idea is that the membrane is attached, in addition to the limbus, to the outermost part of the reticular lamina, stereocilia of the outer hair cells, and supporting cells of the inner hair cells. According to Zwislocki (1974), displacement motion of the organ of Corti toward the scala tympani produces excitation, and motion or displacement toward the scala vestibuli produces inhibition.

2. *Cell Body*

The cuticular plate is composed of a fine homogeneous substance interspersed with randomly distributed patches of dense aggregates (Figs. 5C and 6A). The cuticular plate does not cover the entire cell surface; it is always uncovered in the basal body region and irregularly covers the periphery (Fig. 6A). This cuticle-free zone is also filled with numerous vesicles, and even has a few mitochondria and lysosome granules. The central part of the plate may project deeply into the cytoplasm, thus presenting the shape of an inverted cone or trapezoid in profile. The cuticular substance becomes continuous with the more dense cytoplasmic condensations at the cell junction of the reticular lamina. This cell junction is the zonula occludens; it is seen on the endolymph side, as well as on the Corti's tunnel side (Iurato and Taidelli, 1965). There is asymmetry of the unit membrane at this junction; the inner leaflet of the hair cell is more distinct, has increased electron density, and appears thicker than that of the opposing leaflets of the Deiters' cells.

At the subcuticular zone aggregates of lysosomes, lipofuscin granules, mitochondria, multivesiculated bodies, vesicles, and the Golgi network are the usual constituents (Fig. 6B). From this area to the supranuclear zone is a homogeneous substance within which are scattered small vesicles, RNA particles, varying amounts of glycogen granules, and some mitochondria. The mitochondria are rod-shaped or sometimes very long or branched, and show a distinct distribution pattern. In addition to the subcuticular zone, they are localized in the infranuclear zone where the nerve endings attach, and also at the cell periphery where they lie parallel to the subsurface cisterns. In the bat the mitochondria are not located at the cell periphery; they are very large at the cuticular plate, and small in the infranuclear zone (Fig. 3).

Between the peripheral mitochondria and the plasma membrane of the hair cell are subsurface cisterns (Fig. 7A), which are a rather con-

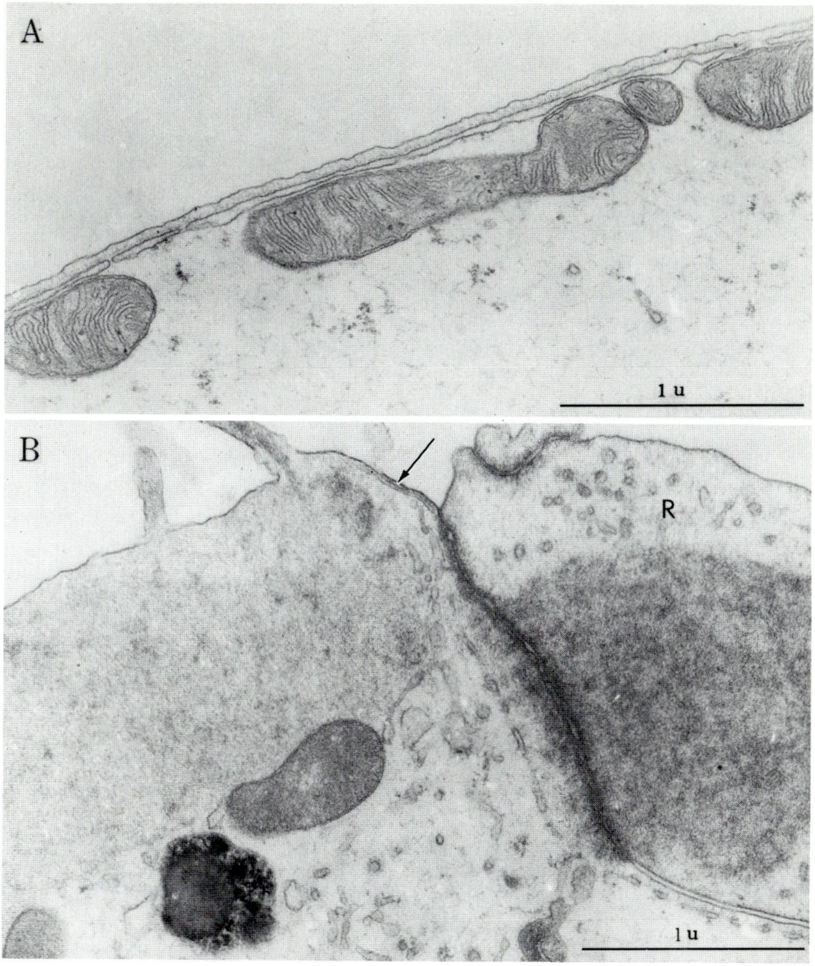

FIG. 7. Peripheral subsurface cisterns of outer hair cell. (A) A dense homogeneous substance is located between the plasma membrane and the subsurface cisterns. Mitochondria lie very close to the cisterns. Squirrel monkey. (B) The subsurface cisterns appear to continue into the cuticle-free zone where other flat cisterns lie parallel (arrow) very close to the apical plasma membrane. Note the dense cytoplasmic condensation at the junction of the hair cell and the reticular lamina (R). Bat.

sistent morphological characteristic in all the species examined including man. They are positioned uniformly at a distance of 300–400 Å from the plasma membrane, and extend from the cuticle-free zone of the cell apex to the infranuclear zone close to the nerve-ending area. In the cuticle-free zone, there are often cisterns which resemble the subsynaptic cisterns located parallel and close to the apical plasma membrane in bat hair cells (Fig. 7B). The peripheral cisterns lie in a single layer in most specimens, including the monkey (Smith, 1968) and man (Kimura *et al.*, 1964), but in the guinea pig and chinchilla they form multiple layers, and the number of lamellae varies according to location in the cell and from cell to cell (Engström, 1967). They resemble smooth endoplasmic reticulum, but sometimes ribosomes are attached to them on their innermost layer (Smith, 1968; Angelborg and Engström, 1973). We found a homogeneous condensation at the outer layer of the cisterns from which filamentous strands extend to the plasma membrane. Filamentous strands may connect the adjacent cisterns (Smith and Dempsey, 1957).

The function of the cisterns is unknown; Engström (1955) thought that they may be involved in the production of cochlear microphonics. Spoendlin (1966) suggested that they may play a part in the maintenance of an electrical potential gradient against the surrounding medium. These subsurface cisterns increase in number or penetrate deeply into the cytoplasm, forming concentrically arranged whorls and laminated bodies, following acoustic trauma (Engström and Ades, 1960). The cisterns increase in number in experimental hydrops specimens also (Kimura, 1967), in which the hair cells undergo gradual degeneration.

Subsynaptic cisterns are located at a distance of 85 Å from the synaptic membrane; the inner and outer layers of the cisterns are separated by a constant gap of 175 Å (Smith and Sjöstrand, 1961), and the outer membrane is more electron-dense. They are always found opposite efferent nerve endings and are totally lacking adjacent to afferent nerve endings. Our observations in the cat, bat, and opossum show that peripheral subsurface cisterns become continuous with subsynaptic cisterns (Fig. 8A). The existence of a connection between peripheral subsurface cisterns and subsynaptic cisterns implies that the peripheral cisterns can be influenced by the efferent nerve or vise versa. Subsynaptic cisterns may either facilitate the effects of efferent nerve activity, or serve as a shield to limit spread of transmitter substance at the local synaptic membrane.

In the hair cells adjacent to the afferent nerve endings, there is often a synaptic bar or a synaptic ring surrounded by synaptic vesicles (Smith and Sjöstrand, 1961) (Figs. 8B and 9A). Sometimes two synaptic bars or rings are found against one nerve ending. The plasma membrane shows presynaptic thickening with or without the presence of synaptic vesicles. The thickening is equal to or greater

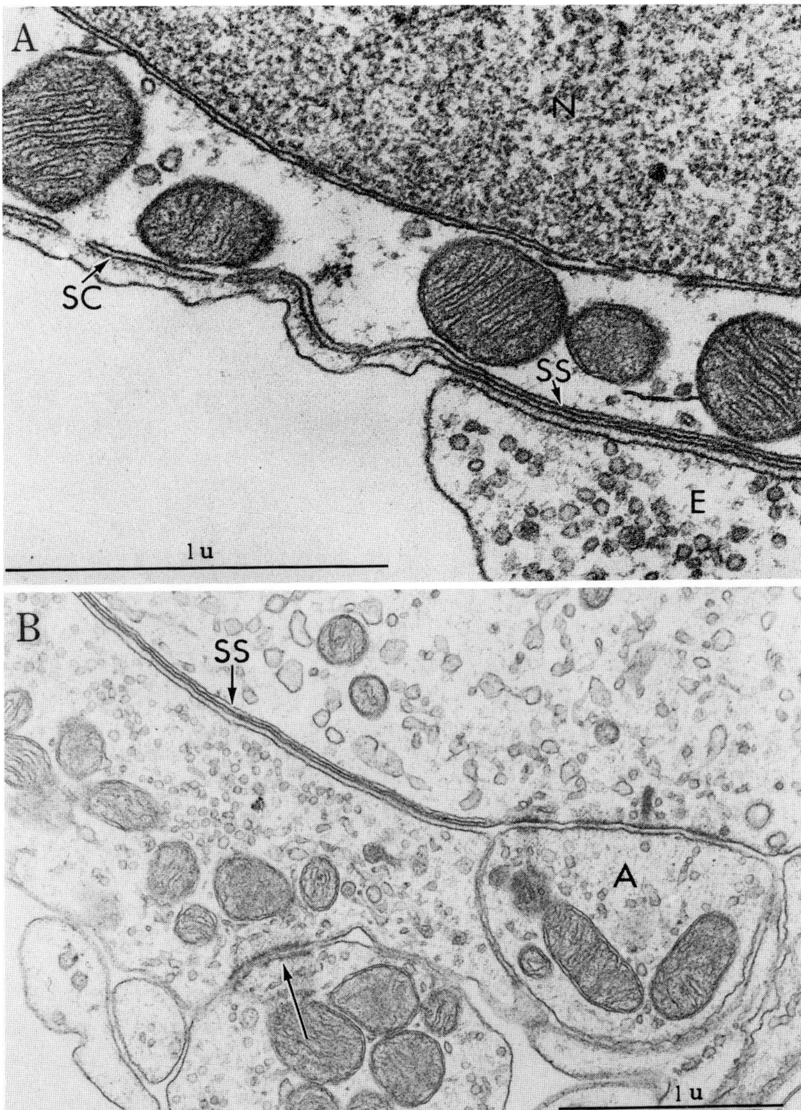

FIG. 8. Synaptic zone of outer hair cell. (A) A subsurface cistern (SC) is continuous with a subsynaptic cistern (SS) which lies adjacent to the efferent nerve ending (E). N, Nucleus. Cat. (B) Synaptic bar with a few vesicles on the hair cell side. A few short thickenings of the synaptic membranes are shown on both sides of the hair cell and the afferent nerve ending (A). The efferent endings contain many vesicles. A desmosome is shown between the efferent nerve ending and efferent varicosity (arrow). SS, Subsynaptic cisterns. Rhesus monkey.

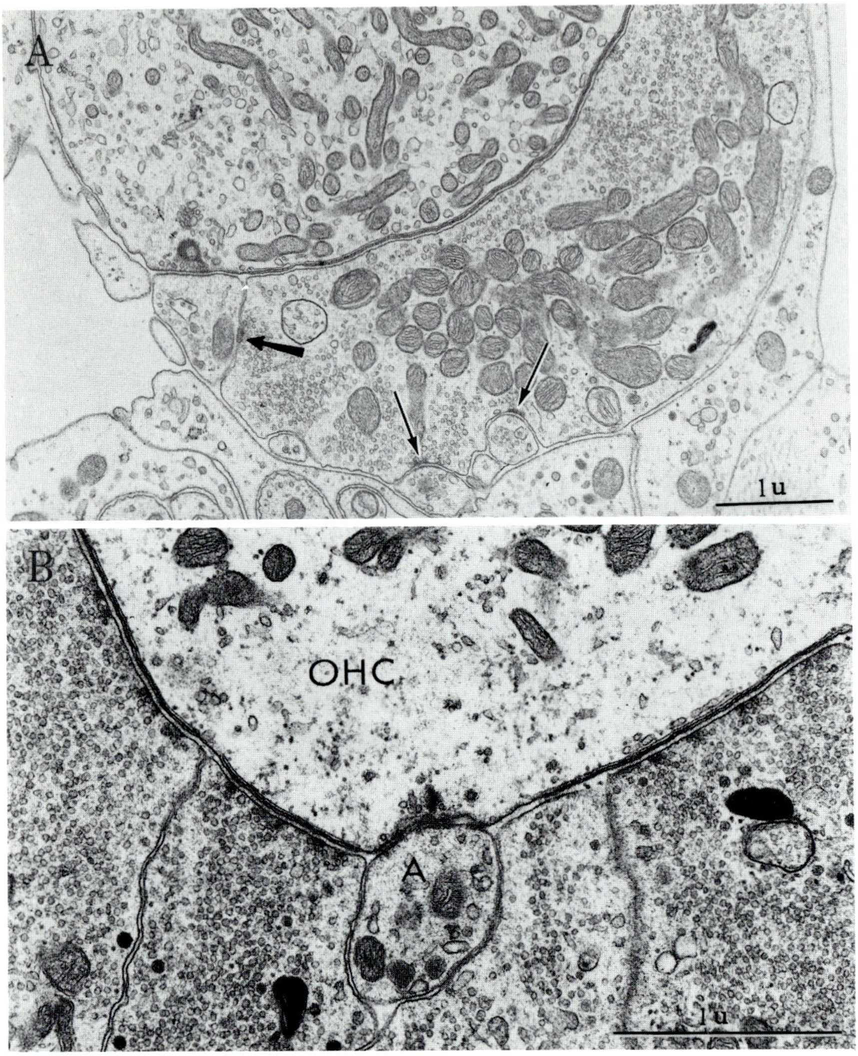

FIG. 9. Synapse at outer hair cell. (A) Loop-form synaptic body in a hair cell against the afferent nerve ending which shows a synapse (large arrow) with the efferent nerve ending. The large efferent nerve ending also shows synapses with two other afferent fibers (small arrows at bottom), Rhesus monkey. (B) Efferent nerve endings show patches of electron-diffuse substance at various parts of the synaptic membrane. A, Afferent nerve ending; OHC, outer hair cell. Guinea pig.

than that of the postsynaptic membrane. Three to four separate thickenings may be seen in a single profile, parallel to those of the afferent nerve endings. The plasma membrane is also invaginated, and synaptic thickenings are shown at the periphery, where the membrane comes closer to the postsynaptic membrane. Fusion of the synaptic membrane at the invaginated portion as shown in the type I vestibular sensory cell by Hamilton (1968) has not been observed yet. Often small pockets resembling pinocytotic vesicles are found in the vicinity of the synaptic zone. Extensive vesicular smooth endoplasmic reticulum, small mitochondria, and sometimes glycogen granules are present in the same area. Tubular filaments are also noted in the apical and basal cytoplasm (von Ilberg, 1969).

3. *Nerve Endings*

There are two types of nerve endings below the outer hair cells, afferent and efferent (Engström and Wersäll, 1953a, 1958; Smith and Dempsey, 1957; Spoendlin, 1959; Iurato, 1961; Smith and Sjöstrand, 1961; Kimura *et al.*, 1964) (Figs. 8B and 9A and B). The tracts of the efferent nerve system were followed from the midbrain into the cochlea by Rasmussen (1946, 1953). Later, their terminals were traced into the organ of Corti by sectioning of the olivocochlear bundle at the midbrain (Iurato, 1962a; Kimura and Wersäll, 1962; Smith and Rasmussen, 1963; Spoendlin and Gacek, 1963).

a. *Afferent Nerve Endings*. The afferent nerve endings abut the hair cells at the basal end. The outstanding features are their small size, about 1 μm, and bouton type. At sites of synapse the afferent nerve endings and outer hair cells show a thickening of the plasma membrane at corresponding points. The width of the postsynaptic membrane is equal to or thinner than that of the presynaptic membrane. The synaptic gap is about 150 Å. The afferent nerve endings are almost completely surrounded by efferent nerve endings, and there is a constant parallel gap between them. According to our observations, there are two types of cell junctions between afferent and efferent nerve endings. One type is an axodendritic junction with aggregates of synaptic vesicles and condensation of an electron-dense substance at the plasma membrane of the efferent nerve ending (Fig. 9A). An electron-diffuse substance is also present in the gap, which has a width identical to that between the hair cell and nerve endings. The second type of junction appears to be the desmosome which has equal thickening on both sides of the membranes; this type is not seen too frequently. The vesicles within the afferent nerve ending are scattered, and their size varies from an equivalent to those within the efferent nerve ending to very large (400–600 Å) (Engström, 1967). In agreement with Smith (1967), we observed that some of the vesicles may contain a dense core. Occasionally, minute electron-dense granules close to the size of glycogen granules are seen, and a few mitochondria are evident.

b. *Efferent Nerve Endings.* Efferent nerve endings are generally larger (3 μm in diameter) than afferent nerve endings. The synaptic surface toward the hair cell is smooth, and at random intervals patches of electron-dense substance accumulate on the inner side of the synaptic membrane (Fig. 9B). The membranes on both sides of the synaptic gap are equal in thickness. The synaptic gap is about 190 Å, which is slightly wider than that of the afferent nerve ending. The efferent nerve endings are located not only at the bottom of the hair cell, but are seen at the level of the nucleus and may extend deeply among the Deiters' cells in an inverted cone shape. Axodendritic synapses are observed not only adjacent to the afferent nerve endings, but also are frequently seen on the Deiters' cell side, where the afferent nerve fibers come very close and touch or invaginate into the efferent nerve endings (Fig. 9A). Another type of junction between the efferent nerve endings is the desmosome which has equal membrane thickening from which filamentous structures protrude vertically (Fig. 8B). Inside the efferent nerve endings are numerous round or ovoid vesicles, about 300–400 Å in diameter (Engström, 1967), and some large vesicles (600–800 Å) with a dense core (Spoendlin, 1972). Many mitochondria are aggregated in the zone away from the synaptic gap junction. Sometimes multivesiculated bodies, an electron-dense irregular body, and an aggregate of neural tubules may be found at the inferior pole. It should be emphasized that the vesicles containing cores are not limited to efferent nerve endings, but also occur in afferent and adrenergic endings.

According to Spoendlin (1966), the number of afferent nerve endings in the cat is 5 to 8, and that of the efferent endings is 6 to 10 per hair cell at a distance of 3 mm from the basal end of the cochlea; there are about 8 afferent and 3 efferent at the apical turn. Smith and Sjöstrand (1961) counted as many as 34 afferent nerve endings in the guinea pig cochlea. A definite distribution pattern of afferent and efferent nerve endings is indicated from the basal to the apical turn. Smith and Sjöstrand (1961) report an equal number at the basal turn; however, at the second and third turns the afferents begin to dominate, particularly in the outermost and middle rows of hair cells. Acetylcholinesterase activity of the efferent nerve endings is high at the basal and second turns and decreases at the apical and extreme basal ends; the activity also decreases from the inner row to the outer row of the outer hair cells (Ishii and Balogh, 1968).

B. Inner Hair Cells

Inner hair cells differ considerably from outer hair cells. They are more-or-less flask-shaped, the cell apex being small in comparison to the main body (Fig. 10). They become slender and taller toward the apex of the cochlea. Their cell body is inclined toward Corti's tunnel, but does not lean in the basal direction as does that of the outer hair cells. Their cuticular surface is oblong in the longitudinal direction (Fig. 11A), and their strongest attachment to the supporting elements is at the reticular lamina. The cell apex is surrounded by the phalanges of the inner pillar cells laterally, the border cells medially, and the phalanges of the inner phalangeal cells between their hair cell tops. At the basal portion they come in contact mostly with the nerve endings and fibers that are supported by the cell processes of the inner phalangeal cells and the border cells. The inner hair cells have fluid spaces toward the inner pillar cells and inferiorly, but they are small in comparison to the spaces of the outer hair cells; the fluid space connects with Corti's tunnel between the inner pillar cells, but such openings are not seen frequently.

1. *Stereocilia*

Stereocilia project from the cuticular plate in a flat "W" pattern at an angle to the main axis of the cell body. The cuticle-free zone is seen at random points, but it is always open toward the pillar cells, where the basal body is often observed. The stereocilia are larger than those of the outer hair cells, about 5200 Å at the top and about 1900 Å at the neck in the squirrel monkey (Kimura, 1966). The taller stereocilia are located toward the pillar cells and appear almost in a straight line along the longitudinal direction (Fig. 11B). There is a notch at the base of the "W", which faces the outer hair cells.

Their morphological characteristics are similar to those of the stereocilia of the outer hair cells. The tops of the stereocilia are not embedded in the tectorial membrane, in contrast to those of the outer hair cells. No imprints are visible on the underside of the tectorial membrane, which suggests that these stereocilia are possibly either not attached to the membrane or adhere very lightly. Between the cilia is a homogeneous substance, as in the outer hair cells. Bredberg *et al.* (1972) reported bridges across these cilia as seen with the scanning electron microscope.

In cross section a total of 75 cilia per hair cell was counted at the

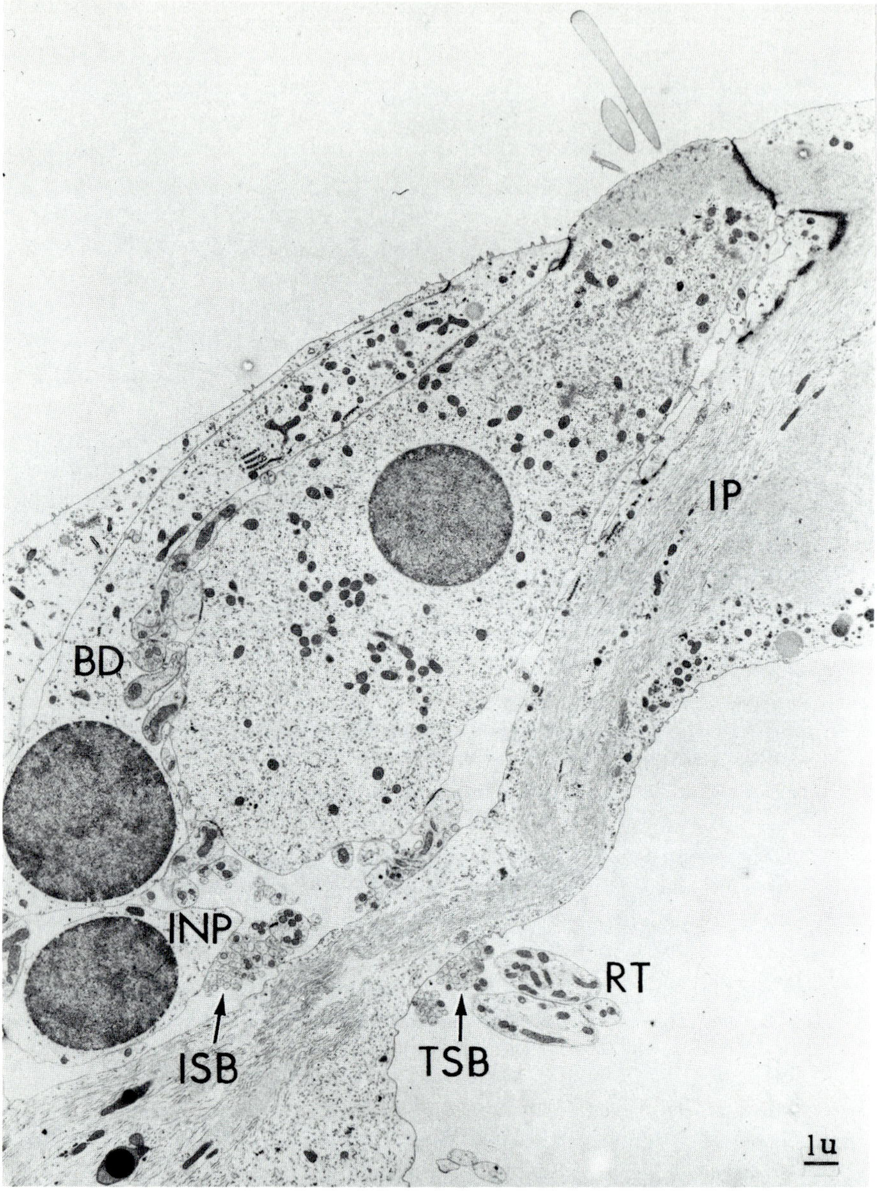

FIG. 10. Inner hair cell of squirrel monkey. BD, Border cell; INP, inner phalangeal cell; ISB, inner spiral bundle; TSB, tunnel spiral bundle; RT, radiating tunnel fibers; IP, inner pillar cell. Note the central position of the nucleus in the inner hair cell.

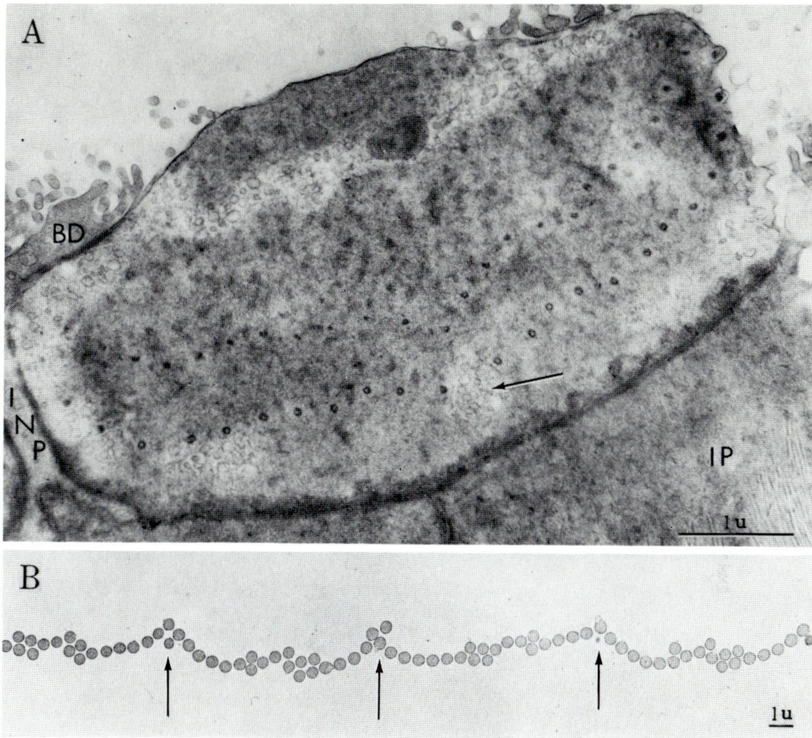

FIG. 11. Inner hair cell top. (A) Cross section of hair cell, showing ciliary roots. Note the absence of a basal body, but numerous vesicles are shown in the corresponding area (arrow). IP, Inner pillar cell; INP, inner phalangeal cell; BD, border cell. Bat. (B) Stereocilia of inner hair cells. Note the closeness of the cilia of adjacent cells. Arrows indicate the border of the hair cells. Basal turn in guinea pig. (From Kimura, 1966.)

cuticular plate in the squirrel monkey, but the number varied considerably from one hair cell to another. The tall cilia of adjacent cells come very close together at the top, and it may be impossible to distinguish which cilia belong to a particular hair cell. Movement of the cilia of one inner hair cell may therefore be transmitted to cilia of adjacent hair cells in a chain reaction in the spiral direction. This spread of ciliary movement is less likely to take place at the cilia of the outer hair cells since the cilia of adjacent cells are more widely separated and are apparently deeply imbedded in the tectorial membrane. However, it should be noted that the inner hair cell bodies come in contact with each other, and also that the nerve fibers are sandwiched between these hair cells in the infranuclear zone.

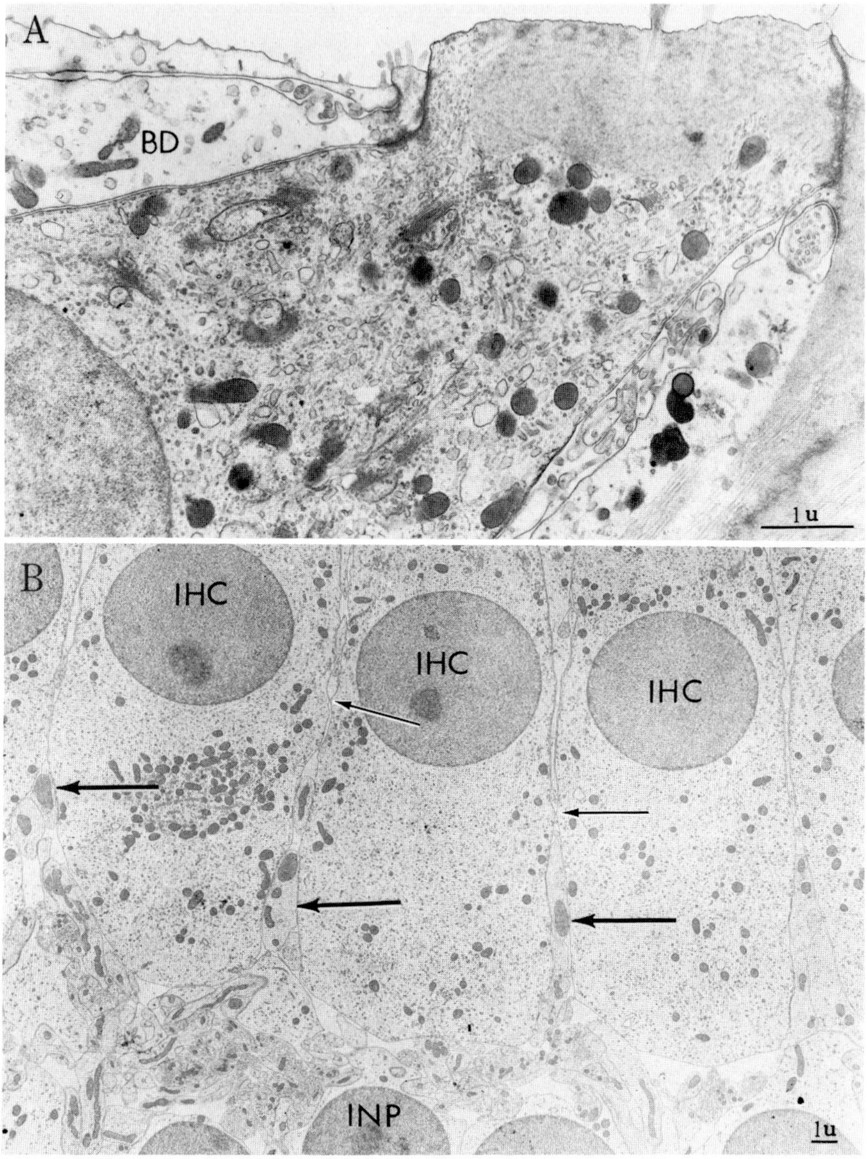

FIG. 12. (A) Apical cytoplasm of inner hair cell, showing the extensive network of the tubular system extending from the cuticle-free zone to the nucleus. Note the small peripheral subsurface cisterns. BD, Border cell. Cat. (B) Longitudinal vertical section of inner hair cells (IHC). The cell bodies lie very close together (small arrows), and the nerve endings touch both hair cells (large arrows). INP, Inner phalangeal cell. Squirrel monkey.

2. *Cell Body*

The nucleus of the inner hair cell is centrally located in comparison to the basal location in the outer hair cell. The subcuticular zone is particularly rich in vesicles, the tubular form of smooth endoplasmic reticulum (Fig. 12A). The Golgi complex is well developed. Some lysosomes, lipofuscin granules, multivesiculated bodies, and coated vesicles are apparent. Mitochondria are round or rod-shaped and are scattered throughout the cytoplasm in contrast to those of the outer hair cells, which tend to localize in specific areas. The plasma membrane of the inner hair cells is underlined by peripheral subsurface cisterns in the supranuclear zone, and they are very short and flat. In the infranuclear zone rough endoplasmic reticula in parallel array, which are virtually lacking in the outer hair cells, are often observed. Mitochondria in this area are more concentrated near the nucleus than in the synaptic zone. The bodies of the inner hair cells come very close together; this is well demonstrated when they are cut in the longitudinal and vertical directions (Fig. 12B) (Smith, 1967). We sometimes noted very small cytoplasmic condensations of equal thickness between the plasma membranes of adjacent hair cells. The gap is about 130 Å.

Between the inner hair cells are usually one to three tightly sandwiched nerve endings (Fig. 12B). The cytoplasm adjacent to the afferent nerve endings shows typical synaptic bodies with vesicles around them (Fig. 13A). The synaptic bodies are more often seen in the inner hair cells than in the outer hair cells. They may be rod-shaped or in loop form, and two or more can be present adjacent to a single nerve ending; in the bat we frequently observed these synaptic bodies evaginating into afferent nerve endings. Subsynaptic cisterns are not generally observed in the inner hair cell opposite the vesiculated nerve endings (Fig. 13B). The basal portion of the inner hair cell differs from that of the outer hair cell in that the bottom is not supported by the cup-shaped arrangement of the Deiters' cells. The basal surface may be round or, more often, extremely irregular with invaginations of the nerve endings, and appears to be resting on a forest of nerve endings and fibers (Figs. 13B and 14A), among which are the small processes of the inner phalangeal and border cells making contact with the hair cells.

3. *Nerve Endings*

There are about 20 nerve endings on each inner hair cell (Spoendlin, 1969). The nerve endings of the inner hair cells are almost exclusively afferent; identification is primarily based on their

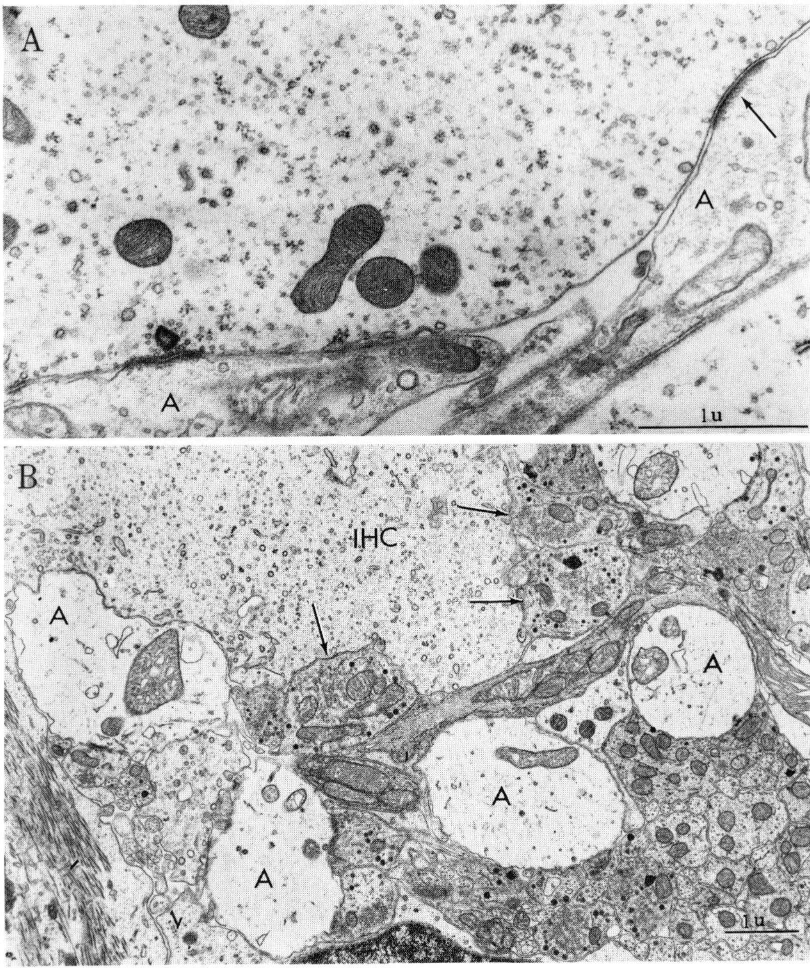

FIG. 13. Nerve endings of inner hair cell. (A) Loop type of synaptic body in the hair cell; a thicker postsynaptic membrane (arrow) is shown in the afferent nerve ending (A). Squirrel monkey. (B) Efferent nerve endings or varicosities (arrows) come in contact with the cell body. Note similar varicosities in the adjacent area. A, Afferent nerve fiber and ending; IHC, inner hair cell. Guinea pig.

abutment against the synaptic bodies of the hair cell. These nerve endings are the terminals of the radial fibers coming from the spiral ganglia (Smith, 1961). They ascend toward the cell apex like fingers; some are as long as 9 μm and sometimes reach above the nucleus. They establish synaptic contact with the hair cell by thickening of the plasma membrane; it is thicker than the presynaptic membrane, at a synaptic gap of about 145 Å (Smith, 1961) (Fig. 13A). The long

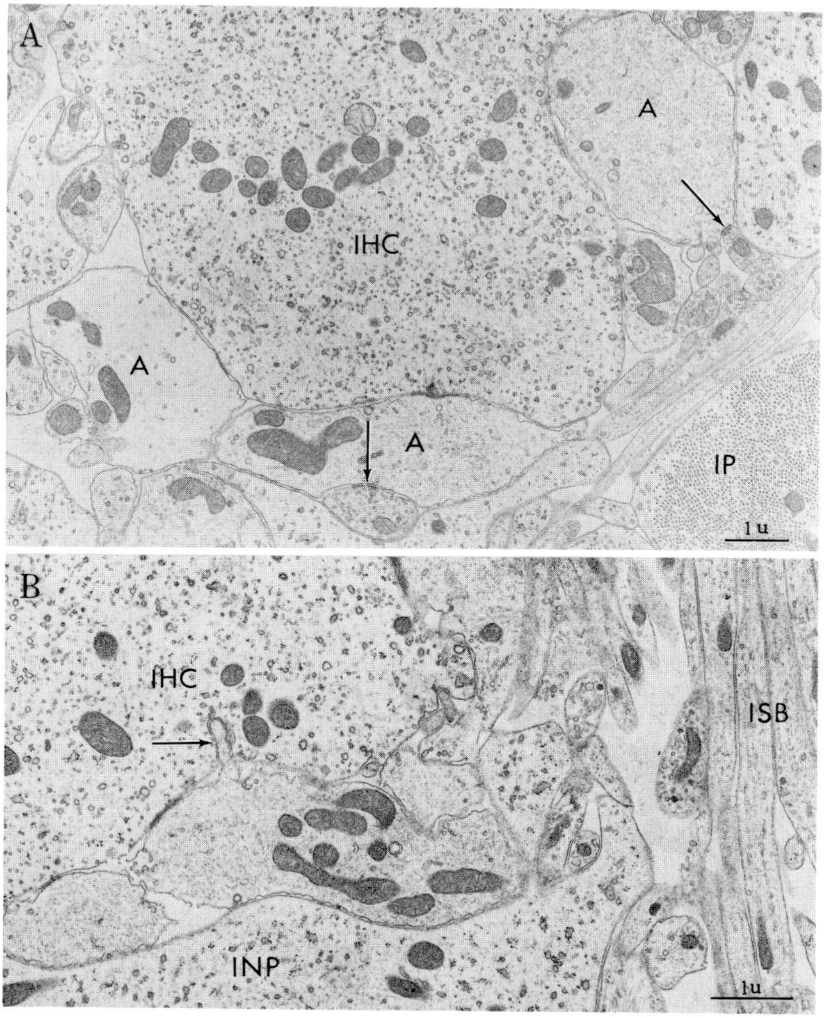

FIG. 14. Nerve endings of the inner hair cell. Squirrel monkey. (A) In one afferent nerve ending (A), a small amount of homogeneous substance is shown; in others there is a large amount of homogeneous substance. Note synapses (arrows) between the afferent nerve endings and efferent fibers. IHC, Inner hair cell; IP, inner pillar cell. (B) Fingerlike extension (arrow) and thickening of the plasma membrane are shown in the same nerve ending. IHC, Inner hair cell; ISB, inner spiral bundle; INP, inner phalangeal cell.

endings may send tiny fingerlike projections into the hair cells (Fig. 14B) or they invaginate to accommodate the presynaptic membrane accompanied by the synaptic body. The nerve endings sandwiched between adjacent hair cells (Fig. 12B) usually have a synaptic membrane thickening on one hair cell, while they lack a conspicuous thickening on other hair cells. It is not determined if the same ending synapses with a second hair cell at another location. The nerve ending is about 1–3 μm in diameter, contains a few mitochondria and some vesicles, and is filled with a homogeneous substance. Some afferent endings appear to contain less homogeneous substance and appear to be almost empty (Fig. 14A).

The nerve endings that contain numerous vesicles resemble the efferent nerve endings seen below the outer hair cells (Fig. 13B), and are small and round (0.4 μm in diameter) (Smith, 1961). It is not certain whether or not these vesiculated nerve endings terminate on the inner hair cell or continue in other directions; the frequency of these contacts with the inner hair cells is low. The efferent, vesiculated portion establishes a typical axodendritic synapse on the opposite pole of the afferent ending, forming a synapse on the hair cell (Fig. 14A).

Thus, in the organ of Corti, the two types of hair cells have a different arrangement of stereocilia and possibly differences in their attachment to the tectorial membrane, as well as differences in distribution of cell organelles and types of nerve endings. This suggests that there must be differences in their responses to mechanical vibration. Recently, Dallos *et al.* (1972) suggested that the electrical potentials produced by the inner hair cells are proportional to the velocity of vibration of the basilar membrane, while those generated by the outer hair cells are proportional to the displacement of the basilar membrane.

IV. Nerve Fibers

The presence of afferent and efferent nerve fibers in the organ of Corti is generally accepted. However, adrenergic nerve fibers are also reported to approach the organ of Corti. Terayama *et al.* (1966) demonstrated their presence along the blood vessels of the tympanic lip of the limbus spiralis, and also at the basilar membrane. Spoendlin (1966), using a similar fluorescence technique, followed the perivascular plexus of the adrenergic fibers up to the cochlear artery, but was unable to demonstrate them further peripherally along the vessels. Instead, he found them distributed among the cochlear nerve fibers, reaching as far as the tympanic lip where they are presumed to form the arcades of the terminal plexus. Densert (1974) likewise noted a rich plexus of adrenergic terminals around the radiating nerve fibers in the region where they lose their myelin sheaths, and also found them around the spiral vessels at the tympanic lip. By experimental procedures he differentiated between adrenergic and cholinergic nerve fibers, both of which contain vesicles with a central core; no adrenergic terminal could be traced into the organ of Corti. The presence of a fourth type of nerve fiber, parasympathetic, was reported by Ross (1969, 1971) in Rosenthal's canal and in the osseous spiral lamina. Since the pathways of the parasympathetic nerve fibers are almost identical to those of the efferent type, doubt has arisen as to whether these investigators are describing the same nerve fibers, or different types running together in the same bundle. The possible existence of autonomic nerve fibers in Rosenthal's canal has been also mentioned by Maw (1973); he demonstrated axodendritic synapses in the intraganglionic bundle.

A. Habenula Perforata

In the osseous spiral lamina, nerve fibers are both myelinated and unmyelinated. Usually the myelinated group is in the majority. These myelinated fibers lose their myelin sheath some distance away from or at the habenula before entering the organ of Corti (Engström and Wersäll, 1958; Smith, 1961) (Figs. 15A and B). In the cat, Spoendlin (1966) observed about 10 to 20 fibers per habenula opening. Our figure from squirrel monkey specimens is slightly higher, averaging about 27, with an extreme variation of 4 to 65 (0.1–0.7 μm in diameter).

In the habenula three types of nerve fibers in various profiles can be seen. One shows exclusively tubules, the second only filaments, and the third both tubules and filaments (Fig. 15B). The nerve fibers with filaments are fewest in number. In the squirrel monkey they account for about 15% of the total fibers in the habenula at the second turn, and range from small to medium size. Attempts to classify the nerve fibers in this area as either afferent or efferent based on the presence of tubules or filaments have not been successful. Sometimes nerve fibers containing many vesicles or vesicles with dense cores are seen in the habenula, but no synapse is observed.

All the nerve fibers in the habenula are surrounded by one to three Schwann cells. It is of interest to note that a single Schwann cell surrounds as many as 20 or more nerve fibers. Usually, the cell pro-

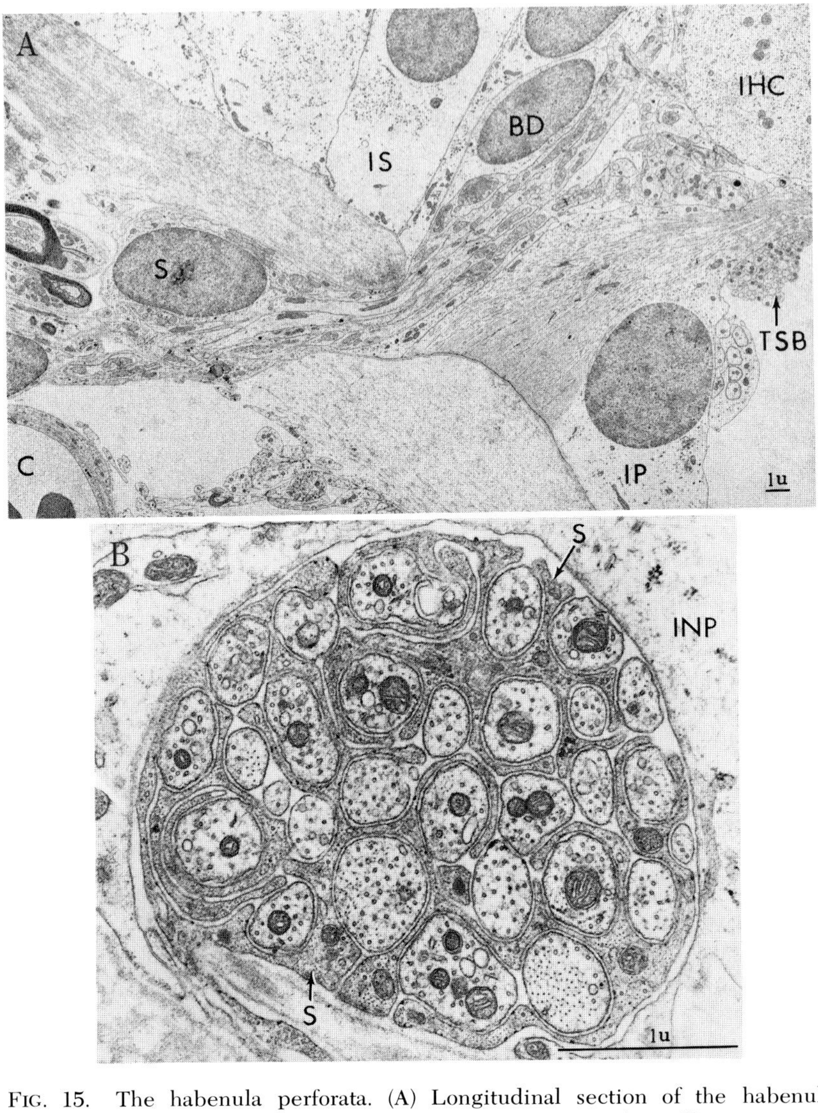

FIG. 15. The habenula perforata. (A) Longitudinal section of the habenula, showing the nerve fibers and the Schwann cell (S). BD, Border cell; TSB, tunnel spiral bundle; IHC, inner hair cell; IP, inner pillar cell; C, capillary; IS, inner sulcus cell. Rhesus monkey. (B) Cross section of the habenula, showing 29 nerve fibers containing tubules, filaments, or both, which are surrounded by only two Schwann cells (S). The entire bundle is further surrounded by the inner phalangeal cell (INP). Squirrel monkey.

cesses of the border cells and the inner phalangeal cells extend deeply downward into the habenula on its peripheral margin. In cross section the supporting cells are located outside, and the Schwann cell inside (Fig. 15B). Above the basilar membrane the nerve fibers lose the Schwann cell processes, but these nerve fibers, singly or in bundles, are partly surrounded by the supporting cells. The wide fluid space noted between the nerve fibers on the tympanic side of the habenula narrows down to the minute intercellular space of the Schwann cells. The gap between the basilar membrane and the Schwann cells of the nerve fibers is filled by the basal portions of the supporting cells. When the Schwann cells meet the supporting cells in the habenula, the basement membranes of each group become continuous; thus there is no free communication between the fluid spaces below and above the habenula.

B. Inner Hair Cell Area

After passing through the habenula, the afferent fibers ascend directly to the inner hair cells, although some of them take a short spiral course before terminating (Smith, 1961; Spoendlin, 1969). According to Spoendlin (1971, 1973), about 20 neurons without branching go directly to every inner hair cell. However, Perkins (1973) reported that the fibers from a single neuron typically end beneath two inner hair cells. The afferent fibers above the habenula are large in size (0.5 μm in diameter) (Smith, 1961; Spoendlin, 1969), in contrast to the efferent fibers.

Smith (1961) described an extensive intertwining of efferent fibers among the radial afferent fibers. They are typically small in diameter (0.1 μm) and show enlargements containing many vesicles and a few mitochondria. The axodendritic synapse is frequently found below the inner hair cells (Rodriguez-Echandia, 1968; Smith, 1968; Spoendlin, 1973). These efferent fibers travel in a spiral direction in bundle form (an inner spiral bundle) beneath the inner hair cells (Fig. 14B). They appear to divide in this area and send a few collateral fibers to the outer hair cells (Perkins, 1973). Iurato (1964) observed that the efferent fibers below the inner hair cells originate homolaterally from the olivocochlear bundle (Rasmussen, 1960), and that about 85% of those traveling to the outer hair cells are of contralateral origin. Spoendlin (1973) stated that about 50% of the inner spiral bundle is part of the contralateral olivocochlear bundle. Our own investigation with R. R. Gacek (unpublished data) suggests that the homolateral fibers also terminate on the outer hair cells, although not as often as the contralateral fibers.

C. Corti's Tunnel Area

From the area below the inner hair cells, nerve fibers pass between the inner pillar cells at their basal portions and enter Corti's tunnel. The many nerve fibers then take a spiral course (tunnel spiral fibers) before moving toward the outer hair cells (Smith and Rasmussen, 1963) (Fig. 16A); the tunnel spiral fibers are mostly efferent. The fibers pass through the tunnel at about four different levels. One group passes high in the tunnel, going directly to the Deiters' cup area; the second group, after taking a high course, descends toward the lower portions of the Deiters' cells; the third takes a direct low route to the inferior portions of the Deiters' cells; and the fourth travels on or between the basal portions of the pillar cells. There are apparently species differences in the distribution of nerve fibers in the tunnel. They pass singly or in bundles with as many as nine fibers. Their size varies from 0.3 to 2.8 μm in diameter.

The efferent fibers are generally acknowledged to pass high in the tunnel (Kimura and Wersäll, 1962; Smith, 1967; Spoendlin, 1969), although Smith indicates that afferent fibers are also mixed among them. In the guinea pig the efferent fibers also pass at the bottom of the tunnel on the pillar cells (Kimura and Wersäll, 1962). In the cat all the fibers passing through Corti's fluid space are efferent (Spoendlin, 1969).

The afferent fibers in the cat travel on or between the pillar cell bases (Spoendlin, 1969). In the squirrel monkey we found that the afferent fibers travel through the tunnel by two routes: one through the fluid space at the lower part of the tunnel, and the other by the route corresponding to the one described by Spoendlin. Nomura and Kirikae (1967) indicated that the afferent fibers in the human tunnel of Corti are directed basalward, whereas the efferent fibers pass diagonally toward the cochlear apex. Branching of the nerve fibers in the tunnel has not yet been observed. These fibers may be partly surrounded by the rod portion of the pillar cells. The fibers located high in the tunnel contain many neurofilaments, while those embedded in the basal cytoplasm of the pillar cells almost exclusively contain neurotubules (Spoendlin, 1966).

Spoendlin (1971, 1973) reported that only 5% of the total afferent fibers pass through the tunnel toward the outer hair cells, and that the other 95% go to the inner hair cells. He also indicated that only a small percentage of the spiral ganglia degenerates when the outer hair cells are completely lost without an obvious decrease in the number of inner hair cells. Our preliminary study in the squirrel monkey supports Spoendlin's contention; about 4–6% of the afferent

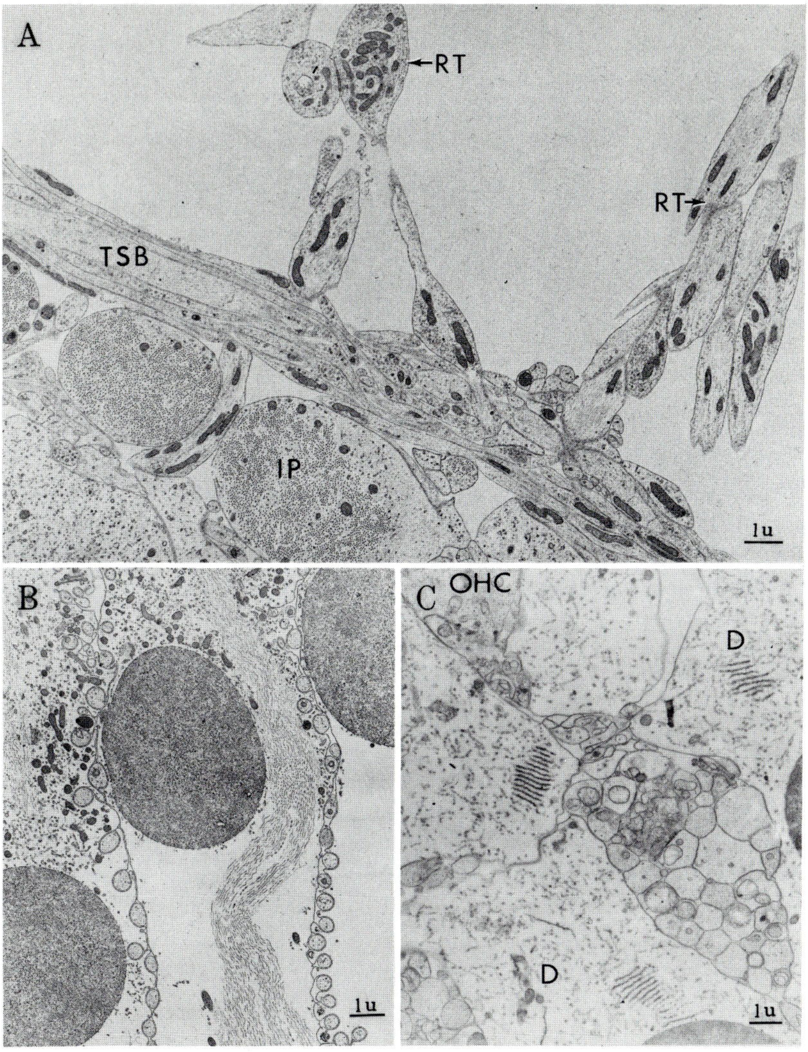

FIG. 16. (A) The tunnel spiral bundle (TSB) and the radiating tunnel fibers (RT) are shown adjacent to the inner pillar cell (IP). The gap between the pillar cells is occupied by nerve fibers and the cell processes of the inner phalangeal cell. Squirrel monkey. (B) The nerve fibers between the Deiters' cells are individually separated in small animals. Cat. (C) The nerve fibers below the outer hair cells (OHC) are shown in bundles comprised of both afferent and efferent fibers which show synapses among them. D, Deiters' cell. Human. (From Kimura *et al.*, 1964.)

fibers pass through Corti's tunnel; when the efferent fibers are not sectioned, about 12–19% of the total fibers entering the habenula cross the tunnel. In human adult specimens, taken from the basal turn of the cochlea, in which no obvious decrease in the nerve fiber population was noted at the osseous spiral lamina, approximately 15–22% of the fibers counted at the habenula passed through the tunnel. A majority of the fibers are surrounded by the basal cytoplasm of the pillar cells, and a small number pass through the fluid space, which is contrary to our observations in the squirrel monkey. Further details will be published at a later date.

1. *Tunnel Fluid*

At present no one knows where the tunnel fluid actually originates. It was presumed to be unlike endolymph, because the high concentration of potassium would interfere with neural transmission in Corti's tunnel. Since no direct communication with endolymph or with the perilymphatic channel could be established in morphological studies, Engström (1960) named the fluid cortilymph. Schuknecht and Seifi (1963) were able to follow acetylcholinesterase precipitates from the scala tympani through the holes in the osseous spiral lamina along the cochlear nerve fibers into Corti's tunnel; therefore they believe the fluid in question is perilymph. The presence of a wide space at the habenula was denied by Engström (1960) and Spoendlin (1966) in their electron microscope studies. The organ of Corti is considered a perilymphatic tissue by Tonndorf *et al.* (1962), on the evidence that it was stained by a vital dye applied in the scala tympani. However, von Békésy (1960) showed that the hair cells were stained by placing a vital dye in the scala vestibuli with an intact Reissner's membrane. Evidently, diffusion of small molecular substances into the organ of Corti occurs from the scala tympani or the scala media side.

The supporting cells of the organ of Corti do not show the morphological characteristics of active fluid transport; however, the Hensen's cells and the border cells reveal numerous tall microvilli on the scala media side. The endolymph may be absorbed through this villous portion into the outer tunnel and into the fluid space around the inner hair cells with a change in chemical composition. The possible nutritional significance of these cells to the hair cells and nerve fibers was mentioned earlier by Engström and Wersäll (1953c), Spoendlin (1966), and Angelborg and Engström (1973). A change in the fluid of Corti's organ would probably produce a profound effect on the mechanism of neural transmission.

D. Outer Hair Cell Area

Below the outer hair cells, the nerve fiber distribution between the afferent and efferent nerves establishes a different pattern. The afferent fibers located at the lower portion of the Deiters' cells travel toward the basal turn direction and gradually rise to the inferior portion of the Deiters' cup. Lorente de No (1937) described the nerve fibers extending basalward as far as one-third of the coil. Fernandez (1951) reported a similar finding, but he also indicated that some of the fibers follow the apical direction. More recently, Smith and Haglan (1973) observed that the afferent fibers travel some distance among the Deiters' cells before branching, and that at the basal turn these fibers either exclusively or predominantly supply a single row of outer hair cells, while at the apical turn they usually supply two or three rows of outer hair cells. They also showed that a single afferent fiber branches and supplies 6 to 12 outer hair cells; the number of nerve endings is greater than the number of hair cells, which suggests that one hair cell may receive one or more nerve endings from the same nerve fiber. According to Spoendlin (1971, 1973), every afferent neuron traveling toward the outer hair cells is branched and innervates about 10 hair cells.

The afferent fibers synapse with the varicose portion of the efferent fibers, or with the inferior pole of the efferent nerve endings, and keep traveling longitudinally until they send nerve stalks out to the base of the hair cells. Axodendritic synapses are common near the hair cells. In small animals synapses are less frequently seen toward the lower portion of the Deiters' cells, where the fibers pass individually (Fig. 16B). However, in human specimens these synapses are often seen, since many of the fibers travel in bundles with both types of fibers intermixed (Fig. 16C). The Deiters' cells develop mesoaxons at the junction where their cytoplasmic processes envelop a single fiber (Engström *et al.*, 1965), and these fibers contain many neural tubules.

E. Afferent Fibers and Spiral Ganglia

The lack of degeneration of the afferent nerve endings and fibers below the outer hair cells after transection of the cochlear nerve trunk was thought to be due to the close relationship of the nerve fibers and the Deiters' cells (Spoendlin and Gacek, 1963). Later, in 1972, Spoendlin indicated that these undegenerated nerve fibers were traceable to the 5% of unmyelinated cochlear neurons that had not degenerated. Ross and Burkel (1973), however, after observing unmyelinated or less myelinated multipolar neurons in Rosenthal's canal, suggested that these neurons are not affected by retrograde degeneration simply because their axons do not extend to the area of the nerve section; they believe that these neurons are postganglionic parasympathetic neurons. However, Spoendlin (1973) states that the unmyelinated neurons are not multipolar. Adrenergic origin for these neurons is ruled out by the fluorescence studies of Spoendlin and Lichtensteiger (1966), Terayama *et al.* (1966), and Spoendlin (1973). Thus a controversy exists over the unmyelinated or thinly myelinated neurons originally described by Rosenbluth (1962). Ross and Burkel (1973) believe that the neurons they described and those described by Rosenbluth are different, although both studies used rat specimens. Kellerhals *et al.* (1967) counted about 10% unmyelinated in the guinea pig; Ross and Burkel (1973) reported 7–8% in the rat; and Spoendlin (1972) counted 5% in the cat.

More recently, Spoendlin (1973) modified his opinion by introducing a third type of neuron (type III), which cytologically resembles the major type (type I) except that the former has a thin myelin sheath. He believes that type III neurons (0.5% of the total neurons) are the origin of the giant fibers, each of which spreads apically and basally to innervate about 10 inner hair cells. Type II (unmyelinated) and type III neurons constitute about 5% of the total neurons that remain after degeneration of type I neurons following cochlear nerve transection. It is also interesting to note that, according to Spoendlin, when the entire organ of Corti including both the inner and outer hair cells is missing following severe acoustic trauma, the unmyelinated neurons (type II) remain even after 14 months. Thus the unmyelinated neurons are resistant to retrograde degeneration from both directions. From his Golgi study, Perkins (1973) reported that the inner hair cells are innervated by two kinds of afferent neurons, radial and spiral; the outer hair cells are innervated by the spiral neurons only. The spiral neurons correspond to the type II neurons of Spoendlin.

F. Efferent Nerves

From the foregoing description it is obvious that distribution patterns of the afferent and efferent nerve fibers are considerably different. It is not understood why the efferent fibers heavily innervate the outer hair cells, whereas only a few contacts are made with the inner hair cells. It is also not clearly understood why so many afferent neurons go to an inner hair cell, whereas a single afferent neuron supplies so many outer hair cells. The synaptic contact of the efferent fibers with the afferent fibers is primarily at the level below the inner hair cells and secondarily at the area below or at the outer hair cells, in addition to their major terminals on the outer hair cells.

It is often noted in the literature that the total number of cochlear efferent fibers in the cat is 500 (Rasmussen, 1960). We presume that this figure is a rough estimate of the myelinated components only. Our study of the nerve fiber count at the vestibulocochlear anastomosis in the guinea pig indicates about 920 myelinated and about 680 unmyelinated fibers. The unmyelinated fibers are numerous at this point (Terayama and Yamamoto, 1971). The main afferent nerve trunk contains only a small number of unmyelinated fibers. According to Warr (1973), the total number of olivocochlear neurons in the cat is sometimes estimated to be more than twice the numerical value provided by Rasmussen.

The function of the efferent nerve system has not been clearly established. Electrical stimulation of the efferent fibers is known to inhibit afferent nerve activities (Galambos, 1956; Desmedt, 1962; Fex, 1962; Wiederhold and Kiang, 1970). However, the role they play in hearing is still in doubt. Galambos (1960) did not find any effect on the hearing of behaviorally conditioned animals. Leibbrandt (1965) thinks that they have an effect on adaptation to sound, while Trahiotis and Elliot (1970) consider their role in the masking effect. Capps and Ades (1968) believe that these fibers are related to frequency discrimination. Igarashi *et al.* (1972) did not observe any change in pure tone audiometry after transection of the efferent fibers at the midline of the fourth ventricle.

V. Supporting Cells

There are two types of supporting cells: one group with intracellular tonofilaments organized in a definite pattern, and another group without these filaments. The group of cells with the tonofilaments establishes the main supporting framework of the organ of Corti. The inner and outer pillar cells and Deiters' cells all contain tonofilaments, and send their filament-containing phalangeal processes from their basal attachment on the basilar membrane to the cell apex where they form the stiff reticular lamina.

A. Pillar Cells

Both inner and outer pillar cells are morphologically similar; the cell apex (head portion) and base (footplate) are larger than the midportion. In the pillar cell the nucleus is located at the basal portion; the cell organelles are not remarkable, and have some mitochondria located close to the tonofilaments and in the area close to the basilar membrane. There are localized electron-dense substances (reticular thickenings) at the border between the pillar cell and hair cell, and at the basal portion of the pillar cells. These substances appear to serve as an anchoring mechanism for the filaments at the top and bottom. The basal condensation was called the basal cone by Angelborg and Engström (1972). According to these investigators, there are two kinds of filaments, a tubular form and microfilaments, which extend from the basal cone to the cell apex or head portion and have diameters of 275 and 60 Å, respectively. The tubular filaments number about 2400 in the guinea pig, and about 1300 in the squirrel monkey (Angelborg and Engström, 1972). Iurato (1961) reported that the number decreases toward the apical turn. Our observation indicates that in the human the tubular filaments often penetrate the basal cone (Fig. 17A) and attach to another more dense serrated part of the cell at the basilar membrane. The filaments of the inner pillar cells extend from the base and attach to the reticular thickenings at the junction of the inner and outer pillar heads, and other filaments continue to the reticular thickenings adjacent to the first row of the

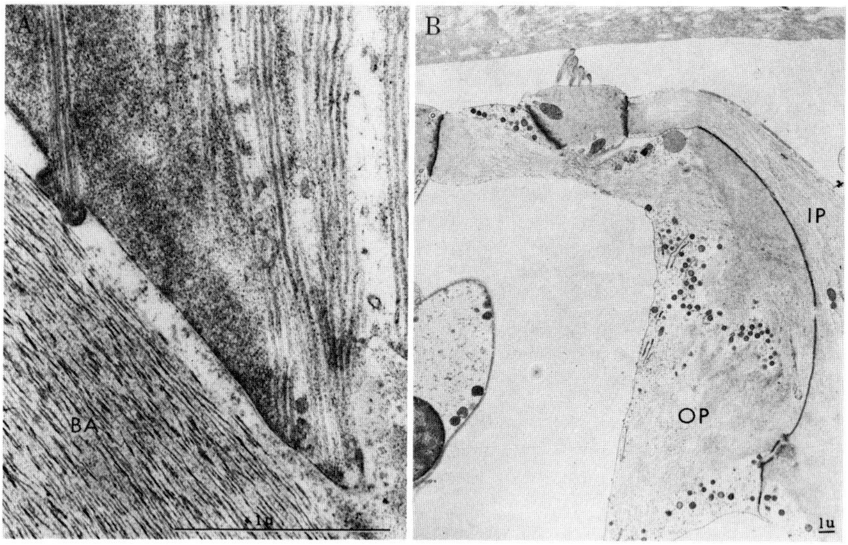

FIG. 17. Pillar cells. (A) The tonofilaments appear to originate from the electron-dense aggregates at the cell base. Note the size of the filaments as compared to those of the basilar membrane (BA). Human. (B) Junction between the inner (IP) and outer (OP) pillar cells. The inner pillar cell abuts only the first row of the outer hair cells, whereas the outer pillar cell abuts both the first and second rows of the hair cells. Squirrel monkey.

outer hair cells (Fig. 17B). At the head portion of the outer pillar cells, filaments also span between the reticular thickenings located at the second row of hair cells and those located at the junctions of the inner and outer pillar cells (Fig. 17B). Thus these filaments at the head portion crisscross at right angles to the filaments coming from the base.

The inner pillar cells establish a wall between the inner hair cell and Corti's tunnel (Fig. 17C). The openings are a slitlike intercellular space often located in the region where the unmyelinated fibers pass toward the outer hair cells. The slits seen in other areas are often closed when followed a little farther along. There is, however, a wide communicating fluid space between the outer pillar cells at their midportion. The apical junctions of the outer and inner pillar cells are long, and have zonulae occludentes at both the scala media and tunnel sides, and zonulae adherentes between them (Iurato and Taidelli, 1965). Desmosomes are seen in the basal portion of the pillar cells.

B. Deiters' Cells

Deiters' cells have an unusual shape. The cell forms a supporting cup for the hair cell at its midportion, and sends a slender stalk from the region below the cup to the top of the cell to form the reticular lamina. The nucleus is located below the cup-shaped part. The cell contains tonofilaments which are similar to those of the pillar cells (Angelborg and Engström, 1972). The bundles of filaments start at the basilar membrane, as in the pillar cells, and from the central position of the basal cytoplasm swing more radially and apicalward around the nucleus (Fig. 18A) and then insert into the opaque cytoplasmic condensation formed at the inner part of the Deiters' cup. Other filaments are diverted below the cup into the stalk ascending in the apical turn direction and insert into the thickening of the reticular lamina which supports the top part of the hair cell adjacent to the one whose base rests on the Deiters' cell itself (Angelborg and Engström, 1972) (Fig. 18B). According to Spoendlin (1966), the insertion is two to three cells apicalward. Since the outer hair cells tilt basalward, the two cell groups crisscross each other. This structural organization may provide some degree of flexibility and protection for the organ of Corti in the outer hair cell area. The number of tonofilaments in the Deiters' cell of the rat is about 600, in contrast to the 3000 to 4000 for the pillar cells (Iurato, 1967). The apical surface of the Deiters' cell is in the shape of a dumbell with the middle section constricted by the abutment of the hair cells.

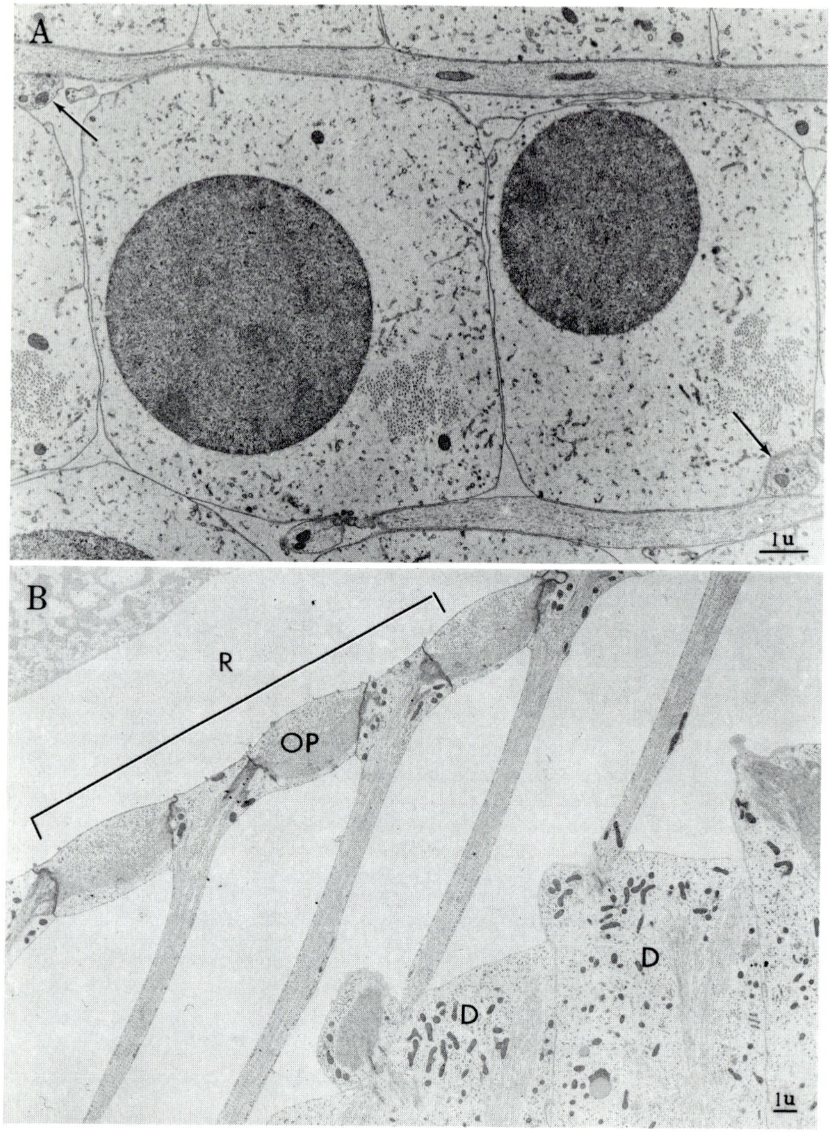

FIG. 18. Deiters' cells. Squirrel monkey. (A) The shape is almost rectangular at the nuclear zone. The tonofilaments are located toward the cochlear apex and also away from Corti's tunnel. Two afferent fibers are touched by the varicosities of the efferent nerve fibers (arrows). (B) Phalangeal processes of the innermost row of Deiters' cells (D) extend to form the reticular lamina (R). The upward slant of the processes is toward the cochlear apex. Between the Deiters' phalanges are the headplates of the outer pillar cell (OP). Compare with Fig. 19B.

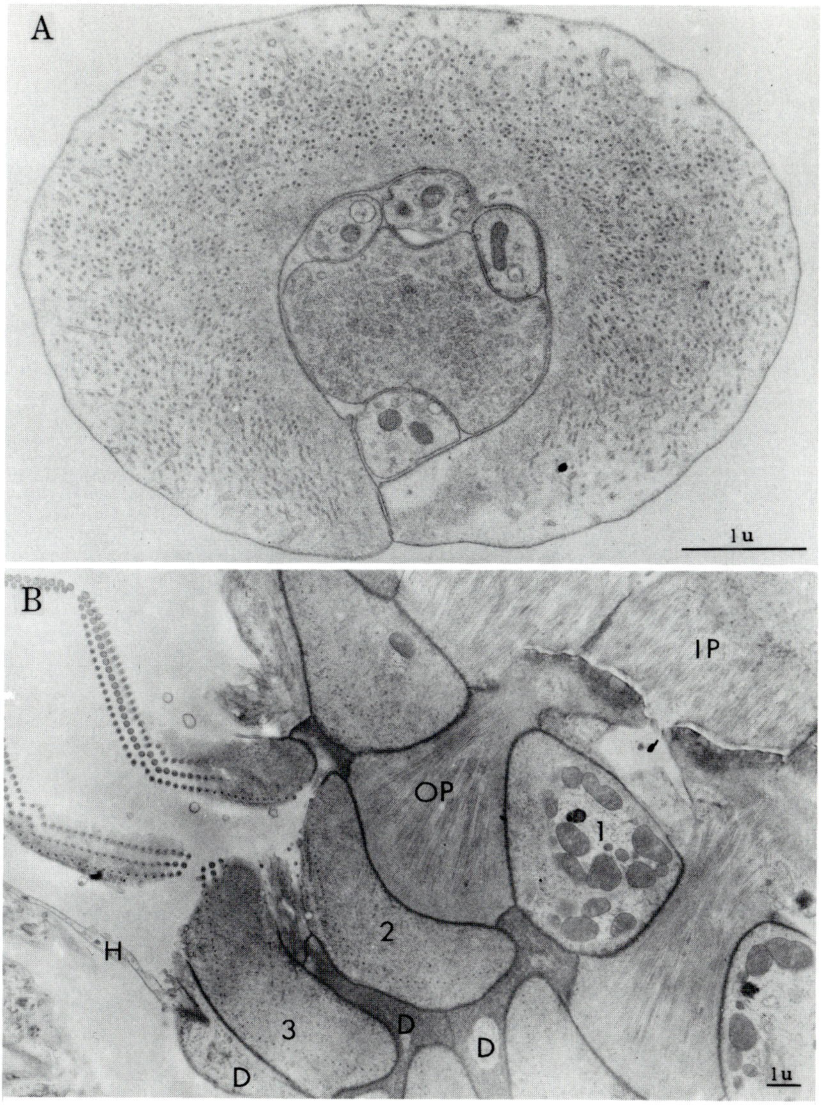

FIG. 19. (A) Cross section of the Deiters' cell at its cup-shaped portion which supports the outer hair cell. In this micrograph four afferent and one large efferent nerve ending are shown. Note the homogeneous substance toward the nerve endings, and the tonofilaments at the periphery of the Deiters' cell. Bat. (B) The reticular lamina. IP, Inner pillar cell; OP, outer pillar cell; D, Deiters' phalanx; H, Hensen's cell. Numbers 1, 2, and 3 represent the first, second, and third rows of the outer hair cells. Bat.

The cell organelles are more numerous in the area between the Deiters' cup and its nucleus. Mitochondria, rough endoplasmic reticulum, lysosomes, and the Golgi apparatus are scattered in this area. The shape of the Deiters' cup is pronounced in the bat (Figs. 3 and 19A), but in other higher species it is less distinct; it looks like the palm of a cupped hand with the side toward Corti's tunnel open, or the cup form is established by a contribution from an adjacent Deiters' cell.

The reticular lamina is the superstructure of the organ of Corti, which very firmly holds the hair cells at their apices (Fig. 19B). It is composed of phalangeal processes of pillar cells and Deiters' cells. The tonofilaments that run horizontally are conspicuous in the phalangeal processes of both pillar cells, while they are lacking or less obvious in the Deiters' processes. According to Iurato and Taidelli (1965), the junction of the reticular lamina, including that of the hair cells, is the zonula occludens; a zonula adherens is less frequently observed, and desmosomes are rare. At the cell junctions an electron-dense substance (reticular thickening) accumulates, more pronounced on the phalangeal processes than on the hair cell side. The unique aspect of this junction is that the zonula occludens extends almost the full thickness of the reticular lamina, although it is intermittent along its length.

C. Hensen's Cells

The Hensen's cells are attached to the outermost margin of the reticular lamina which is formed by the phalangeal processes of the third row of Deiters' cells. The Hensen's cells in this area are extremely thin, having short microvilli on the endolymph side. However, farther out adjacent to the outer tunnel, the endolymphatic surface of the Hensen's cells show remarkable development of microvilli which are numerous and very tall. Some of them are taller than the thickness of the cell processes from which they project (Fig. 20A). Toward the outer tunnel the Hensen's cells show clublike or polypous projections. The cytoplasm in this area contains some mitochondria, scattered RNA particles, lysosome granules, numerous tubules, and vesicles. Most of the other parts of the Hensen's cells are characterized by the paucity of cell organelles; often the cytoplasm appears almost empty. The nucleus is located high, near the cell apex and close to the endolymph surface. It is not known whether all the Hensen's cells reach the basilar membrane. The cell junction between the reticular lamina and the Hensen's cell is the zonula occludens (Beagley, 1965), which is reported to be the weakest part and is subject to tearing from acoustic trauma. Neural ele-

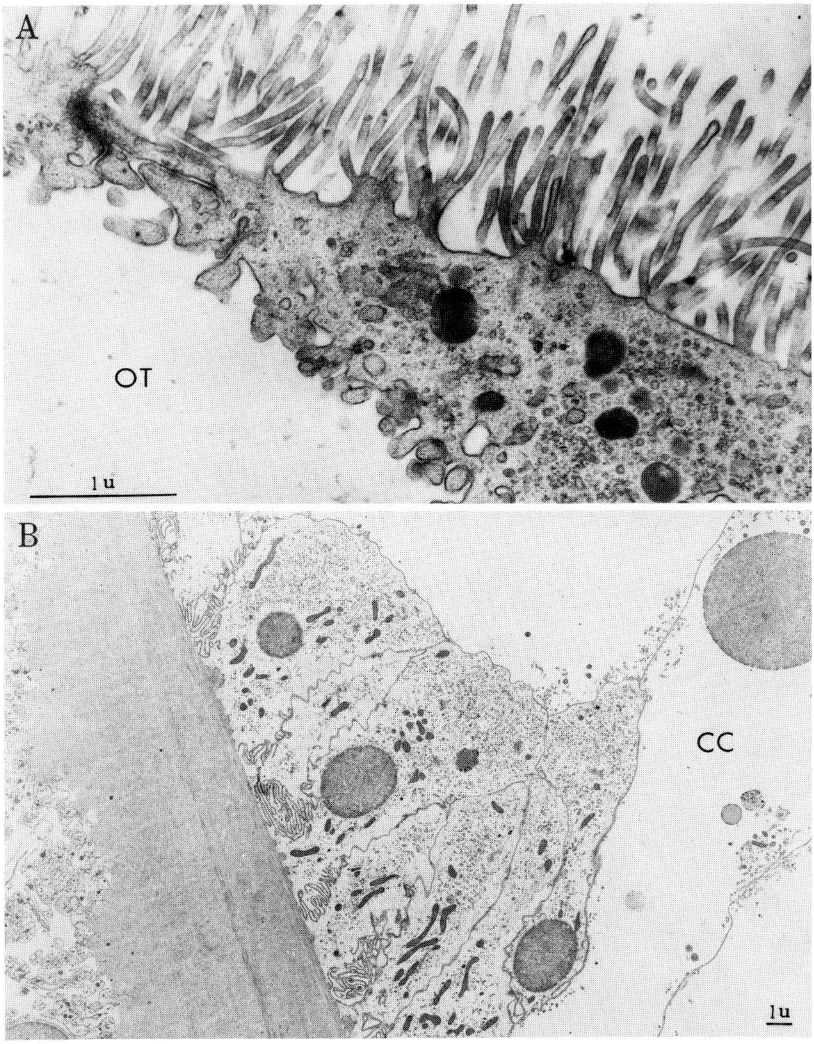

FIG. 20. (A) The Hensen's cell, showing an enormous number of microvilli on the endolymph side. OT, Outer tunnel. Bat. (B) Boettcher cells. They form a cluster over the basilar membrane and show plasma membrane infoldings at their basal portions. The cytoplasm of the Claudius cell (CC) appears almost empty. Squirrel monkey.

ments are generally not observed among or near the Hensen's cells, although they are sometimes seen, according to Engström *et al.* (1966) and Wright and Preston (1973). The cell junction of Hensen's cells at the basal portion is the occludens type, which may be seen at the basilar membrane area or at a slightly higher level. Such occludens are intermittent along the cell borders.

D. Claudius Cells and Inner Sulcus Cells

The Claudius cells and inner sulcus cells are morphologically similar to the Hensen's cells, except that the former are shorter in height. They have short microvilli in varying numbers at the apical surfaces where mitochondria are found in moderate numbers. The cell organelles are extremely scarce and have poorly developed Golgi apparatuses (Figs. 15A and 20B).

E. Boettcher Cells

Boettcher cells are usually limited to the basal turn area between the basilar membrane and the Claudius cells. These cells do not reach the endolymphatic fluid space, but rest on the basilar membrane in a cluster (Fig. 20B). The cytoplasm is rich in vesicles, tubules, and free ribosomes, and shows a moderate number of scattered mitochondria. The cell junction between the Claudius cell and the Boettcher cell is the fascia occludens. Boettcher cells frequently show interdigitations among themselves at their basal cytoplasm (Ishiyama *et al.*, 1970). In the bat tonofilament bundles in tubular form, similar to those seen in the Deiters' cells, are sometimes observed extending from the apical portion around the nucleus toward the inferior portion. In the guinea pig close to or continuous with the basilar membrane and between the cells is a mazelike or fingerprint structure which has a very thick electron-dense line alternating in parallel with an equally thick light zone. No reasonable explanation is offered for Boettcher cells being limited to the basal turn.

F. Inner Phalangeal Cells and Border Cells

The inner phalangeal cells and the border cells are similar in morphology; only their locations differ (Figs. 10 and 12A). The inner phalangeal cells are located between the inner hair cells and the inner pillar cells, and between the inner hair cells on the pillar cell side. These cells correspond to the Deiters' cells; however, there is no cytological resemblance between these two types of supporting cells. The inner phalangeal cell is very irregular in shape. The cell surface has some microvilli; the apex is very small, and in the shape of an I beam between the inner hair cell tops (Fig. 11A). The border cells are located more toward the inner sulcus cell side of the inner hair cells. Their cell surfaces are very small, but show numerous tall microvilli and often a basal body below. The cells are thin and follow closely the contour of the inner hair cells toward the base, where they show an irregular form of cytoplasm.

Both the inner phalangeal and border cells contain nuclei in their basal cytoplasm where they often partly surround unmyelinated nerve fibers individually or in bundle form. At the base the cytoplasm often extends into the habenula perforata and partly surrounds it. The cell organelles are not particularly remarkable in either type of cell. There are slightly more mitochondria in the apical portion. Some rough endoplasmic reticulum, free ribosomes, lysosome granules, vesicles, and tubules are scattered in the cytoplasm. The inner phalangeal cells and border cells both lack tonofilaments. While they form reticular thickenings at their cell tops, the basal portions have no resemblance to the Deiters' cells. No cytoplasmic dense substance accumulates toward the inner hair cell, in contrast to the Deiters' cell cup. The space between the hair cell and these supporting cells is generally wide, except near the cell apex area. All the supporting cells on the basilar membrane are underlined by the basement membrane throughout, except at the habenula perforata.

VI. Basilar Membrane

The basilar membrane underlines the organ of Corti, extending from the basilar crista of the spiral ligament to the tympanic lip of the limbus spiralis. The membrane is composed of a fibrous portion, a homogeneous ground substance, mesothelial cells, and capillaries. Many details were provided earlier by Spoendlin (1957), Engström and Wersäll (1958), Iurato and de Petris (1961), and Iurato (1962b). The reader is particularly referred to Iurato's work (1967) for major coverage of this subject. According to Iurato, collagen and elastic materials are missing from the basilar membrane and other perilymphatic connective tissues.

The basilar membrane is divided into two parts, the pars tecta which extends from the outer pillar cells to the tympanic lip, and the pars pectinata which extends from the outer pillar cells to the basilar crista. The pars tecta contains radially arranged filaments 180–250 Å in diameter, which form a compact and uniform layer embedded in a homogeneous ground substance (Fig. 21A). This part also contains capillaries (vas spiralis) running longitudinally. Our studies (Kimura and Ota, 1974) and others show that these vessels are not fenestrated; sometimes nerve fibers are observed in this area (Terayama *et al.*, 1966). In the pars pectinata the filaments are grouped in bundles which are observed in two strata separated by the homogeneous ground substance (Fig. 21B). The bundles at the upper stratum are of medium size and show anastomoses. The lower bundles are large in size and are separated by the homogeneous substance (Iurato, 1967).

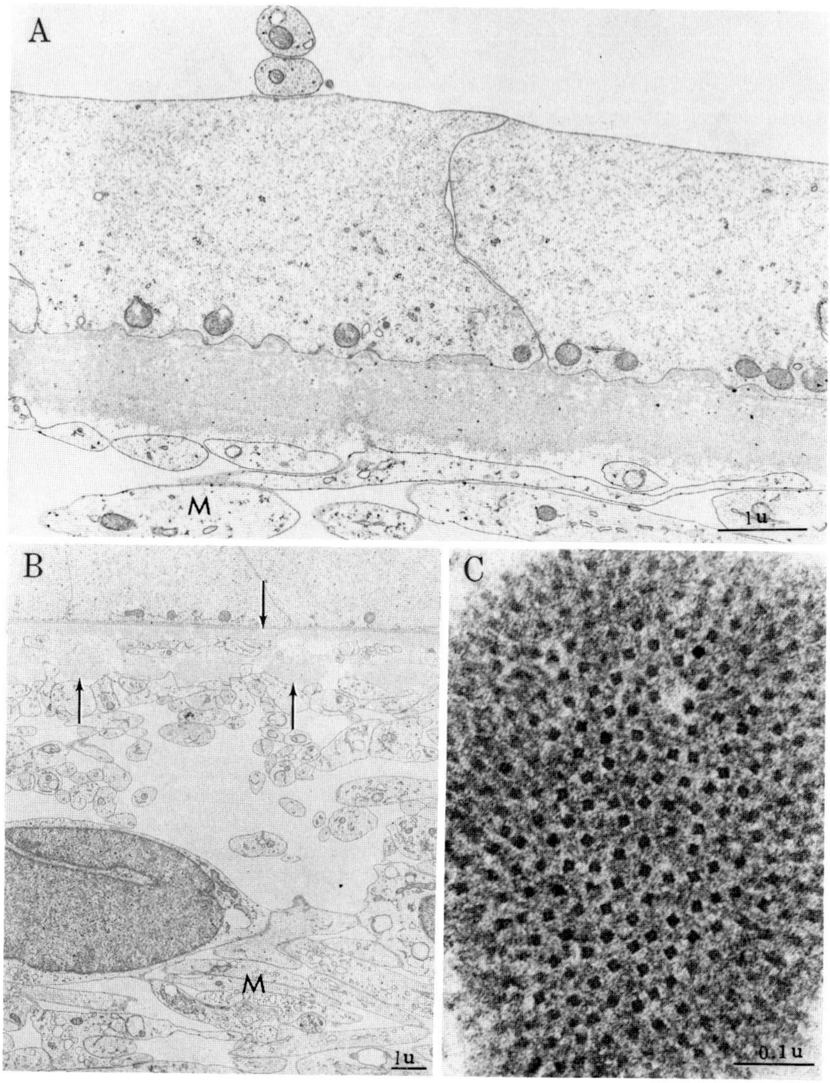

FIG. 21. The basilar membrane. (A) The pars tecta, showing a single stratum of the fibrous layer. Note two nerve fibers on the pillar cells. The junction between the pillar cells is the zonula occludens. M, Mesothelial cells. Squirrel monkey. (B) The pars pectinata, showing a double strata of the fibrous layer (arrows). The lower layer is in bundle form. The mesothelial cell layer (M) is very thick. Squirrel monkey. (C) A high magnification of the basilar fibers, showing a rectangular shape. Each fiber appears to be composed of subfibrils. Guinea pig.

The space between the two major strata is randomly filled by filaments in small bundles. The filaments of the basilar membrane are more often rectangular than cylindrical (Fig. 21C) in cross section. Hamilton (1967) earlier described fibers (110 Å) of rectangular or diamond shape with four subfibrils (20–50 Å) in the vestibular perilymphatic space. He also noted a smaller fiber (50 Å) in diffuse form near the endolymph epithelial cells. The fibrous layer adjacent to the basement membrane forms a continuous layer from the basal to apical turns; however, loose areas are sometimes observed among these fibers at the pars pectinata.

Mesothelial cells are longitudinally arranged in the lower turns, but at the apical turn they may also be observed in the radial direction (Angelborg and Engström, 1974). These cells often show kinocilia protruding a short distance. There are many cell layers in the lower turn in the squirrel monkey (Fig. 21B), but the number of layers decreases toward the apex (Angelborg and Engström, 1974). The cytoplasm of the mesothelial cells shows no unusual organization. It contains scattered mitochondria, free ribosomes, vesicles containing a diffuse substance, and a small Golgi apparatus. In the longitudinally arranged cell processes are numerous filaments, vesicles, and long mitochondria. The cell bodies establish the macula occludens. The mesothelial cells with long cytoplasmic extensions, the homogeneous substance, and the fibrous layer appear to interfere with the penetration of some of the submicroscopic particles from the scala tympani into Corti's tunnel (Angelborg, 1974). Morphological evidence indicates that there is no free communication between the perilymphatic space and Corti's tunnel through the basilar membrane.

References

Angelborg, C. (1974). *Acta Oto-Laryngol., Suppl.* **319,** 19.

Angelborg, C., and Engström, H. (1972). *Acta Oto-Laryngol., Suppl.* **301,** 49.

Angelborg, C., and Engström, H. (1973). *In* "Basic Mechanisms in Hearing" (A. R. Møller, ed.), pp. 125–183. Academic Press, New York.

Angelborg, C., and Engström, B. (1974). *Acta Oto-Laryngol., Suppl.* **319,** 43.

Beagley, H. A. (1965). *Acta Oto-Laryngol.* **60,** 479.

Bélanger, L. F. (1953). *Science* **118,** 520.

Boettcher, A. (1869). *Nova Acta Phys.-Med. Acad. Nat. Curios.* **35,** 1.

Bredberg, G. (1968). *Acta Oto-Laryngol., Suppl.* **236,** 1.

Bredberg, G., Lindeman, H. H., Ades, H. W., and West, R. (1970). *Science* **170,** 861.

Bredberg, G., Ades, H. W., and Engström, H. (1972). *Acta Oto-Laryngol., Suppl.* **301,** 3.

Capps, M. J., and Ades, H. W. (1968). *Exp. Neurol.* **21,** 147.

Corti, A. (1851). *Z. Wiss. Zool.* **3,** 109.
Dallos, P., Billone, M. C., Durrant, J. D., Wang, C. -Y., and Raynor, S. (1972). *Science* **177,** 356.
Davis, H. (1959). *In* "Handbook of Physiology" (Amer. Physiol. Soc., J. Field, ed.), Sect. 1, Vol. I, pp. 565–584. Williams & Wilkins, Baltimore, Maryland.
Densert, O. (1974). *In* "The Adrenergic Innervation of the Inner Ear." Malmö General Hospital, Malmö, Sweden.
Desmedt, J. E. (1962). *J. Acoust. Soc. Amer.* **34,** 1478.
de Vries, H. (1949). *Acta Oto-Laryngol.* **37,** 334.
Duvall, A. J., and Wersäll, J. (1964). *Acta Oto-Laryngol.* **57,** 581.
Engström, H. (1955). *Rev. Laryngol. Otol. Rhinol.* **11,** 808.
Engström, H. (1960). *Acta Morphol. Neer. Scand.* **3,** 195.
Engström, H. (1967). *J. Laryngol. Otol.* **81,** 687.
Engström, H., and Ades, H. W. (1960). *Acta Oto-Laryngol., Suppl.* **158,** 219.
Engström, H., and Wersäll, J. (1953a). *Acta Oto-Laryngol.* **43,** 1.
Engström, H., and Wersäll, J. (1953b). *Acta Oto-Laryngol.* **43,** 323.
Engström, H., and Wersäll, J. (1953c). *Ann. Otol., Rhinol., & Laryngol.* **62,** 507.
Engström, H., and Wersäll, J. (1958). *Int. Rev. Cytol.* **7,** 535.
Engström, H., Sjöstrand, F. S., and Wersäll, J. (1955). *Proc. Int. Cong. Otorhinolaryngol., 5th, 1953* pp. 563–568.
Engström, H., Ades, H. W., and Hawkins, J. E., Jr. (1962). *J. Acoust. Soc. Amer.* **34,** 1356.
Engström, H., Ades, H. W., and Hawkins, J. E., Jr. (1965). *In* "Contributions to Sensory Physiology" (W. D. Neff, ed.), Vol. I, 1–34. Academic Press, New York.
Engström, H., Ades, H. W., and Andersson, A. (1966). *In* "Structural Pattern of the Organ of Corti," pp. 1–172. Williams & Wilkins, Baltimore, Maryland.
Engström, H., Ades, H. W., and Bredberg, G. (1970). *In* "Sensorineural Hearing Loss" (G. E. W. Wolstenholme and J. Knight, eds.), pp. 127–156. Churchill, London.
Fernandez, C. (1951). *Laryngoscope* **61,** 1152.
Fex, J. (1962). *Acta Physiol. Scand.* **55,** Suppl. 189.
Flock, Å. (1965). *In* "Electron Microscopic and Electrophysiological Studies on the Lateral Line Canal Organ and on the Macula Utriculi of a Teleost Fish," Almqvist & Wiksell, Stockholm.
Flock, Å. (1971). *In* "Handbook of Sensory Physiology" (W. R. Lowenstein, ed.), p. 396. Springer-Verlag, Berlin and New York.
Flock, Å., Kimura, R. S., Lundquist, P. -G., and Wersäll, J. (1962). *J. Acoust. Soc. Amer.* **34,** 1351.
Friberg, U., and Ringertz, N. R. (1956). *J. Embryol. Exp. Morphol.* **4,** 313.
Friedmann, I. (1959). *J. Biophys. Biochem. Cytol.* **5,** 263.
Galambos, R. (1956). *J. Neurophysiol.* **19,** 424.
Galambos, R. (1960). *In* "Neural Mechanisms of the Auditory and Vestibular Systems" (G. L. Rasmussen and W. F. Windle, eds.), pp. 137–151. Thomas, Springfield, Illinois.
Hamilton, D. W. (1967). *Anat. Rec.* **157,** 627.
Hamilton, D. W. (1968). *J. Ultrastruct. Res.* **23,** 98.
Held, H. (1926). *In* "Handbuch der normalen und pathologischen Physiologie" (A. Bethe, ed.), Vol. II, pp. 467–526. Springer-Verlag, Berlin and New York.
Hensen, V. (1863). *Z. Wiss. Zool.* **13,** 481.
Hilding, A. G. (1952). *Ann. Otol., Rhinol., & Laryngol.* **61,** 354.
Hilding, D. A., and Wersäll, J. (1962). *Acta Oto-Laryngol.* **55,** 205.

Igarashi, M., and Alford, B. R. (1969). *Acta Oto-Laryngol.* **68**, 420.
Igarashi, M., Alford, B. R., Nakai, Y., and Gordon, W. P. (1972). *Acta Oto-Laryngol.* **73**, 455.
Ishii, D., and Balogh, K., Jr. (1968). *Acta Oto-Laryngol.* **66**, 282.
Ishiyama, E., Cutt, R. A., and Keels, E. W. (1970). *Ann. Otol., Rhinol., & Laryngol.* **79**, 54.
Iurato, S. (1960). *Z. Zellforsch. Mikrosk. Anat.* **52**, 105.
Iurato, S. (1961). *Z. Zellforsch. Mikrosk. Anat.* **53**, 259.
Iurato, S. (1962a). *Z. Zellforsch. Mikrosk. Anat.* **56**, 40.
Iurato, S. (1962b). *Exp. Cell Res.* **27**, 162.
Iurato, S. (1964). *Monit. Zool. Ital., Suppl.* **72**, 60.
Iurato, S. (1967). *In* "Submicroscopic Structure of the Inner Ear" (S. Iurato, ed.), pp. 59–168. Pergamon, Oxford.
Iurato, S., and Bairati, A., Jr. (1959). *Monit. Zool. Ital., Suppl.* **67**, 242.
Iurato, S., and de Petris, S. (1961). *Electron Microsc. 1960, Proc. Eur. Reg. Conf., 2nd, 1960* Vol. II, pp. 814–817.
Iurato, S., and Taidelli, G. (1965). *Atti Cong. Ital. Microsc. Elettronica, 5th, 1965*, pp. 167–170.
Kellerhals, B., Engström, H., and Ades, H. W. (1967). *Acta Oto-Laryngol., Suppl.* **226**, 1.
Kiang, N. Y. S. (1965). *In* "Discharge Patterns of Single Fibers in the Cat's Auditory Nerve," Res. Monogr. No. 35, pp. 84–92. MIT Press, Cambridge, Massachusetts.
Kikuchi, K., and Hilding, D. (1965). *Acta Oto-Laryngol.* **60**, 207.
Kimura, R. S. (1966). *Acta Oto-Laryngol.* **61**, 55.
Kimura, R. S. (1967). *Ann. Otol., Rhinol., & Laryngol.* **76**, 664.
Kimura, R. S., and Ota, C. Y. (1974). *Acta Oto-Laryngol.* **77**, 231.
Kimura, R., and Wersäll, J. (1962). *Acta Oto-Laryngol.* **55**, 11.
Kimura, R. S., Schuknecht, H. F., and Sando, I. (1964). *Acta Oto-Laryngol.* **58**, 390.
Kolmer, W. (1927). *In* "Handbuch der mikroskopischen Anatomie des Menschen" (W. von Möllendorff, ed.), pp. 250–478. Springer-Verlag, Berlin and New York.
Leibbrandt, C. C. (1965). *Acta Oto-Laryngol.* **59**, 124.
Lim, D. J. (1969). *Acta Oto-Laryngol., Suppl.* **255**, 1.
Lim, D. J. (1972). *Arch. Otolaryngol.* **96**, 199.
Lim, D. J., and Melnick, W. (1971). *Arch. Otolaryngol.* **94**, 294.
Lindeman, H. H., Ades, H. W., Bredberg, G., and Engström, H. (1971). *Acta Oto-Laryngol.* **72**, 229.
Lorente de No, R. (1937). *Laryngoscope* **47**, 373.
Lowenstein, O., and Wersäll, J. (1959). *Nature (London)* **184**, 1807.
Maw, A. R. (1973). *J. Laryngol. Otol.* **87**, 619.
Nomura, Y., and Kirikae, I. (1967). *Ann. Otol., Rhinol., & Laryngol.* **76**, 57.
Perkins, R. W. (1973). *Anat. Rec.* **175**, 410 (abstr.).
Plotz, E., and Perlman, H. B. (1955). *Laryngoscope* **65**, 291.
Rasmussen, A. T. (1940). *Laryngoscope* **50**, 67.
Rasmussen, G. L. (1946). *J. Comp. Neurol.* **84**, 141.
Rasmussen, G. L. (1953). *J. Comp. Neurol.* **99**, 61.
Rasmussen, G. L. (1960). *In* "Neural Mechanisms of the Auditory and Vestibular Systems" (G. L. Rasmussen and W. F. Windle, eds.), pp. 105–115. Thomas, Springfield, Illinois.
Retzius, G. (1881). *In* "Das Gehörorgan der Wirbeltiere," Samson & Wallin, Stockholm.

Retzius, G. (1884). *In* "Das Gehörgan der Wirbeltiere," Samson & Wallin, Stockholm.
Rodriguez-Echandia, E. L. (1967). *Z. Zellforsch. Mikrosk. Anat.* **78**, 30.
Rodriguez-Echandia, E. L. (1968). *Z. Zellforsch. Mikrosk. Anat.* **85**, 183.
Rosenbluth, J. (1962). *J. Cell Biol.* **12**, 329.
Ross, M. D. (1969). *J. Comp. Neurol.* **135**, 453.
Ross, M. D. (1971). *Acta Oto-Laryngol., Suppl.* **286**, 1.
Ross, M. D., and Burkel, W. (1973). *Acta Oto-Laryngol.* **76**, 381.
Schuknecht, H. F., and Seifi, A. E. (1963). *Ann. Otol., Rhinol., & Laryngol.* **72**, 687.
Smith, C. A. (1955). *Anat. Rec.* **121**, 451.
Smith, C. A. (1961). *Ann. Otol., Rhinol., & Laryngol.* **70**, 504.
Smith, C. A. (1967). *Rev. Panamer. Otorrinolaringol. Broncoesofagol.* **1**, 77.
Smith, C. A. (1968). *Advan. Sci.* **24**, 419–433.
Smith, C. A., and Dempsey, E. W. (1957). *Amer. J. Anat.* **100**, 337.
Smith, C. A., and Haglan, B. J. (1973). *Acta Oto-Laryngol.* **75**, 203.
Smith, C. A., and Rasmussen, G. L. (1963). *Ann. Otol., Rhinol., & Laryngol.* **72**, 489.
Smith, C. A., and Sjöstrand, F. S. (1961). *J. Ultrastruct. Res.* **5**, 523.
Spoendlin, H. (1957). *Pract. Oto-Rhino-Laryngol.* **19**, 192.
Spoendlin, H. (1959). *Pract. Oto-Rhino-Laryngol.* **21**, 34.
Spoendlin, H. (1966). *In* "The Organization of the Cochlear Receptor," p. 1. Karger, Basel.
Spoendlin, H. (1969). *Acta Oto-Laryngol.* **67**, 239.
Spoendlin, H. (1971). *Arch. Klin. Exp. Ohren Nasen Kehlkopfheilkd.* **200**, 275.
Spoendlin, H. (1972). *Acta Oto-Laryngol.* **73**, 235.
Spoendlin, H. (1973). *In* "Basic Mechanisms in Hearing" (A. Møller, ed.), pp. 185–234. Academic Press, New York.
Spoendlin, H., and Gacek, R. R. (1963). *Ann. Otol., Rhinol., & Laryngol.* **72**, 660.
Spoendlin, H., and Lichtensteiger, W. (1966). *Acta Oto-Laryngol.* **61**, 423.
Tanaka, T., Kosaka, N., Takiguchi, T., Aoki, T., and Takahara, S. (1973). *In* "Scanning Electron Microscopy," Part III. pp. 427–433 (Abstr.). Proceedings of the Workshop on Scanning Electron Microscopy in Pathology. April 1973, ITT Research Institute, Chicago, Illinois.
Terayama, Y., and Yamamoto, K. (1971). *Acta Oto-Laryngol.* **72**, 385.
Terayama, Y., Holz, E., and Beck, C. (1966). *Ann. Otol., Rhinol., & Laryngol.* **75**, 1.
Tonndorf, J. (1970). *In* "Foundations of Modern Auditory Theory" (J. V. Tobias, ed.), Vol. 1, pp. 205–254. Academic Press, New York.
Tonndorf, J., Duvall, A. J., and Reneau, J. P. (1962). *Ann. Otol., Rhinol., & Laryngol.* **71**, 801.
Trahiotis, C., and Elliot, D. N. (1970). *J. Acoust. Soc. Amer.* **47**, 592.
von Békésy, G. (1960). *In* "Experiments in Hearing," p. 1–745. McGraw-Hill, New York.
von Ilberg, C. (1969). *Arch. Klin. Exp. Ohren Nasen Kehlkopfheilkd.* **194**, 408.
Warr, W. B. (1973). *Anat. Rec.* **175**, 464 (abstr.).
Wersäll, J. (1973). *In* "Basic Mechanisms of Hearing" (A. R. Møller, ed.), p. 235–256. Academic Press, New York.
Wersäll, J., and Flock, Å. (1967). *In* "Sensorineural Hearing Processes and Disorders," (H. F. Schuknecht, ed.), Henry Ford Hosp. Int. Symp., p. 3. Little, Brown, Boston, Massachusetts.
Wiederhold, M. L., and Kiang, N. Y. S. (1970). *J. Acoust. Soc. Amer.* **48**, 950.
Wislocki, G., and Ladman, A. (1955). *J. Anat.* **89**, 3.
Wright, C. G., and Preston, R. E. (1973). *Brain Res.* **58**, 37.
Zwislocki, J. J. (1974). *J. Acoust. Soc. Amer.* **55**, 578.

Endocrine Cells of the Gastric Mucosa

Enrico Solcia[1,2] Carlo Capella[1,2] Gabriele Vassallo[1]
and Roberto Buffa[1,2]

I. Introduction

Chromaffin cells were first observed by Heidenhain (1870) in dog gastric mucosa; *osmiophilic* cells were noted in the same area by Nussbaum (1879), Grützner and Menzel (1879), and Toldt (1880). In the intestinal mucosa granular *yellow* cells were described by Nicolas (1891), and *basigranular acidophil* cells by Kultschitzky (1897). Very likely, these cells were related to the basigranular yellow cells found by Schmidt (1905) in unstained sections of intestinal mucosa fixed in Orth's Muller-formol fluid. Schmidt correctly attributed the yellow staining of the granules to their interaction with the chromium salt of the fixative, hence the name *enterochromaffin* (EC) cells, which was introduced by Ciaccio (1907).

The endocrine nature of the EC cells and their independence of adrenal chromaffin cells were first recognized by Masson (1914), who also discovered the silver-reducing power or *argentaffinity* of such cells. Their azo-coupling or diazonium reaction was reported by Cordier and Lison (1930), who suggested a diphenol structure for the chromaffin, argentaffin, and diazo-reactive substance stored in EC cell granules. A series of histochemical, biochemical, and pharmacological researches allowed Vialli and Erspamer (1937a,b,c,

[1] Institute of Pathological Anatomy, The University of Pavia, Pavia, Italy.
[2] Histopathology, Histochemistry and Ultrastructure Center, The University of Pavia-Varese, Varese, Italy. Supported by C.N.R. Grant 74.00272.04.

1942) to extract and partially purify this substance, which was found to be an amine, the so-called enteramine. Later Erspamer and Asero (1952) identified enteramine as 5-hydroxytryptamine (5HT).

The presence in the gastrointestinal mucosa of basigranular acidophilic cells resembling EC cells, but lacking chromaffinity and argentaffinity, was first reported by Kull (1913). Then argyrophilic, nonargentaffin cells were described, which were reputed by several investigators to be morphologically and functionally related to EC cells (Erspamer, 1939; Hamperl, 1952), although the hypothesis of an independent endocrine function was also considered (Clara, 1957). (Argyrophilic structures take up silver from silver solutions without reducing it; their blackening is obtained only by means of some exogenous reducer.) More recently, with the help of new histological, histochemical, and ultrastructural techniques, many endocrine cells unequivocally differing from EC cells and reproducing morphological features of peptide-producing cells were demonstrated (Solcia and Sampietro, 1965a,b; Solcia *et al.*, 1967; Carvalheira *et al.*, 1968; Orci *et al.*, 1968). The existence of non-EC peptide-producing cells was soon confirmed by the immunohistochemical identification of the gastrin cell in the pyloric mucosa (McGuigan, 1968).

On ultrastructural and histochemical grounds, non-EC endocrine cells of the gastrointestinal mucosa were classified as four (Forssmann *et al.*, 1969), five (Capella *et al.*, 1969), or six independent cell types (Vassallo *et al.*, 1969). Then, at the Wiesbaden Symposium (October, 1969), a classification was agreed on by the Pavia, Geneva, and London groups. Six distinct types of endocrine cells were added to the well-known EC cells (Solcia *et al.*, 1970b). The immunohis-

TABLE I
REVISED WIESBADEN CLASSIFICATION OF ENDOCRINE CELLS OF THE GASTROINTESTINAL MUCOSA AND PANCREAS

Stomach		Intestine		Pancreas	
Cell type	Hormone	Cell type	Hormone	Cell type	Hormone
EC	5HT	EC	5HT	EC	5HT
A-like	Gastroglucagon	L (EG)	Enteroglucagon	A	Glucagon
D	?	D	?	D	?
D_1	?	D_1	?	D_1	?
G	Gastrin	G	Gastrin	—	—
ECL	Histamine	—	—	—	—
		S	Secretin	—	—
		I	CCK ?	—	—
				B	Insulin

TABLE II
GUT ENDOCRINE CELLS

Cell type	Distribution	Hormone
EC	Gastrointestinal tract, pancreas	5HT
G	Antrum, pylorus, duodenum	Gastrin
ECL	Stomach: body, fundus	?(Histamine)
A	Stomach: body, fundus	Gastroglucagon
D	Stomach, upper small intestine, pancreas	?
D_1	Stomach, intestine, pancreas	?
X	Stomach; duodenum?	?
S	Duodenum; jejunum	Secretin
I	Duodenum, jejunum	CCK ?
K	Small intestine	GIP
L	Small and large intestine	Intestinal GLI

tochemical and ultrastructural identification of new cell types (Kobayashi *et al.*, 1970; Sasagawa *et al.*, 1970; Bussolati *et al.*, 1971; Ferreira 1971; Vassallo *et al.*, 1971b; Capella and Solcia, 1972) suggested a revision of the Wiesbaden classification. The revised Wiesbaden classification of gut endocrine cells, which was agreed on in 1973 at the Bologna Symposium by several groups (Solcia *et al.*, 1973), is reported in Table I, together with the proposed function of each cell type. In light of the immunohistochemical detection of gastric inhibitory peptide (GIP) cells (Polak *et al.*, 1973), and of further ultrastructural investigations (Osaka *et al.*, 1973b; Lechago and Bencosme, 1973; Solcia *et al.*, 1974c), we believe that two cell types must be added to those reported in Table I: the K cell of the small intestine, probably related to the GIP, and the gastric X cell (Capella *et al.*, 1969: Fig. 4), now to be distinguished from gut D cells (Table II).

Only the endocrine cells represented in the gastric mucosa are extensively discussed in this article; however, reference to the endocrine cells of the small intestine and pancreas is made, given their close interrelationship with gastric cells.

II. Gastric Endocrine Cells

A. General Features

In routine histological specimens of the gastric mucosa, the endocrine cells appear as single, small to medium-sized cells with clear cytoplasm, scattered here and there inside the epithelium lining the glands. However, not all the "clear" cells (Feyrter, 1953) found in routinely stained sections proved to be endocrine in nature when tested with granule-staining techniques or immunohistochemistry. Electron microscopy also showed that some of the clear cells found with light microscopy of semithin sections correspond to agranular poorly differentiated cells.

As a rule the endocrine cells lie directly on the basal lamina of the glands. In the pyloric mucosa most of them reach the lumen in small areas, as a result of thin cytoplasmic extensions; most cells of the fundic mucosa lack luminal contacts. Secretory granules are preferentially grouped at the base of the cell, while the Golgi complex is supranuclear; this pattern, which is more evident in cells showing luminal endings, suggests some functional polarity of the cell. Mitochondria, lysosomes, endoplasmic reticulum, and free ribosomes are moderately represented.

As in studies dealing with other endocrine structures, the ultrastructural identification of different cell types has been based mostly on the shape, size, and inner texture of their secretory granules (Orci *et al.*, 1968). The application of silver techniques to electron microscopy has been of special help for some cell types (Vassallo *et al.*, 1971a). The ultrastructural peculiarities of their luminal contacts and of their cytoplasmic organelles also proved useful.

The attribution of each endocrine function of the gastrointestinal mucosa to the appropriate cell type has been achieved: (1) by comparing in the various tracts of the gut mucosa the distribution of each cell with that of the candidate hormone as assayed in mucosal or whole gut extracts (Vassallo *et al.*, 1969; Solcia *et al.*, 1970b); and (2) by detecting in light microscopy—using immunohistochemistry or amine histochemistry—the cell producing a given hormone and then (a) by comparing the morphology, distribution, and staining patterns of such cell with that of the various ultrastructurally identified endocrine cell types (Bussolati and Pearse, 1970; Bussolati *et al.*, 1971; Polak *et al.*, 1971a,b,c), and (b) by directly identifying at the electron microscopy level the histochemically stained cell, either by applying electron microscopy and the histochemical technique to consecutive sections of the same cell (Solcia *et al.*, 1970b; Bussolati and Canese, 1972), or by studying in the electron microscope thin sections stained with the appropriate histochemical technique (Solcia *et al.*, 1970b; Greider *et al.*, 1972).

Only the attribution of 5HT to EC cells and that of gastrin to G cells fullfil requirement 2b; the function of other cells, such as S, L (EG), and K cells is suggested by evidence of type 2a; the evidence supporting the attribution of CCK to the I cell remains substantially at stage 1. For other cells no functional evidence at all is available.

B. Argentaffin or EC Cells

EC cells react with all the histochemical methods detecting 5HT, including those related to its phenol function and its reducing power, such as chromaffin, diazonium thioindoxyl, hematoxylin (catechol dye condensation), argentaffin, and ferric ferricyanide reactions, and those depending on its pyrrole ring, such as dimethylaminobenzaldehyde–nitrite and xanthydrol reactions (Fig. 1). Apart from the chromaffin reaction, which requires fixation in dichromate and chromic acid mixtures (Orth's or Helly's fluids), the reactions of the first group are successfully applied to specimens fixed in formaldehyde, glutaraldehyde, acrolein, and glyoxal, or to mixtures of these substances with picric acid, Hg^{2+} and Ca^{2+} salts, and so on. Reactions depending on the pyrrole ring require fixation in glutaraldehyde or acrolein solutions; formaldehyde-fixed material is unsuitable for such indole tests, because formaldehyde (and glyoxal) blocks C-2 (α) of 5HT, which must be free to allow this substance to react with dimethylaminobenzaldehyde or xanthydrol. In addition to C-2, formaldehyde also reacts with the amino group, thus giving a new condensation ring involving the lateral chain of 5HT. This new

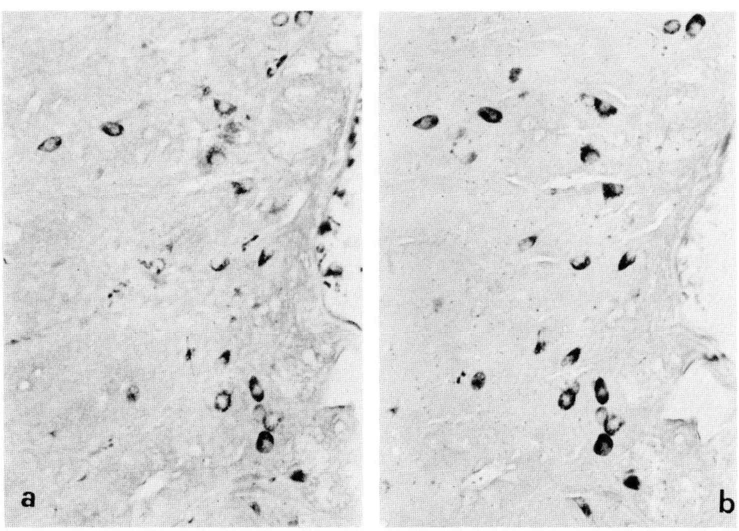

FIG. 1. EC cells of guinea pig duodenal mucosa fixed in glutaraldehyde, stained with xanthydrol (a), and restained by the diazonium reaction (b). ×230.

ring accounts for the strong yellow fluorescence and the Holcenberg-Benditt ninhydrin reaction of formaldehyde-treated EC cells and 5HT; the last-mentioned reactions are not shown by native 5HT or unfixed EC cells (Solcia *et al.*, 1966, 1969a; Geyer, 1968a).

Some doubts concerning the reactivity of the chromaffin, argentaffin, and diazo-reactive substance of EC cells in indole tests, as well as its identification with 5HT (Lillie *et al.*, 1973), have been overcome by restaining indole-reactive cells by the diazonium technique (Solcia and Buffa, 1974). Moreover, spectrofluorometric analysis of formaldehyde-treated 5HT and formaldehyde-treated EC cells gave identical spectra (Håkanson *et al.*, 1970). Thus identification of the chromaffin substance as 5HT is now fully supported by histochemical tests. Minor discrepancies between the behavior of 5HT and that of EC cell granules in respect to the diazonium reaction at acid pH values (Lillie *et al.*, 1973) could be due to the fact that in EC granules 5HT is closely linked to other substances (ATP, proteins; see below this section) which may interfere with its reactivity.

It must be noted that different amounts of cells are detected with different 5HT methods, mostly depending on the different sensitivity of such methods. In particular, gastric and pancreatic EC cells, which often store less 5HT than intestinal EC cells, are more prone to escape detection with chromaffin, xanthydrol, Pearse's thioindoxyl, or Clara's hematoxylin reactions than with other techniques. More cells are detected with the argentaffin reaction, the diazonium reaction following glutaraldehyde fixation, or formaldehyde-induced fluorescence, particularly with the Falck-Hillarp technique.

It is well known that, in addition to EC cells, other 5HT-storing cells react with the above methods under appropriate technical conditions. Among these there are some mast cells of rat, mouse, *Mastomys natalensis*, horse, and dog, as well as thyroid calcitonin (C) cells of horse, sheep, and goat, A cells of pig pancreas, some carotid body cells, some ECL cells of cat and rabbit, and stomach and bronchial endocrine cells of cat and rabbit (Solcia *et al.*, 1969a). All these cells also show intense yellow fluorescence with the very sensitive Falck-Hillarp technique (Falck *et al.*, 1964; Enerback, 1966; Hamberger *et al.*, 1966; Cegrell *et al.*, 1968). Thus cells reacting with 5HT methods are not necessarily the same as EC cells; not even in the gut mucosa where, in some species, 5HT-storing mast cells or ECL cells occur inside the epithelium. Some unspecific staining methods, mostly unrelated to 5HT (Solcia *et al.*, 1974c), proved useful for more appropriate identification of EC cells when coupled with 5HT methods. Grimelius' silver method (Grimelius, 1968), using dilute silver nitrate solutions at acid pH (under these condi-

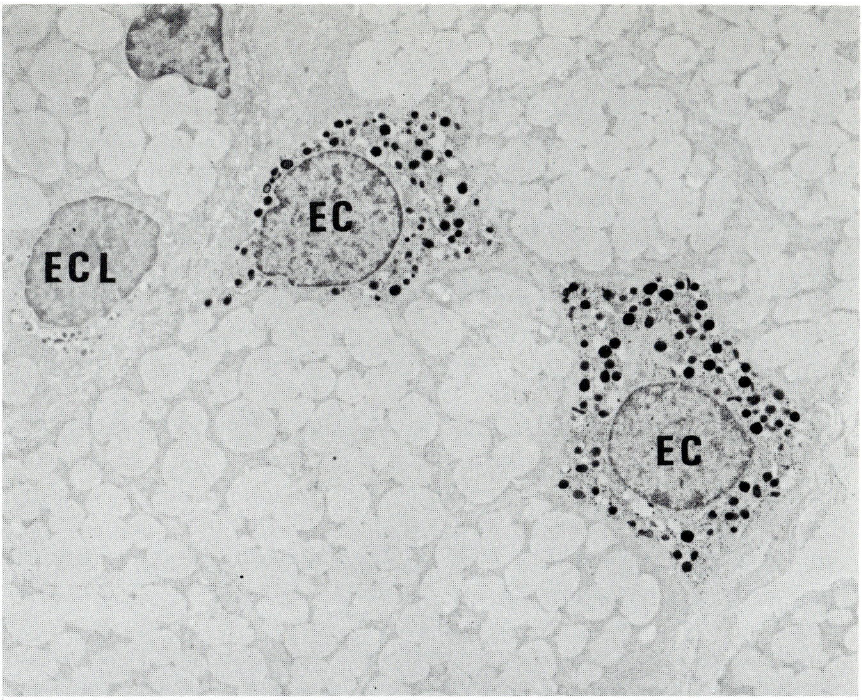

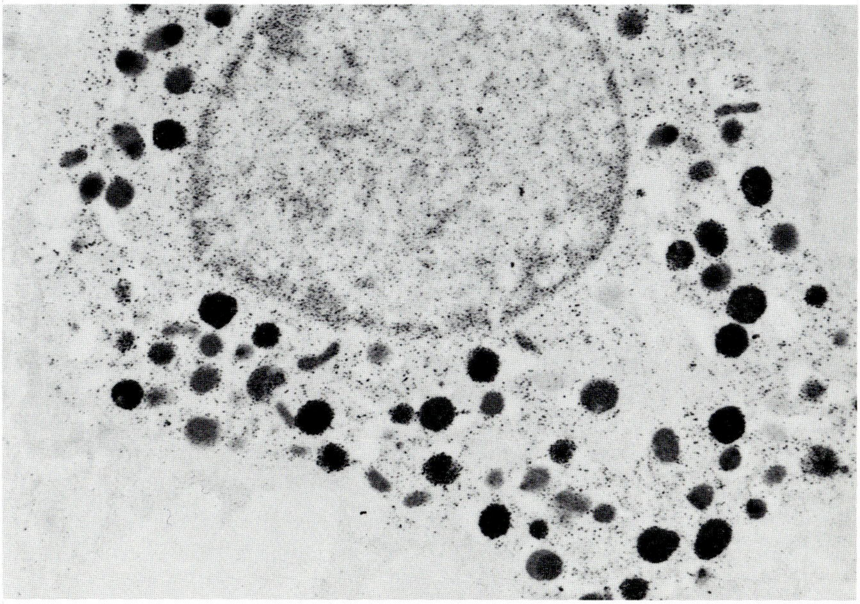

FIG. 2. Dog fundic mucosa fixed with 2% paraformaldehyde plus 2.5% glutaraldehyde in 0.1 *M* phosphate buffer (pH 7.3) and stained by the argentaffin reaction; section counterstained with uranyl acetate. Note two EC cells with argentaffin granules and an ECL cell with small, nonargentaffin granules. ×4300.

FIG. 3. Details of an EC cell in Fig. 2. ×11500.

tions 5HT fails to reduce silver), basic dyes after hot hydrochloric acid treatment (Solcia *et al.*, 1968), and lead–hematoxylin (Solcia *et al.*, 1969b) have been used for this purpose. The staining of EC cell granules by these methods is prevented by pepsin, trypsin, and papain, but is not affected by fixation in formol–alcohol–acid mixtures, which remove 5HT. These findings, together with those of a series of histochemical tests (Solcia and Sampietro, 1966; Solcia *et al.*, 1974c), suggest that some protein component (argyrophilic glycoproteins? hormonal peptides?) is stored in EC cell granules. Further histochemical studies made by Geyer (1968b) confirm this hypothesis.

A more reliable and selective demonstration of EC cells is obtained by directly applying argentaffin or chromaffin reactions in electron microscopy (Figs. 2 and 3). In this way EC cells have been directly identified at the ultrastructural level and shown to display granules of peculiar shape, clearly different from those of the non-EC 5HT-storing cells reported above (Vassallo *et al.*, 1969, 1971b; Bus-

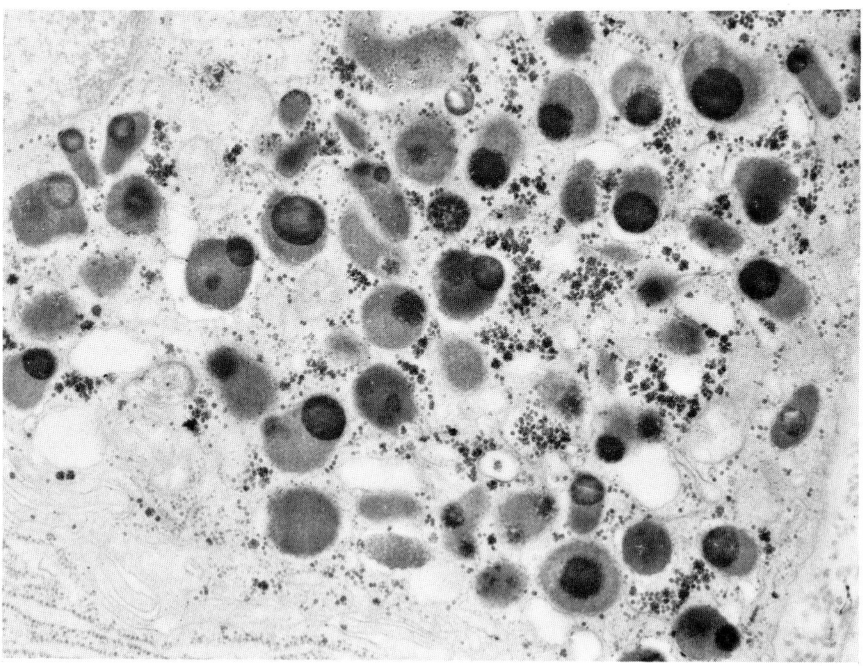

FIG. 4. Granules of an EC cell of dog fundic mucosa fixed in formaldehyde–glutaraldehyde mixture and postfixed in osmium tetroxide; section stained with uranyl acetate and lead citrate. Note dense bodies and particles inside granules. ×25,000.

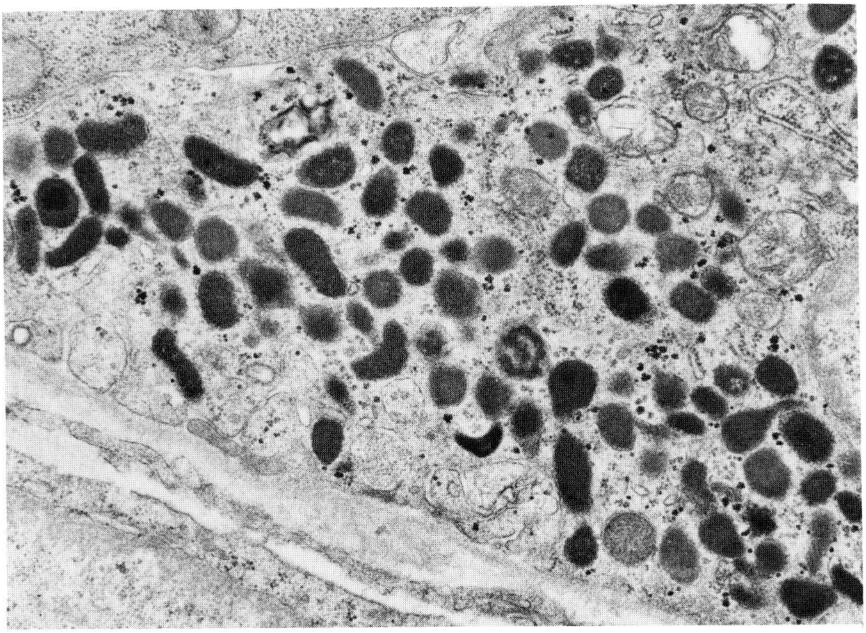

FIG. 5. EC cells of rabbit duodenum. ×18,000.

solati *et al.*, 1971). In thin sections EC granules appear pleomorphic: round, ovoid, oblonged, pear-shaped, triangular, kidney-shaped, U-shaped, and so on. According to Ferreira (1971), these various shapes of EC granules in sections are related to the biconcave structure of most of them. At least in the intestine, the morphology of the granule seems partly dependent on its 5HT content. In fact, Golgi-associated EC granules and granules of reserpine- or *p*-chlorophenylalanine-treated EC cells, which show poor reactivity with 5HT methods, are mostly round in shape (Capella *et al.*, 1969; Solcia *et al.*, 1974c). Sometimes, 5HT-poor, round, slightly osmiophilic granules have been found scattered everywhere in the cytoplasm of untreated EC cells, admixed with 5HT-rich, pleomorphic, strongly osmiophilic granules, and seldom outnumbering the latter. Such cells, which may escape detection with 5HT methods in light microscopy, are better detected with electron microscopy.

EC granules often show heterogeneous internal structure, with highly osmiophilic bodies or particles immersed in a less osmiophilic matrix (Figs. 4 and 5). Both the highly osmiophilic component and the matrix of the granules have been found to react with Masson's argentaffin reaction and other silver techniques, although the former is usually more effective. In some granules the osmiophilic bodies show central cavitation and emptying, and corresponding areas of unreactivity to silver techniques. In most granules the membrane enveloping EC granules adheres closely to the matrix; however, in gastric EC cells of the dog, rat, and mouse, a wide space is often found between the eccentrically located core and the enveloping membrane. This granule pattern must be carefully evaluated to avoid confusion with granules of ECL cells (see Section II,D).

p-Chlorophenylalanine has been found to inhibit selectively tryptophan-hydroxylating enzymes, thus blocking 5HT synthesis (Koe and Weissman, 1966). Reserpine is known to block 5HT uptake (Carlsson, 1966). EC granules of animals treated with *p*-chlorophenylalanine (Fig. 6) or reserpine lost most of their 5HT content—as shown by their lack of reactivity with 5HT methods with both light and electron microscopy—but retained some of their osmiophilia and most of their Grimelius reactivity (Wetzstein *et al.*,

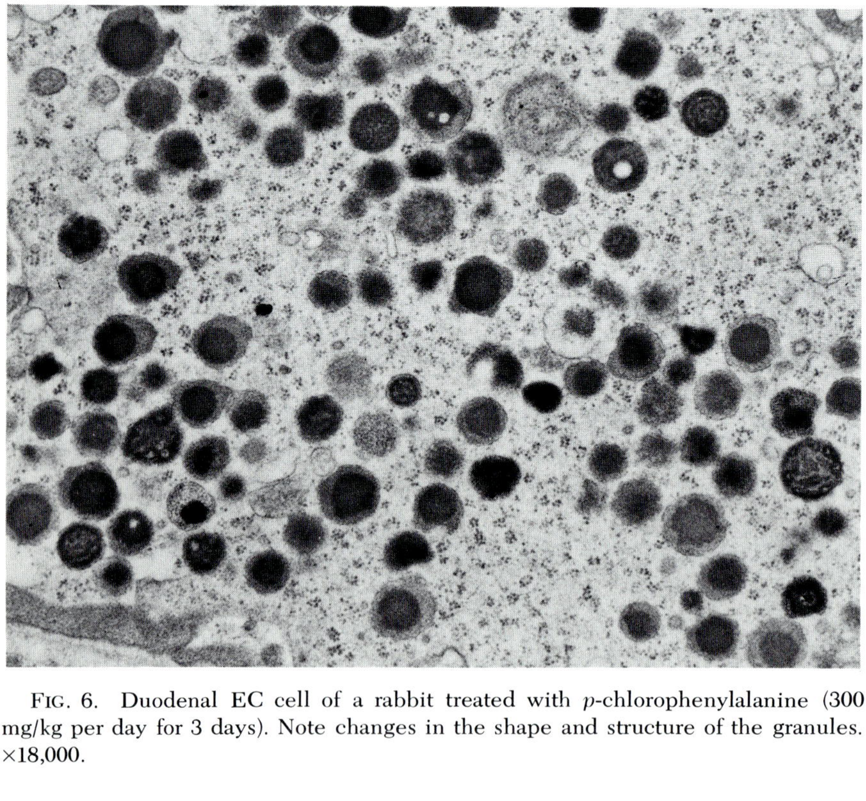

FIG. 6. Duodenal EC cell of a rabbit treated with *p*-chlorophenylalanine (300 mg/kg per day for 3 days). Note changes in the shape and structure of the granules. ×18,000.

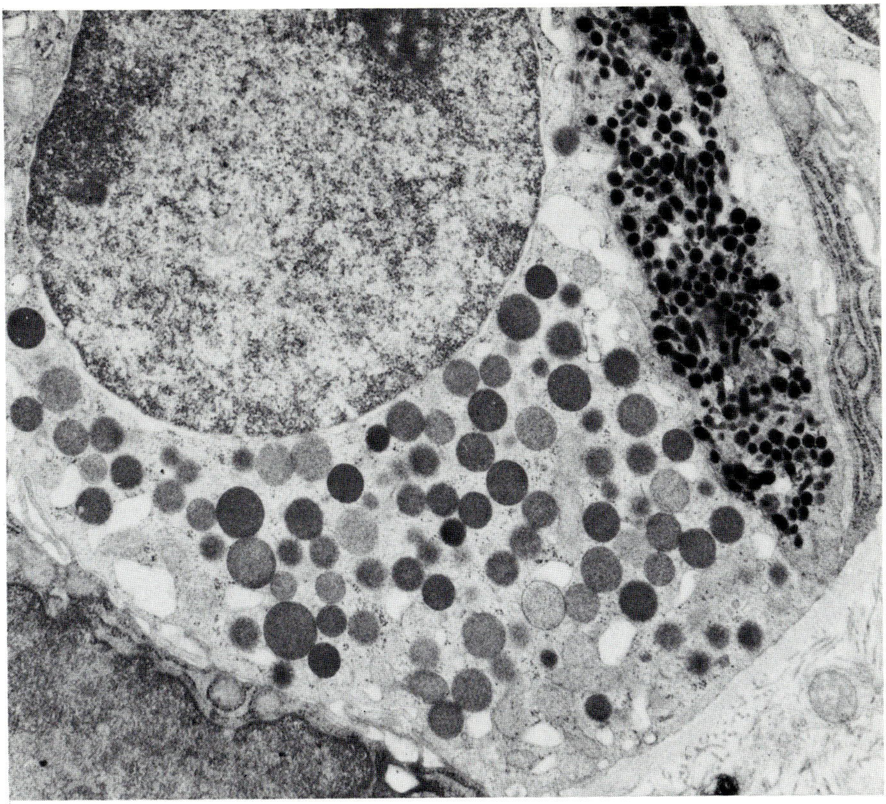

FIG. 7. Human pyloric gland showing a thin extension of an EC cell with small, elongated, strongly osmiophilic granules, as well as a D cell with large, round, weakly osmiophilic granules. ×14,000. (From Vassalo *et al.*, 1971b.)

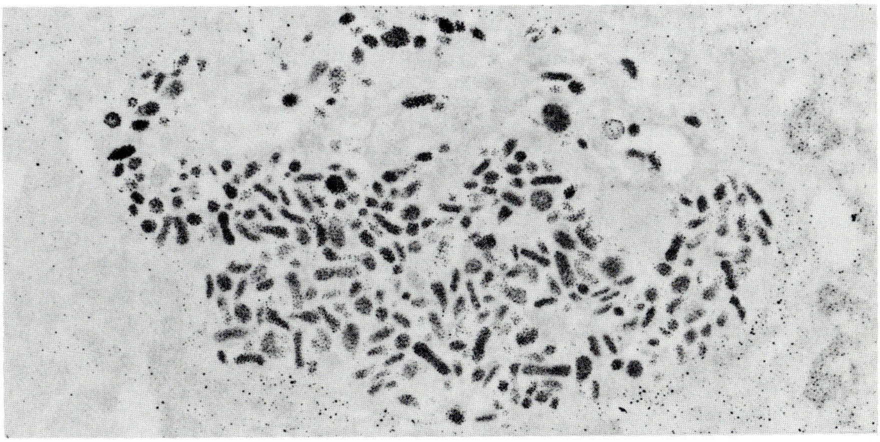

FIG. 8. Granules of a human pyloric EC cell stained by the argentaffin reaction. ×13,000.

1964; Capella *et al.*, 1969, 1972; Solcia *et al.*, 1974c). This confirms that, besides 5HT, nonamine components are also present in EC granules. Sometimes large doses of reserpine given to sensitive animals (guinea pig, dog) resulted in dissolution of granule contents. Intestinal or pyloric EC cells are much more sensitive to reserpine and *p*-chlorophenylalanine than EC cells of gastric fundic mucosa. Unlike *p*-chlorophenylalanine, reserpine affects mitochondria (which are often swollen and disrupted) and increases the glycogen stores of intestinal EC cells (Capella *et al.*, 1969). Granule exocytosis seems to be unaffected by both drugs.

Apart from a relatively few cells occurring in the pancreas and biliary tree, EC cells are mostly distributed in the gastrointestinal tract, especially in the mucosa of the small and large intestine and the pyloric mucosa. In fundic gastric mucosa, the number of EC cells differs markedly according to the animal species; they are numerous in pig, dog, rabbit, and monkey, relatively few occur in guinea pig, cat, and man, and they are practically absent in rat, mouse, and *Mastomys*. EC cells are fairly well represented in the glands of human cardiae.

EC cells of the fundic mucosa, like other endocrine cells in this area, lack luminal contacts; these are found regularly in intestinal EC cells and less frequently in pyloric EC cells. In man gastric EC cells are small, lengthened, and have small secretory granules (Figs. 7 and 8). They are obviously different from the large, pyramidal or flask-shaped EC cells with large granules, which are typical of the human duodenum. Gastric EC granules are smaller than duodenal EC granules in several animal species also, including guinea pig, cat, and rabbit. EC cells with small granules also appear in the rabbit ileum. These differences in granule size, cell shape and size, luminal endings, glycogen content (much more evident in intestinal EC cells), and drug sensitivity are likely related to the functional specialization of EC cells in different areas, perhaps involving their nonamine secretory products.

Hypertonic glucose in the duodenum has been shown to release 5HT and kallikreins (Drapanas *et al.*, 1962; Reichle *et al.*, 1967; Zeitlin and Smith, 1970), and to stimulate emiocytosis of EC granules (Kobayashi and Fujita, 1973). The luminal endings of intestinal EC cells may be involved in this response. Besides 5HT, both kallikreins (Oates *et al.*, 1964) and prostaglandinlike substances (Sandler *et al.*, 1968) are released from argentaffin carcinoid tumors or "argentaffinomas" which reproduce closely the histochemical and ultrastructural patterns of EC cells (Fig. 9).

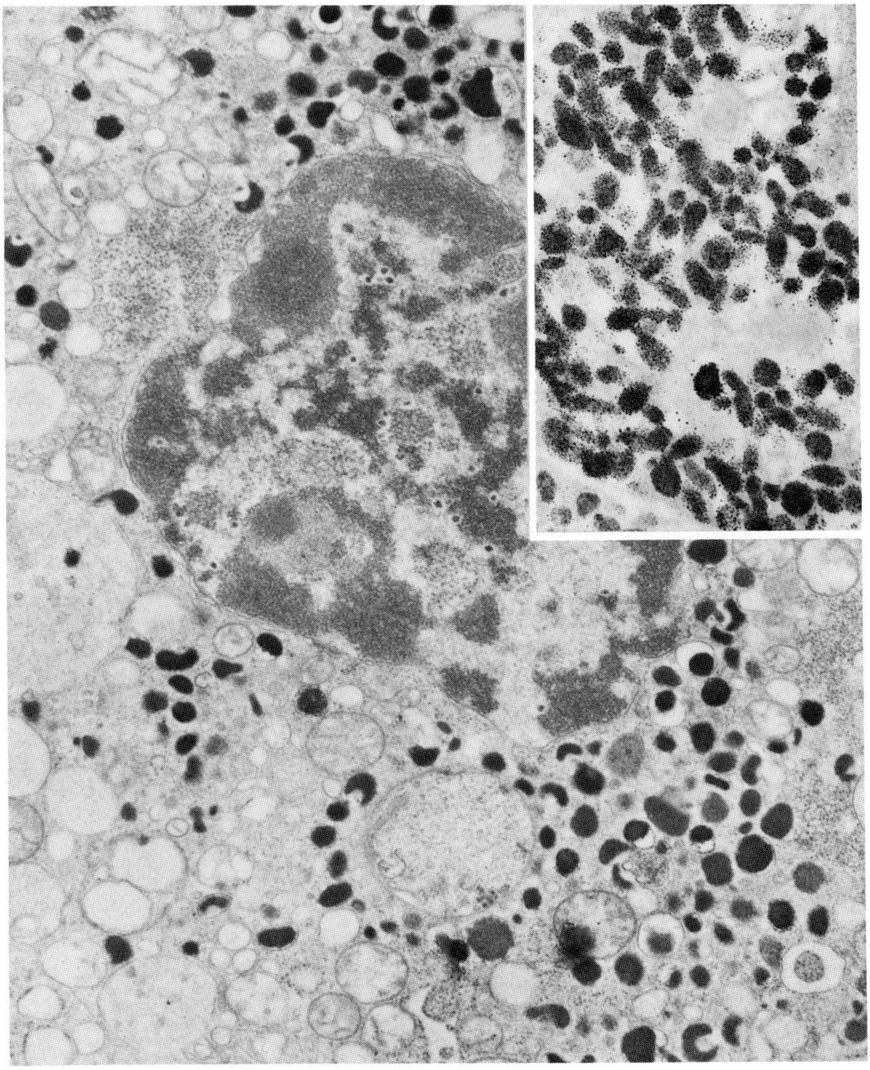

FIG. 9. Ultrastructure of EC granules in a carcinoid tumor of the ileum. ×16,500. Inset: Granules of the same tumor stained by the argentaffin reaction. ×21,000.

A possible role of intestinal EC cells in the regulation of gut motility has been repeatedly suggested (Bülbring and Crema, 1959; Kaltiala, 1971). They should be able to respond to chemical and mechanical stimuli by releasing substances affecting gut motility, such as 5HT, kallikreins (known kinin releasers), substance P, prostaglandins, and motilin (Brown *et al.*, 1973), which has been recently shown to be stored in EC cells of the duodeno-jejunal mucosa (Pearse *et al.*, 1972). Some intramucosal interaction of gastric EC cells with gastric secretions should also be considered, given the ability of 5HT to inhibit hydrochloric acid secretion (Haverback and Davidson, 1958) and to stimulate mucous secretion (White and Magee, 1958).

C. G Cells

Until recently, the cellular site of gastrin production was not known. Only nonmorphological data were available, and these indicated as a possible gastrin cell a nerve cell of Meissner's Plexus (Baugh *et al.*, 1958). Then light (Solcia and Sampietro, 1965a; Carvalheira *et al.*, 1968; Solcia *et al.*, 1968) and electron microscopy studies (Solcia *et al.*, 1967; Orci *et al.*, 1968) demonstrated in pyloric glands of mammals many intraepithelial endocrinelike cells different from EC cells and with characteristics of peptide-secreting cells. Such cells were supposed to be related to gastrin, the only peptide hormone known to come from the pyloric area, and were labeled G cells Further ultrastructural studies (Capella *et al.*, 1969; Forssmann *et al.*, 1969; Solcia *et al.*, 1969c; Vassallo *et al.*, 1969) identified among these non-EC pyloric endocrine cells a well-defined cell type of fairly large size, showing vesicular granules ranging from 150 to 400 nm, with a relatively loose core of floccular, dotted, or filamentous appearance and moderate osmiophilia (Figs. 10 and 11). It was assumed that this type of cell, which was found to be mostly restricted to pyloric gastric mucosa, corresponded better to the requisites of the supposed gastrin cell, and therefore among the various types of endocrine cells of the gastric mucosa, the term G cell was used for only this type of cell.

In the meantime, McGuigan (1968) succeeded in staining gastrin cells by the immunofluorescence method, using antigastrin antibodies. McGuigan and Greider (1971) demonstrated that these cells were distinct from the EC cells; however, they failed to obtain staining of the immunofluorescent cells with the selective techniques for endocrine cell granules that had already been shown to stain the supposed gastrin cells. This discrepancy has been ruled out by Bussolati and Pearse (1970), Larsson *et al.* (1973b), and Solcia *et al.* (1974a), who repeatedly showed immunofluorescent gastrin cells to react with Grimelius' silver nitrate technique (Fig. 12), a method known to stain ultrastructurally defined G cells (Solcia *et al.*, 1969b, 1970b; Vassallo *et al.*, 1971b). The reactivity of G cells with lead–hematoxylin (Solcia *et al.*, 1969b; Vassallo *et al.*, 1969, 1971b) has also been reproduced for gastrin immunofluorescent cells (Pearse and Bussolati, 1972; Beltrami *et al.* 1975).

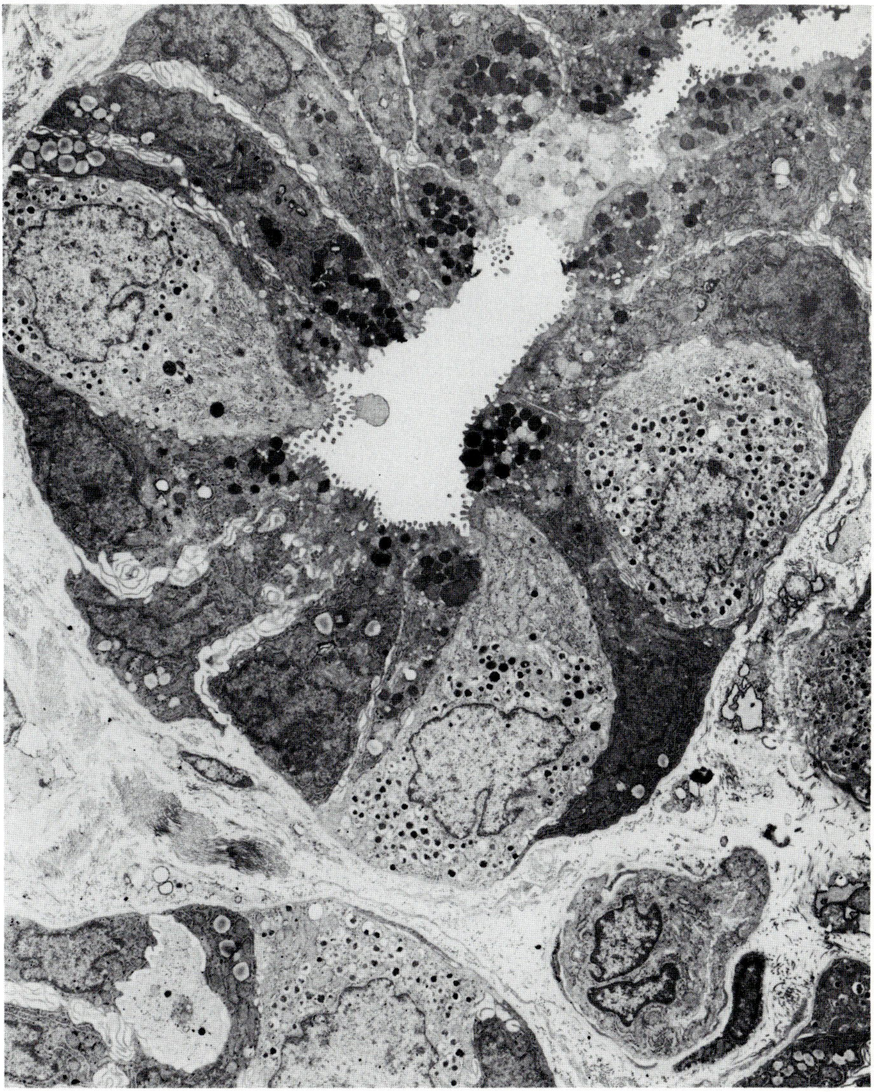

FIG. 10. Cat pyloric mucosa showing several G cells. Note juxtaluminal secretory granules of mucous neck cells. ×3000.

The correspondence of gastrin immunofluorescent cells to ultrastructurally defined G cells has also been ascertained by applying immunofluorescence and electron microscopy to adjacent sections of the same cell (Bussolati and Canese, 1972; Canese and Bussolati, 1974). It was deduced that G cells of cat and man, already identified ultrastructurally (Forssmann and Orci, 1969; Vassallo *et al.*, 1969,

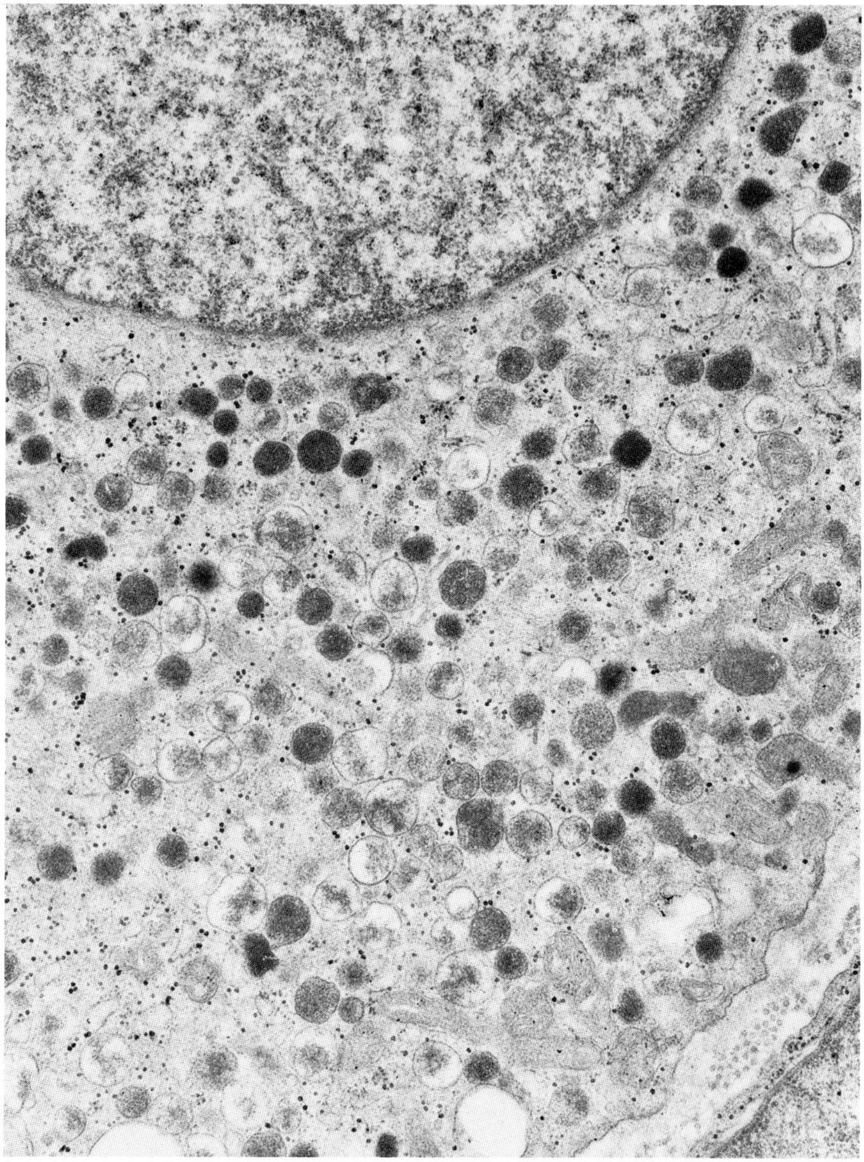

FIG. 11. Human pyloric G cell. Note several vesicular-type granules storing filamentous and dotted material of relatively weak osmiophilia. ×20,000.

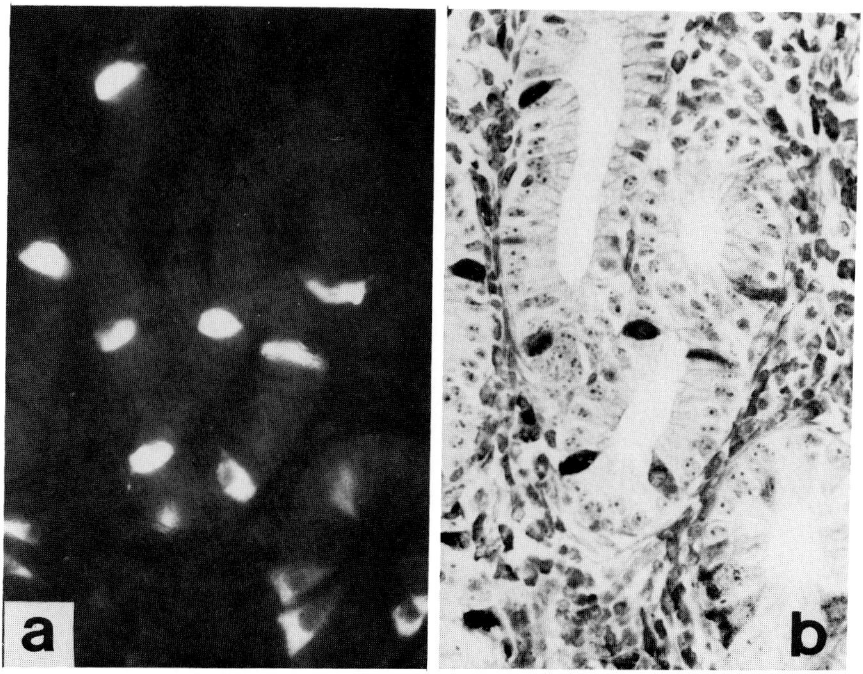

FIG. 12. Bouin-fixed paraffin section of human pyloric mucosa showing G cells stained by the indirect immunofluorescence test using antibodies against synthetic human gastrin I (a), and restained by Grimelius' silver technique (b). ×500.

1971b; Sasagawa *et al.*, 1970), reacted to antigastrin antibodies; other types of ultrastructurally defined endocrine cells of the same pyloric glands (EC and D cells) failed to react. Thus it is now generally agreed that G cells contain gastrin (Solcia *et al.*, 1970b; Forssmann, 1970; Pearse *et al.*, 1970; Sasagawa *et al.*, 1970; Creutzfeldt *et al.*, 1971; Lefranc and Pradal, 1971; Frexinos *et al.*, 1972; Mitschke and Becker, 1973; Bencosme and Lechago, 1973; Lechago and Bencosme, 1973). The apparently opposite conclusions of Greider *et al.* (1972) are likely due to misinterpretation of the cells they found to react with the immunoperoxidase technique at the ultrastructural level; in fact, several of the "gastrin" cells they show in their article seem to have the ultrastructural features of G cells.

In addition to Grimelius' silver and lead–hematoxylin, G cells are also stained with acid dyes such as Evans' blue (Bussolati *et al.*, 1972) and erythrosin in Herlant's tetrachrome (Bencosme and Lechago, 1973). They take up amine precursors (Figs. 13 and 15a), hence show amine precursors uptake and decarboxylation (APUD)

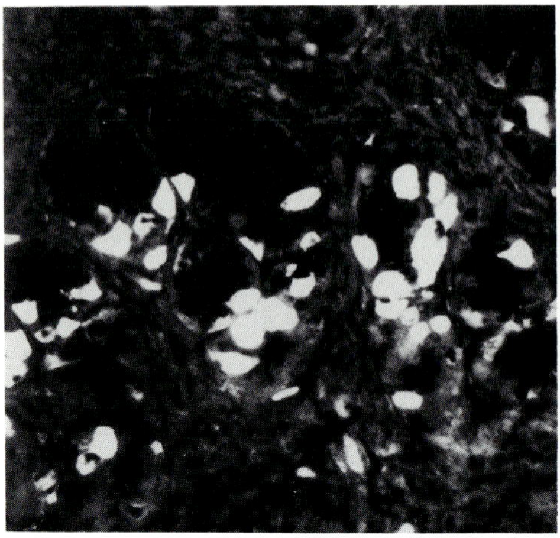

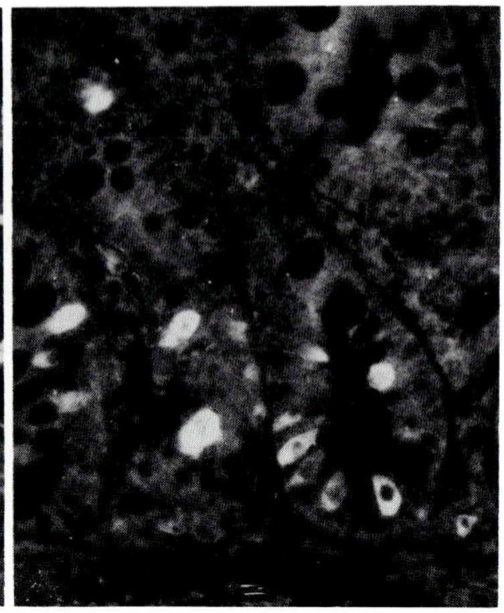

FIG. 13 (left). Pyloric mucosa of dog injected with 50 mg/kg of L-dopa 4 hours before sacrifice. Fixed in formaldehyde–glutaraldehyde buffered solution; cryostat section observed under ultraviolet light. Large G cells with intense (blue-green) fluorescence are scattered together with small (yellowish) EC cells. ×250.

FIG. 14 (right). Gastrin immunofluorescent cells in crypts of human duodenal mucosa. Bouin fixed paraffin section. ×250.

characteristics (Pearse, 1969). Moreover, they show selective fluorescence with the formaldehyde-ozone technique, because of a tryptophilic peptide (Håkanson *et al.*, 1972). In man and cat, G cells also show fluorescence with the hydrochloric acid–formaldehyde technique, whose histochemical meaning remains obscure (Håkanson *et al.*, 1973; Larsson *et al.*, 1973b). A moderate reactivity with methods known to stain protein-bound tryptophan (our unpublished observations) and intense α-glycerophosphate dehydrogenase activity (Carvalheira *et al.*, 1968; Becker, 1969) have also been noted in G cells.

In the pyloric glands G cells are preferentially distributed in the junctional area between mucous neck cells and mucoid cells (Solcia *et al.*, 1969c). In animals that, like the rat, mouse, and rabbit, have a thin or scarcely developed mucoid zone, G cells are scattered in the deep half of the mucosa; in species with a more developed mucoid zone, such as man, pig, dog, guinea pig, and cat, they are preferentially distributed in the intermediate third of the glands, with more (cat) or less (pig, dog, man) substantial amounts of cells in the deep third and only a few of them in the superficial third (McGuigan, 1968; Vassallo *et al.*, 1969, 1971b; Solcia *et al.*, 1970b; Capella and Solcia, 1972; Rubin, 1972b). In cat and dog the distribution of G cells is well in accordance with that of gastrin activity in tangentially cut sections of pyloric mucosa (Bromé *et al.*, 1968).

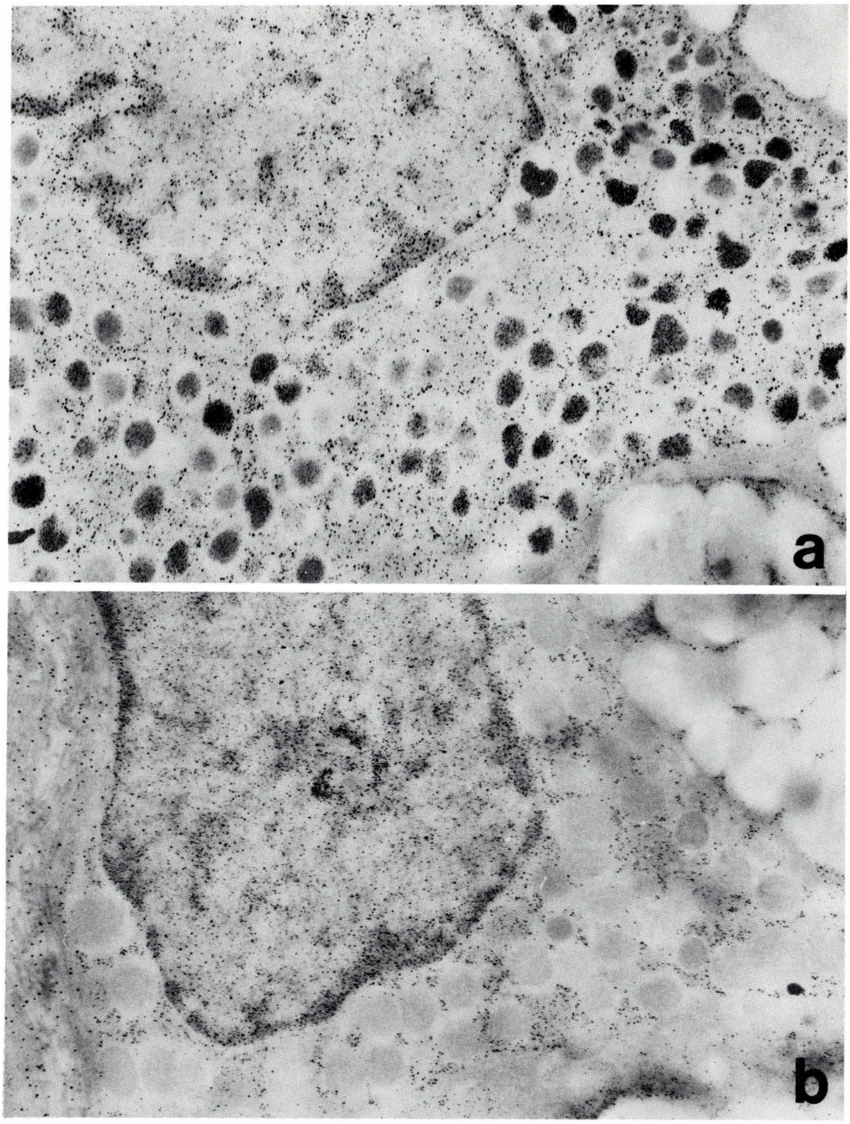

FIG. 15. Pyloric mucosa of dog as in Fig. 13, fixed in the same aldehyde solution, stained by the argentaffin reaction and embedded in Epon–Araldite; ultramicrotome sections stained with uranyl acetate. (a) G cell showing deposits of reduced silver, particularly over secretory granules; G cells of untreated dogs failed to show silver deposits following the same procedure. ×13,000. (b) Unreactive granules of a D cell. ×13,000.

Immunohistochemical (Polak *et al.*, 1972; Solcia *et al.*, 1974d) and ultrastructural findings (Cavallero *et al.*, 1972; Frexinos *et al.*, 1973) suggest the presence of G cells in the duodenal mucosa, especially in the crypts of the proximal human duodenum (Fig. 14). This is in keeping with extraction from the proximal intestine of gastrinlike substances, as shown by both bioassay (Emås and Fyrö, 1968; Emås *et al.*, 1971) and radioimmunoassay (Nilsson *et al.*, 1970; Berson and Jalow, 1971). However, in the intestinal mucosa ultrastructurally proved G cells seem to be less well represented than gastrin immunofluorescent cells. The possibility of gastrin coming from intestinal non-G cells, although unproved so far, cannot be excluded. This hypothesis could explain some chemical and functional peculiarities of intestinal gastrin, for instance, the prevalence of "big gastrin" over heptadecapeptide gastrin (Berson and Jalow, 1971), the presence of "big big gastrin" (Jalow and Berson, 1972), and the failure of hydrochloric acid to suppress gastrin release (Stern and Walsh, 1973). However, a possible crossreactivity of cholecystokinin cells with gastrin antisera must also be considered, given the close chemical similarities known to exist between cholecystokinin and gastrin (Mutt and Jorpes, 1967).

No G cells have been found ultrastructurally or immunohistochemically in unaltered fundic gastric mucosa. Some ultrastructural findings suggesting the presence of G cells in cardial mucosa (Forssmann *et al.*, 1969; type V cells) await immunohistochemical confirmation. The presence of gastrin activity in cardiae has been shown by Gregory (1962).

No reliable ultrastructural evidence supporting the presence of G cells in normal pancreas has been obtained so far. F cells (Bencosme and Liepa, 1955; Munger *et al.*, 1965), which have been interpreted as the pancreatic equivalent of G cells (Forssmann, 1970), differ in several ultrastructural and histochemical aspects from pyloric G cells (Solcia *et al.*, 1974b) and fail to react with antigastrin antibodies. Several attempts to detect gastrin cells in the pancreas by immunofluorescence have been unsuccessful, despite simultaneous staining of pyloric G cells (Creutzfeldt *et al.*, 1971; Lotstra *et al.*, 1974; Solcia *et al.*, 1974d). However, islet cells reacting with antigastrin antibodies have been reported by some investigators (Lomsky *et al.*, 1969; Greider and McGuigan, 1971; Polak *et al.*, 1972); they have been identified as D cells (see Section II,E). Immunochemical and bioassay evidence for gastrin in normal pancreas is quite conflicting (Hallenbeck *et al.*, 1963; Nilsson *et al.*, 1970; Creutzfeldt *et al.*, 1971; Greider and McGuigan, 1971; Omole *et al.*, 1972; Rehfeld and Iversen, 1973; Lotstra *et al.*, 1974).

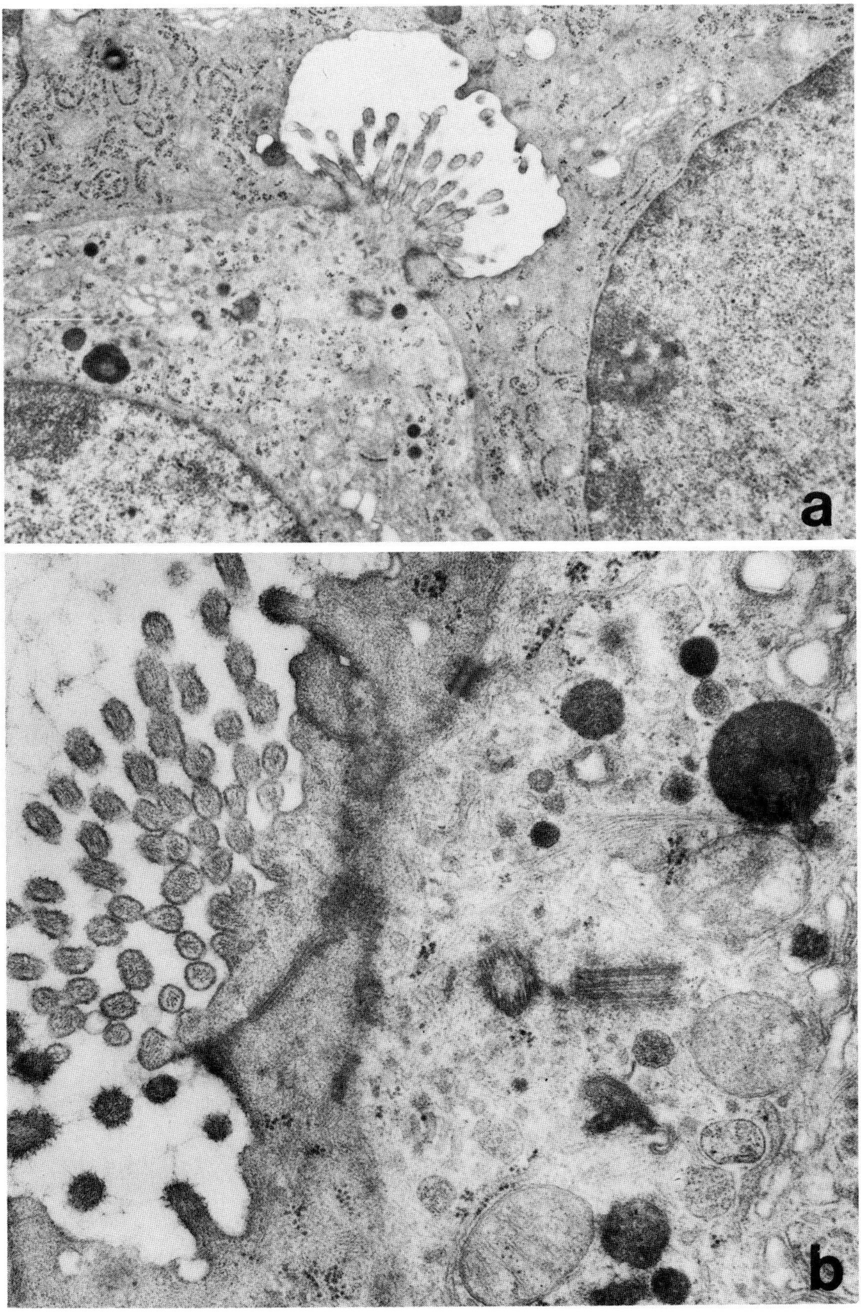

FIG. 16. Luminal endings of G cells in guinea pig pyloric mucosa. Note microvilli, tight junctions with neighboring mucous cells, centrosomes, supranuclear Golgi complex, and lysosomes. (a) ×10,000. (b) ×23,000.

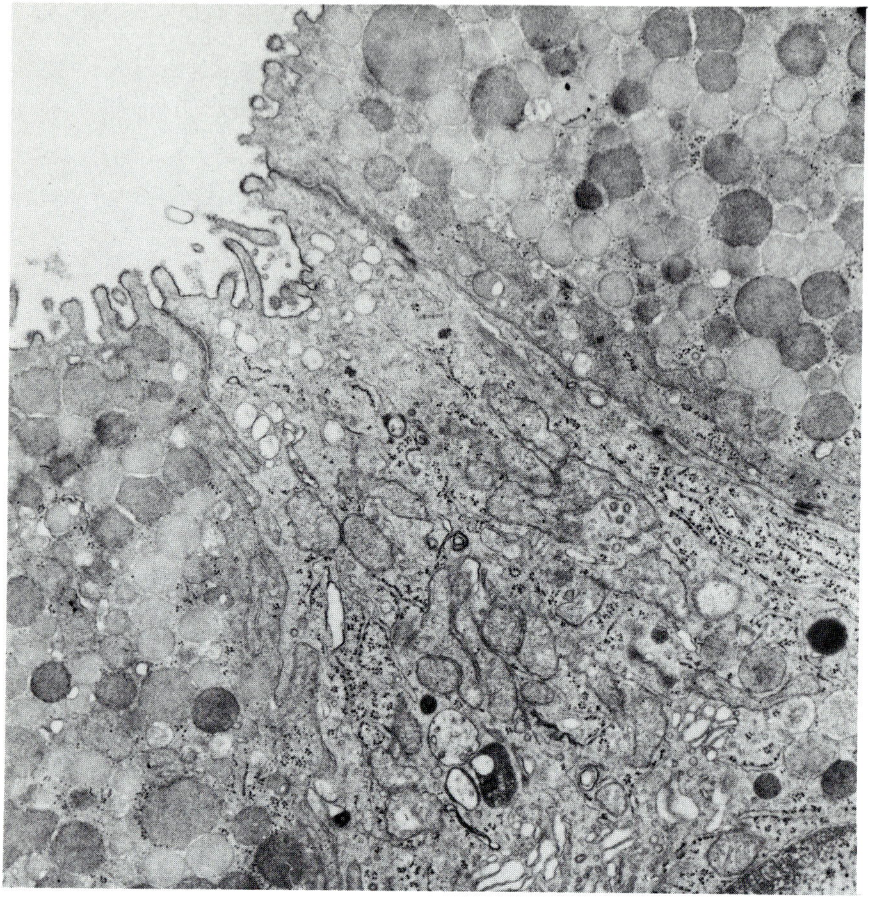

FIG. 17. Supranuclear part of pig G cell showing many pinocytoticlike vesicles just beneath the luminal surface. ×13,500. (From Capella and Solcia, 1972.)

Various stimuli are known to enhance gastrin release from the pyloric mucosa; among these are food, direct vagal stimulation, hypoglycemia, calcium, acetylcholine, choline, glycine, alanine, ethanol, catecholamines, reserpine, some anesthetics, distention of the antrum, and increased pH values at the luminal surface of the pyloric mucosa. Ultrastructural changes in G cells have been shown to occur after feeding or administration of ethanol (Forssmann, 1970), anesthetics, or reserpine (Lechago and Bencosme, 1973); loss of granule content with increase in the number of vesicular "empty" granules has been usually found. Similar changes have been observed in G cells of dog antral pouches transplanted onto the colon (Lechago and

Bencosme, 1973), and in human G cells of hypergastrinemic achlorhydric patients suffering from pernicious anemia (Creutzfeldt *et al.*, 1971; Polak *et al.*, 1972). Both intracytoplasmic solubilization of granular secretory products (Forssmann and Orci, 1969) and granule exocytosis (Osaka *et al.*, 1973a) have been considered as possible subcellular mechanisms of gastrin release from G cells.

As a rule, G cells contact both the basal lamina and the lumen of pyloric glands. The luminal endings often show microvilli, pinocytoticlike vesicles, cilia, and centrioles (Figs. 16 and 17). Likely, these structures play some role in the mechanism of cell stimulation by luminal stimuli, either chemical or mechanical (Solcia *et al.*, 1967, 1970b). It was recently shown that the gastrin-releasing power of some gastric secretatogues is related to the rate of their absorption by the pyloric mucosa (Andersson and Elwin, 1972). Morphological features are compatible with direct absorption of secretagogues by the G cells at their luminal endings. Alternatively, interaction of luminal contents with some receptor located on the luminal surface of the G cell should be considered. Nerve and blood supplies, which are well developed in the lamina propria of the pyloric mucosa, should also modulate G-cell function, as suggested by many physiological and pharmacological experiments.

G cells have been shown to undergo hyperplastic proliferation in the "retained excluded antrum" following unappropriate gastric surgery (Solcia *et al.*, 1970b) in the pyloric and duodenal mucosa of patients with recurrent duodenal ulcer, hyperchlorhydria, and hypergastrinemia (Figs. 18 and 19) (Solcia *et al.* 1970a; Polak *et al.*, 1972; Cowley *et al.*, 1973; Ganguli *et al.*, 1973), as well as in the pyloric mucosa of hypergastrinemic achlorhydric patients suffering from long-standing pernicious anemia (Creutzfeldt *et al.*, 1971; Polak *et al.*, 1972; C. Capella, F. Miglio, G. Gasbarrini, and E. Solcia, unpublished data, 1973). Patients with uncomplicated duodenal ulcer, gastric hypersecretion, and roughly unchanged fasting blood gastrin, often show numerous pyloric and duodenal G cells with a high gastrin content, but no extensive G-cell hyperplasia (Solcia *et al.*, 1970a; Polak *et al.*, 1972; Mitschke and Becker, 1973). Increased amounts of gastrin-rich G cells have also been found in the pyloric mucosa of hyperparathyroid and acromegalic patients (Creutzfeldt *et al.*, 1971; Polak *et al.*, 1971c). The low gastrin content of G cells in pernicious anemia, as shown by immunohistochemistry, has been correlated with the empty granules of such G cells and attributed to accelerated continuous release of gastrin and defective hormone storage (Creutzfeldt *et al.*, 1971; Polak *et al.*, 1972).

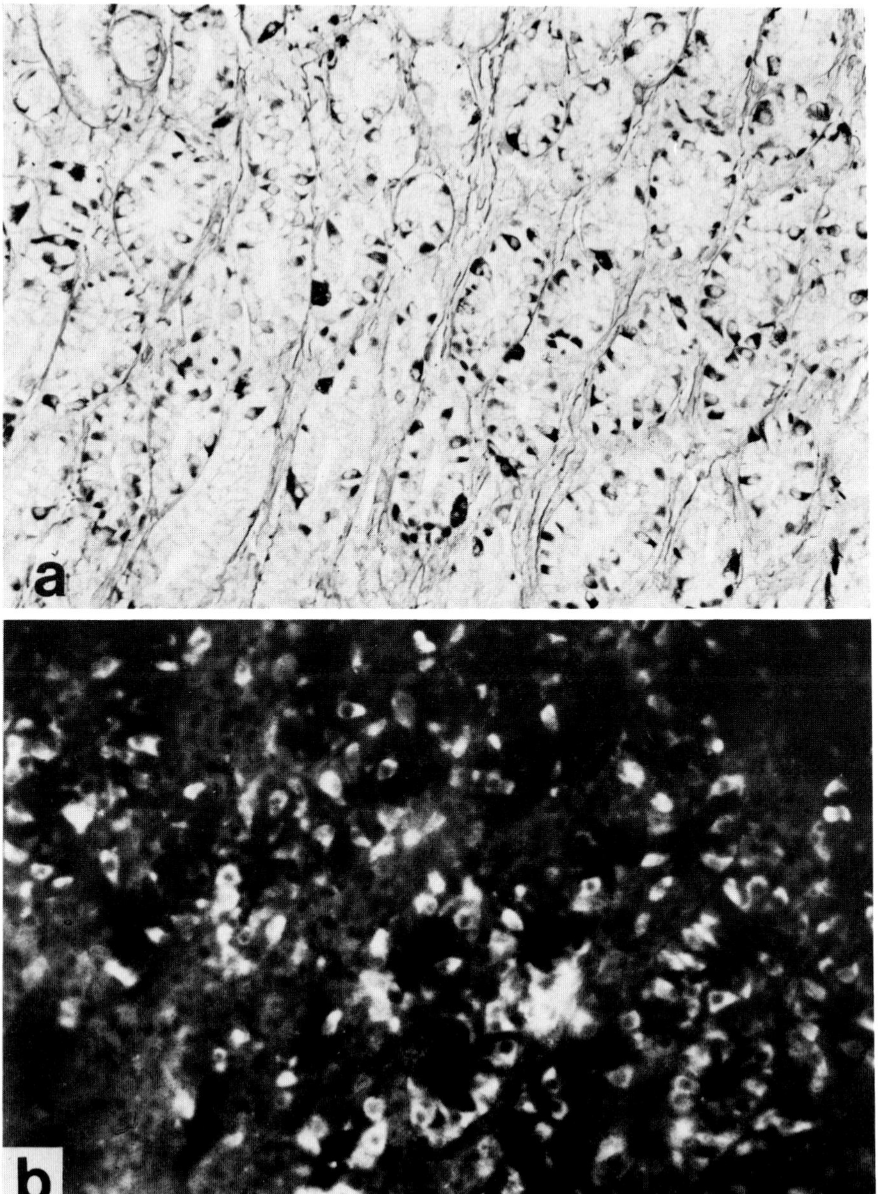

FIG. 18. Extensive hyperplasia of pyloric G cells in a woman with severe gastric hypersecretion, multiple peptic ulcer involving the distal duodenum, and gastrinlike activity in the urine on rat bioassay. The patient was cured following gastrectomy; 5 years later her blood gastrin was constantly below 50 pg/ml (normal values are below 100 pg/ml). (a) Grimelius' silver. ×240. (From Solcia *et al.*, 1970a.) (b) Gastrin immunofluorescence. ×280.

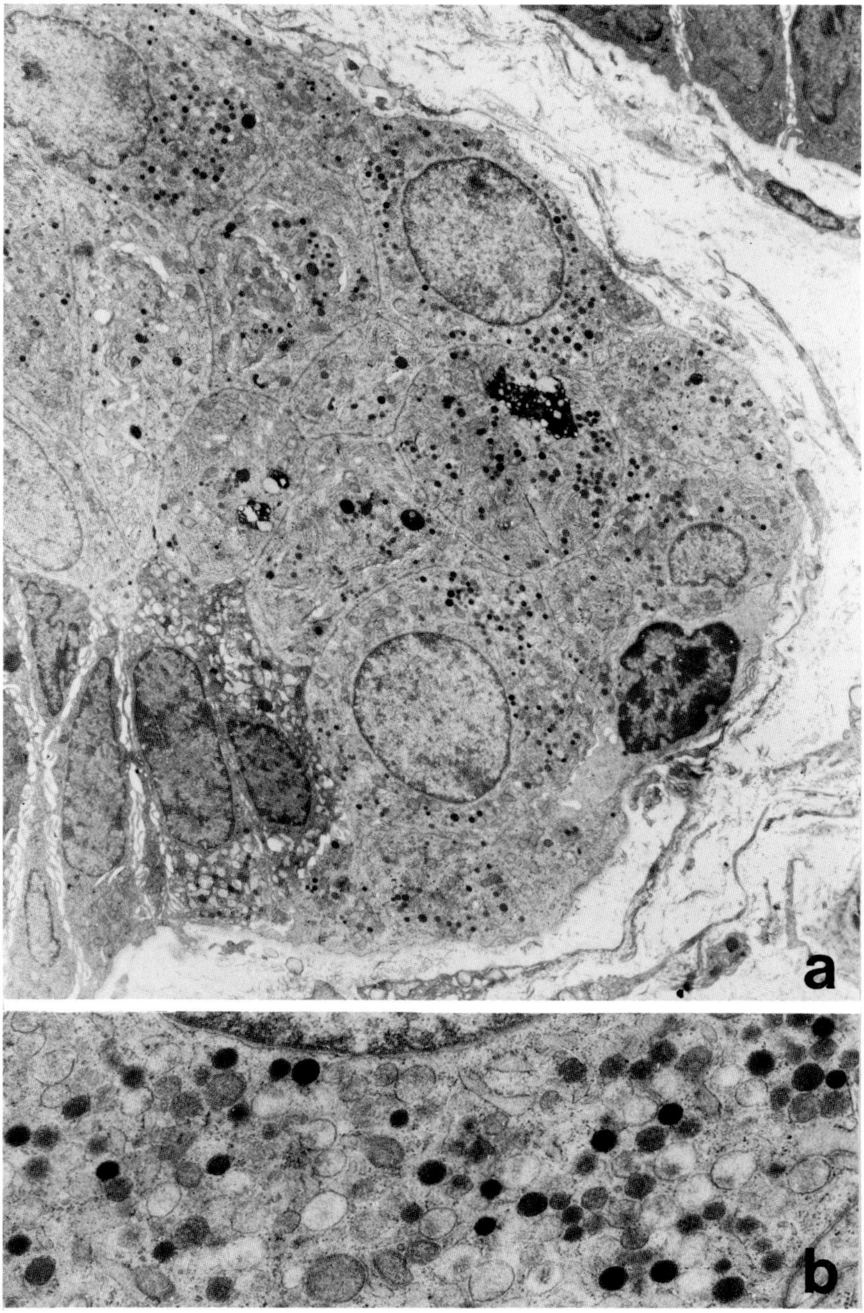

FIG. 19. (a) Micronodular G-cell growth in pyloric gland of a patient with the Zollinger-Ellison syndrome and hypergastrinemia (case kindly supplied by Prof. A. G. E. Pearse and Dr. J. M. Polak). ×4000. (b) Details of (a) showing G-type granules. ×17,500.

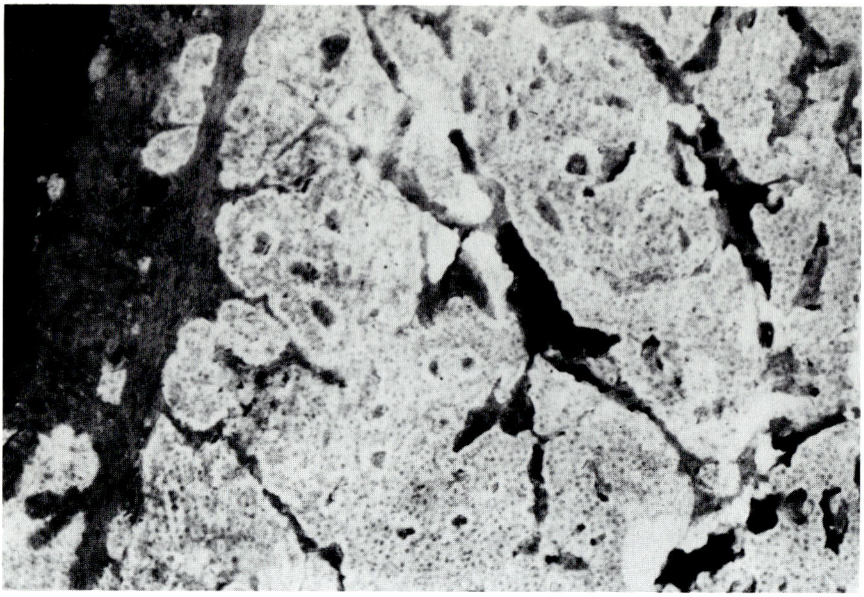

FIG. 20. Duodenal gastrinoma stained by gastrin immunofluorescence. Note nests of reactive tumor cells in ulcerated mucosa (left) and extensive growth of the tumor in submucosa. ×120.

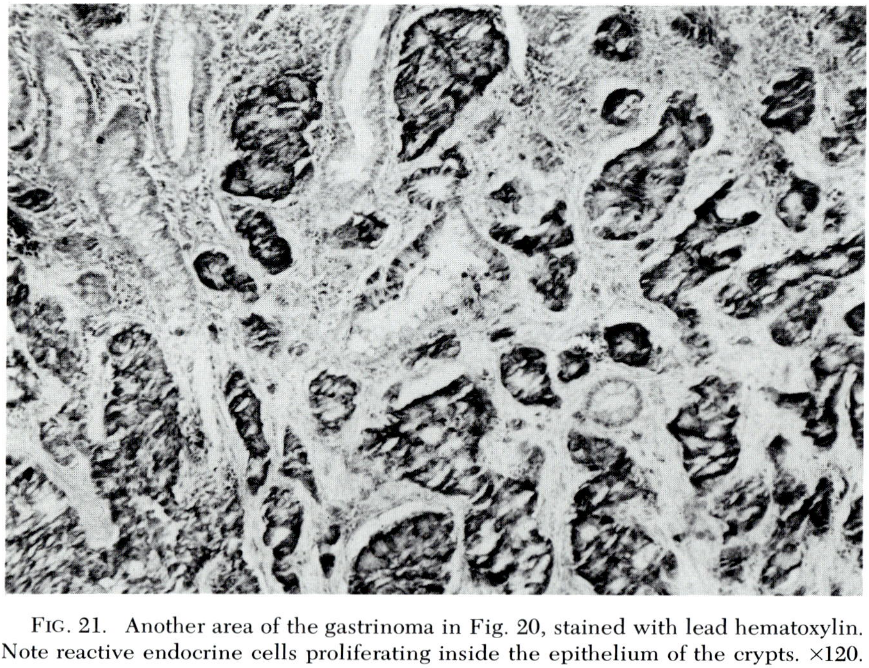

FIG. 21. Another area of the gastrinoma in Fig. 20, stained with lead hematoxylin. Note reactive endocrine cells proliferating inside the epithelium of the crypts. ×120.

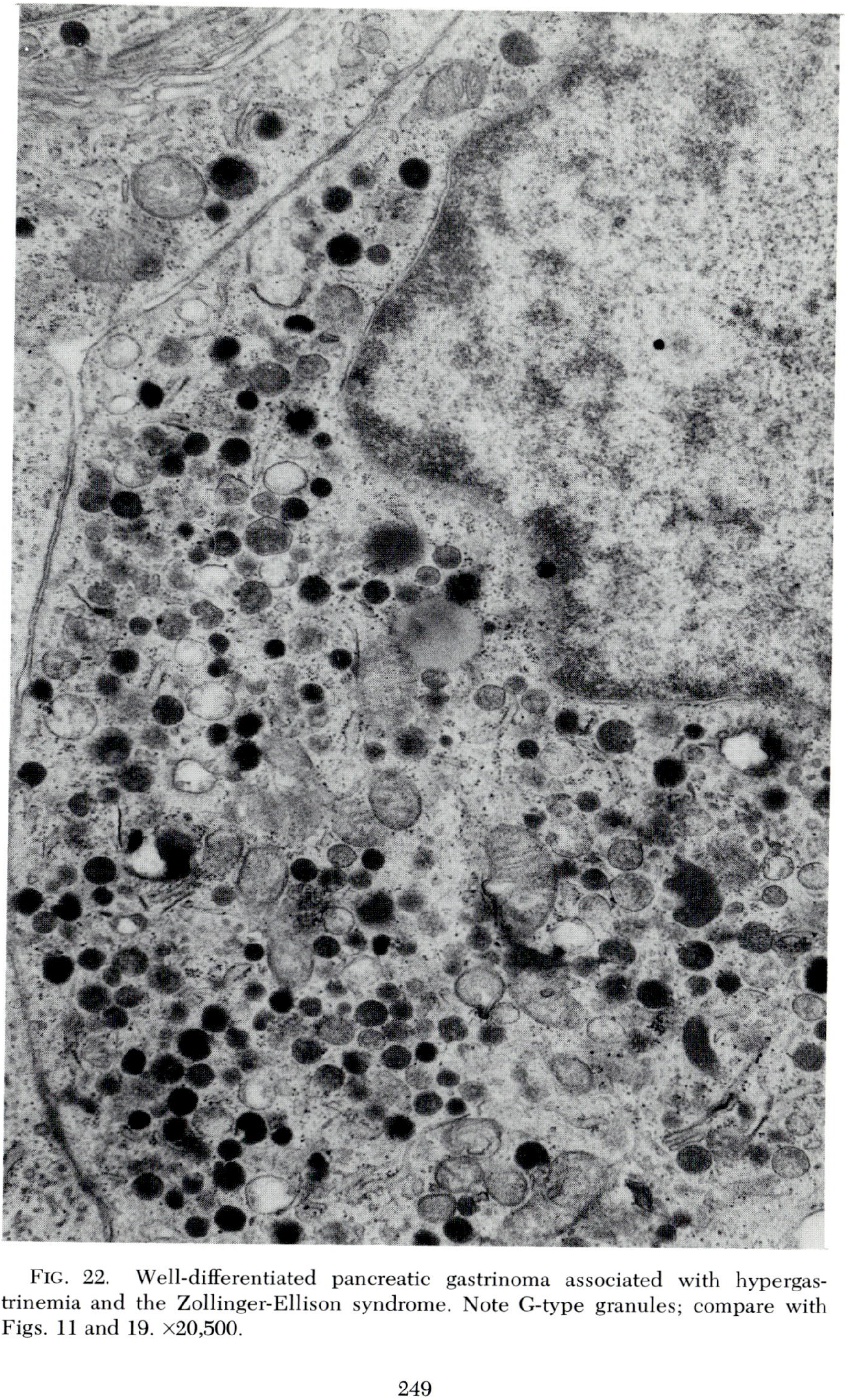

FIG. 22. Well-differentiated pancreatic gastrinoma associated with hypergastrinemia and the Zollinger-Ellison syndrome. Note G-type granules; compare with Figs. 11 and 19. ×20,500.

Only two gastrin-producing G-cell tumors have been described so far in the stomach (Royston *et al.*, 1972; Larsson *et al.*, 1973a). Duodenal gastrinomas (Figs. 20 and 21) occur more frequently (Oberhelman *et al.*, 1961; Weichert *et al.*, 1967); they are probably related to duodenal G cells (Solcia *et al.*, 1974d). The origin of gastrin-producing tumors arising in the pancreas (Zollinger and Ellison, 1955) remains uncertain, given the absence of G cells in normal human pancreas (see above). Some of such pancreatic tumors show unequivocal evidence of G cells on ultrastructural (Fig. 22) and histochemical examination (Creutzfeldt *et al.*, 1971; Solcia *et al.*, 1974d). Many of them are quite undifferentiated, having cells with few nondiagnostic granules and some features resembling those of the ductular islet cells occurring in fetal pancreas (Vassallo *et al.*, 1972; Greider *et al.*, 1974; Solcia *et al.*, 1974d). In addition to gastrin, these malignant tumors often produce insulin glucagon, amines, ACTH and other "inappropriate" hormones (Law *et al.*, 1965; O'Neal *et al.*, 1968; Block *et al.*, 1969; Sircus, 1969). They are probably to be interpreted as "stem cell" tumors undergoing partial, incomplete and

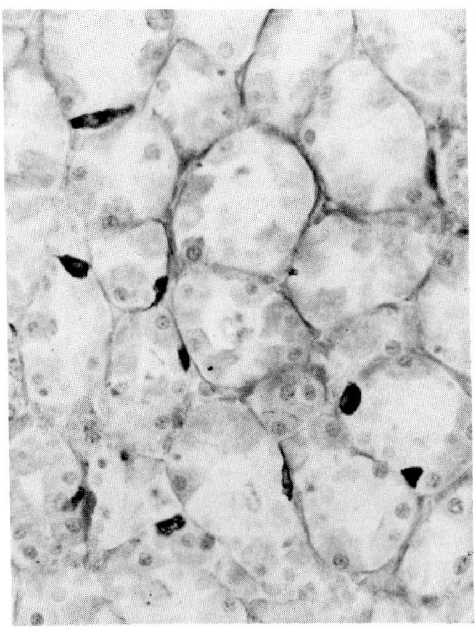

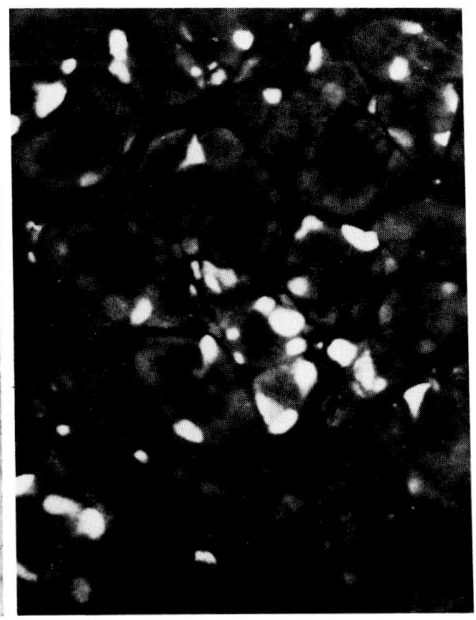

FIG. 23 (left). Argyrophilic cells of human fundic mucosa stained by the Sevier-Munger technique. Most of these are nonargentaffin and correspond to ECL cells. ×300.

FIG. 24 (right). Fundic mucosa of dog treated with L-dopa (same as in Figs. 13 and 15); formaldehyde-induced fluorescence according to the Falck-Hillarp technique. In addition to yellow EC cells, many blue-green cells are present, partly corresponding to ECL cells. ×300.

sometimes unappropriate differentiation toward multiple endocine lines.

D. ECL Cells

ECL cells are small, irregularly shaped, heavily argyrophilic (with Grimelius', Bodian's, and Sevier-Munger's methods) which are scattered in fairly large numbers in fundic gastric glands of cells (Fig. 23). Ultrastructurally (Figs. 25 and 39), they show either vesicular granules with an irregular argyrophilic core eccentrically located in a wide space, or a round, relatively compact (or coarsely granular), argyrophilic core surrounded by a membrane often of wavy appearance and forming a thin, clear space (Capella *et al.*, 1969, 1971; Solcia *et al.*, 1970b; Vassallo *et al.*, 1971b; Capella and Solcia, 1972). Unlike other endocrine cells of the fundic mucosa, ECL cells show high pseudocholinesterase activity which is mostly localized in the endoplasmic reticulum and secretory granules (Capella *et al.*, 1973; Monga *et al.*, 1974).

Only part of the nonargentaffin argyrophilic "enterochromaffinlike" cells described by Håkanson *et al.* (1970) resemble the ECL cells described above. In fact, in addition to ECL cells, A-like, G, D_1, and X cells also show argyrophilia and take up amine precursors (L-5HTP and L-dopa), developing amine-related fluorescence (Figs. 13 and 24) (Carvalheira *et al.*, 1968; Solcia *et al.*, 1970b, 1974c; Capella *et al.*, 1971; Rubin *et al.*, 1971; Ericson *et al.*, 1972). The argyrophilic histamine-storing cells peculiar to murine fundic mucosa (Håkanson and Owman, 1967, 1969) are more closely related to ECL cells, although X or A-like cells could also store histamine in these species (Håkanson *et al.*, 1971; Tjälve, 1971; Bussolati and Monga, 1973). Many of the argyrophilic nonargentaffin cells found by Dawson (1948) in rat fundic mucosa should also be interpreted as ECL cells, which in this tissue are largely prevalent over other types.

As a rule, ECL cells of untreated animals are not argentaffin; like other argyrophilic cells, they become argentaffin only after treatment with amine precursors (Håkanson *et al.*, 1971; Solcia *et al.*, 1974c). However, ultrastructural findings suggest that some ECL cells of cat and rabbit can display argentaffin granules even under basal conditions (Vassallo *et al.*, 1969). They differ from argentaffin EC cells in being poorly reactive with the hydrochloric acid–basic dye technique and with lead–hematoxylin.

ECL cells are scattered in the fundic glands, especially in their

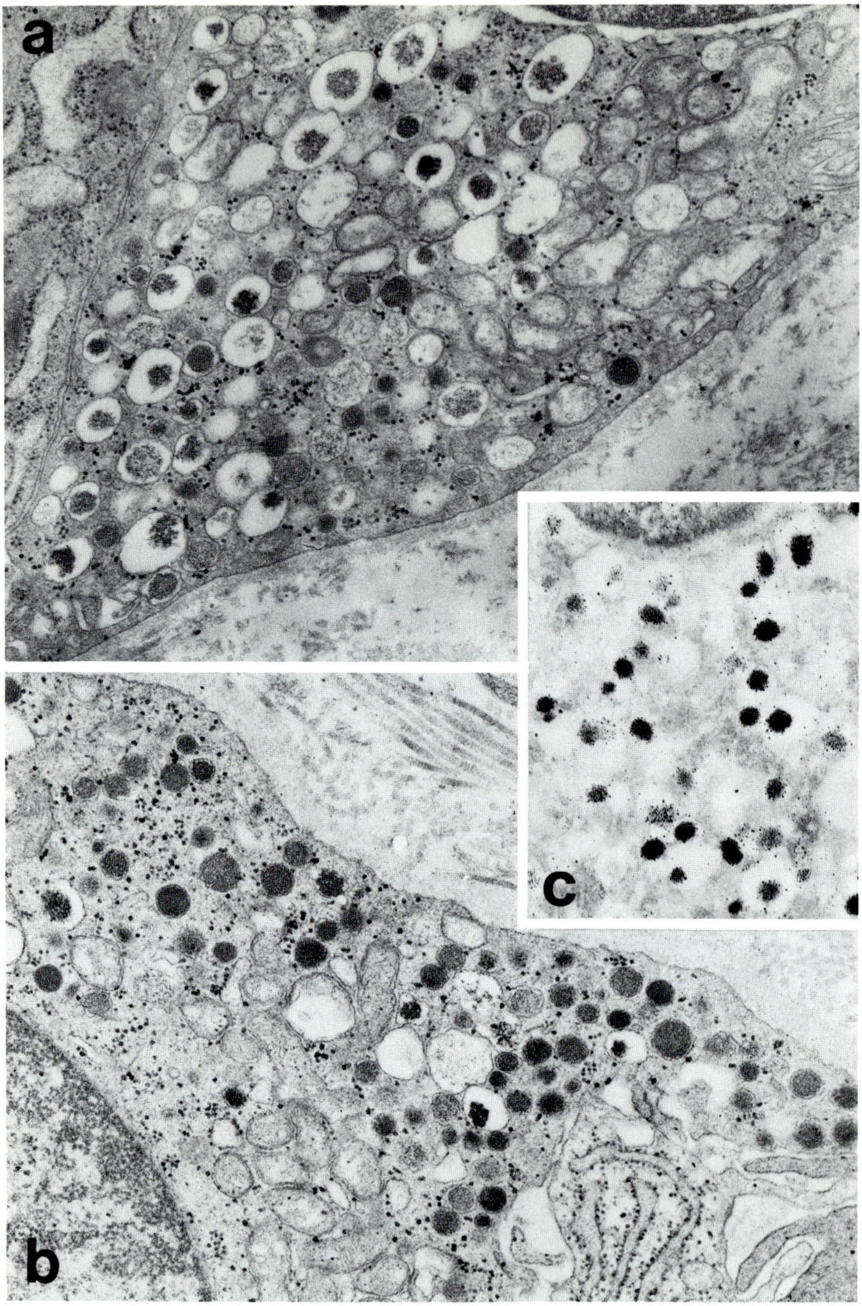

FIG. 25. ECL cells of human fundic mucosa showing prevalence of vesicular granules (a) or more compact granules (b), and intense reactivity of the granule core with Sevier-Munger silver (c). (a and b) 21,000. (c) 19,500. (From Vassallo *et al.*, 1971b.)

deep and intermediate regions; very few ECL cells have been found in the neck of the glands, and none in the epithelium covering the luminal surface of the mucosa and the gastric pits. Like other endocrine cells of the fundic mucosa, ECL cells lack any contact with the lumen of the glands, usually being covered by oxyntic or peptic cells (Capella *et al.*, 1969, 1971; Vassallo *et al.*, 1969; Solcia *et al.*, 1970b). According to Kobayashi *et al.* (1971), fundic endocrine cells are of the "closed" type, in contrast to most pyloric and intestinal cells which, being in direct contact with the lumen, are of the "open" type.

ECL cells are in close contact with the basal lamina of the glands; just beneath this, nerve endings and blood capillaries are often found. It seems likely that the activity of ECL cells is mostly dependent on stimuli coming from these structures. In fact, Bussolati and Monga (1973) noted stimulation of ECL cells with discharge of their histamine content in pylorus-ligated Shay rats. As shown by Brodie (1966), severe gastric hypersecretion secondary to intense vagal stimulation occurs in these animals. Moreover, blood gastrin, either endogenous or exogenous, is known to regulate specific histidine decarboxylase which, according to all available evidence, is the key enzyme for the synthesis of histamine inside ECL cells (Aures *et al.*, 1968, 1970).

In keeping with these data, we found hypertrophy of ECL cells, with signs of hyperfunction (prominent Golgi and reticulum, highly vesicular granules), in rats treated for months with high doses of caerulein, a known gastrin analog (Fig. 26). Hypertrophy and hyperplasia of ECL cells have been also found in patients suffering from the Zollinger-Ellison syndrome with long-standing severe hypergastrinemia (Figs. 27 and 28). Human ECL cells lack histochemically detectable histamine (Håkanson *et al.*, 1970), despite the fact that the inhibition by burimamide and related antihistaminic drugs of gastrin-induced secretion in the human stomach (Willie *et al.*, 1972) suggests that some histamine-dependent mechanism might also operate in the human stomach. It seems pertinent to recall here that some Zollinger-Ellison patients show large amounts of histamine in the blood and urine (Dotevall and Walan, 1970; Block *et al.*, 1969), and that some argyrophilic gastric "carcinoid" tumors (probably related with ECL cells) produce large amounts of histamine, causing severe gastric hypersecretion (Sandler and Snow, 1958; Oates and Sjoerdsma, 1962; Campbell *et al.*, 1963; Granérus *et al.*, 1966; Noll and Levine, 1970). Thus it seems likely that ECL cells are under the control of gastrin and vagus, and that part of the hormone and nerve

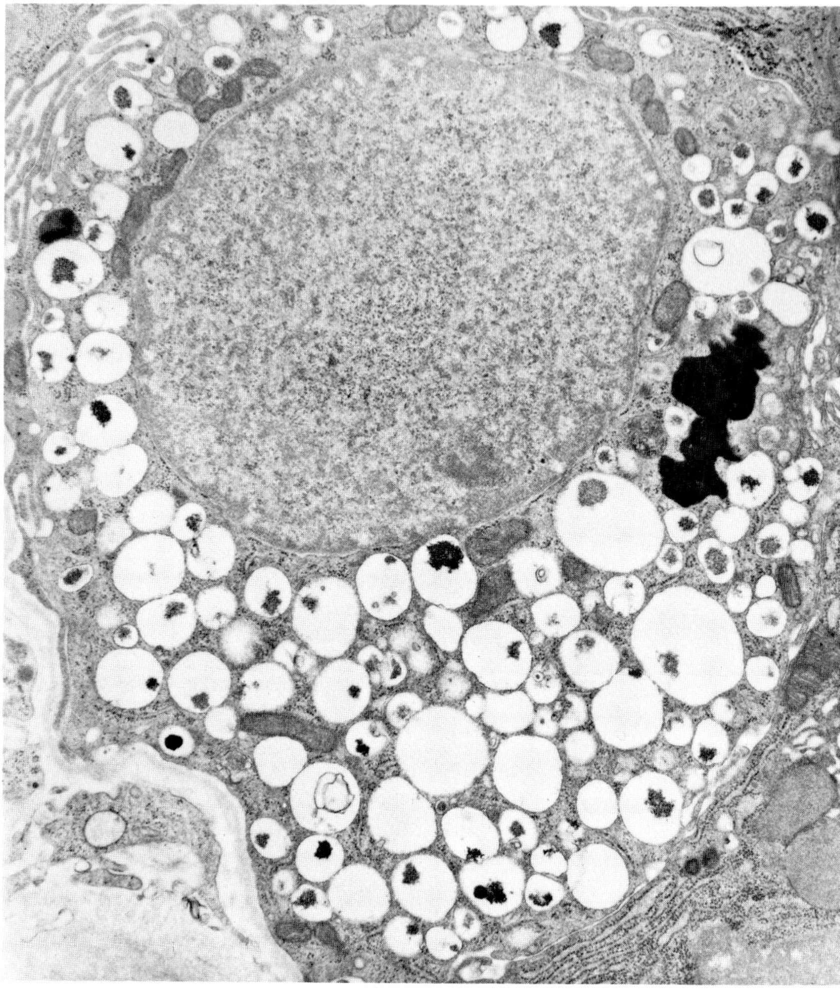

FIG. 26. ECL cell of a rat given 250 μg of caerulein per day for 6 months. Note very large, vesicular granules. ×12,000.

influence on gastric secretion is mediated by ECL cells. In ECL cells of caerulein-treated rats, we found evidence of granule exocytosis at the surface confronting oxyntic and peptic cells. Possible release of active agents other than histamine should be investigated; ECL cells of man and several mammals, although lacking histochemically detectable histamine stores (Håkanson *et al.*, 1970), still show many secretory granules, whose content is presently unknown.

The involvement of ECL cells in histamine-producing carcinoid

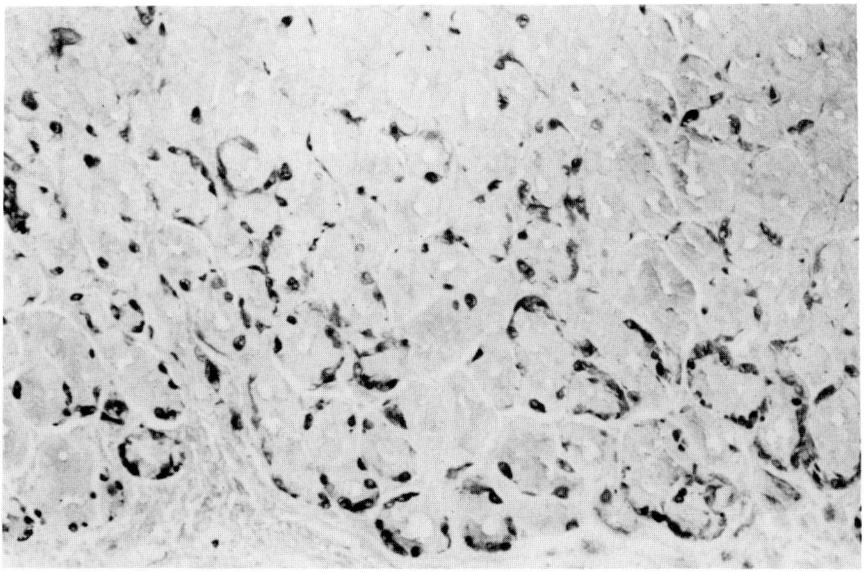

FIG. 27. Extensive hyperplasia of argyrophilic cells—mostly ECL cells—in the fundic mucosa of a patient with the Zollinger-Ellison syndrome and long-standing severe hypergastrinemia. Compare with Fig. 23. ×120.

tumors arising in the human stomach (Solcia *et al.*, 1970a) (Fig. 29) is supported by recent studies on the histamine-producing argyrophilic carcinoids arising in the stomach of *M. natalensis* (Hosoda *et al.*, 1970). Many cells of such tumors have been found to reproduce some ultrastructural and histochemical features of nontumor ECL cells (Capella *et al.*, 1973). Gastric hypersecretion and duodenal ulcers develop in animals bearing these histamine-producing carcinoids, just as in man.

E. D CELLS

Non-EC endocrine cells blackened with Davenport's alcoholic silver nitrate method and showing metachromatic basophilia at slightly acid pH values, were first noted in the pyloric mucosa and compared with D cells of the pancreatic islets, known to display similar staining patterns (Solcia and Sampietro, 1965a). Then cells resembling ultrastructurally pancreatic D cells, while clearly differing from both EC and gastrin cells, were described (Forssmann *et al.*, 1968; Cavallero *et al.*, 1969; Forssmann, 1970; Kobayashi *et al.*, 1970, 1971; Sasagawa *et al.*, 1970; Solcia *et al.*, 1970b; Vassallo *et al.*, 1971b; Rubin, 1972b). In some articles D cells were reported to be X

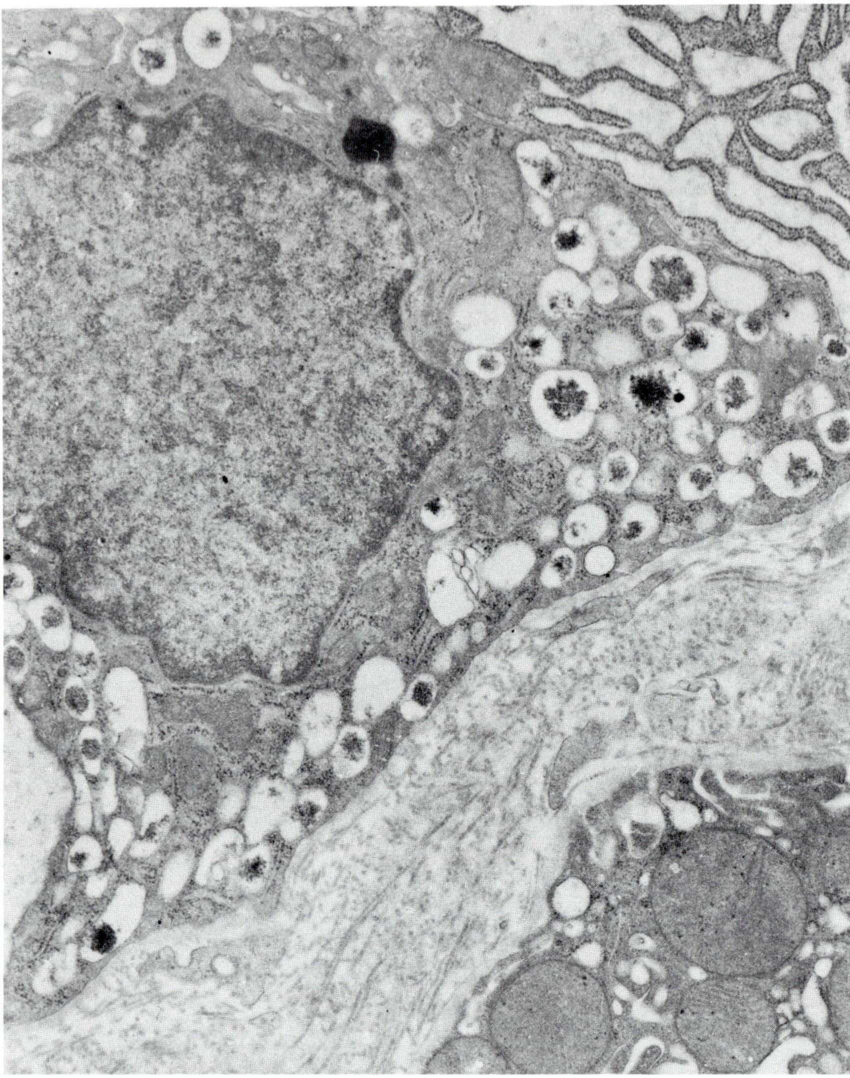

FIG. 28. Ultrastructure of an ECL cell of the same patient as in Fig. 27. Note large vesicular granules. Compare with Fig. 25. ×15,000.

cells, from which they were not distinguished (Vassallo *et al.*, 1969; Solcia *et al.*, 1970b); in other articles the cells reported as D cells were in fact D_1 cells (Forssmann *et al.*, 1969). Recent progress in the ultrastructural detection of pancreatic D cells, as well as in the identification of pancreatic D_1 (IV type) cells (Munger *et al.*, 1965; Like, 1967; Shibasaki and Ito, 1969; Greider *et al.*, 1970; Misugi *et al.*,

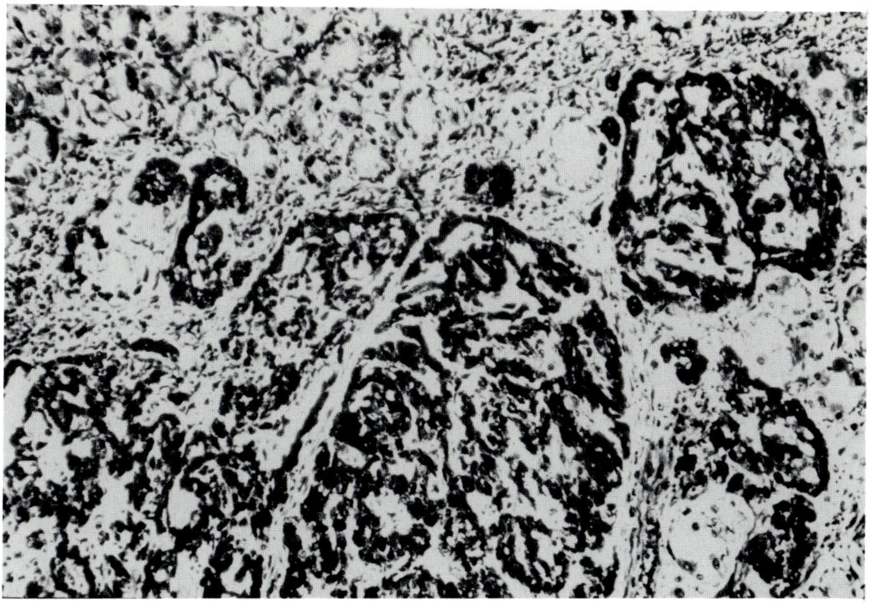

FIG. 29. Intramucosal multiple carcinoid tumor strongly reacting with Sevier-Munger silver. The tumor, which failed to react with 5HT methods and gastrin immunofluorescence, reproduced the staining pattern of ECL cells; it was associated with high histamine levels in the urine. ×120.

1970; Deconinck *et al.*, 1972; Like and Orci, 1972; Munger, 1972; Vassallo *et al.*, 1972), allowed D cells to be more easily recognized even in the gastrointestinal mucosa. They are characterized by round, weakly osmiophilic, relatively large granules (about 200–400 nm, with marked differences according to the animal species) with a homogeneous core and a closely applied membrane (Figs. 7 and 30). They react with Davenport's silver, lead–hematoxylin, and hydrochloric acid–basic dye technique, while being consistently unreactive in light and electron microscopy with Grimelius' silver (Figs. 31a and 40b), Sevier-Munger's silver (Capella and Solcia, 1972), and phosphotungstic acid (Fig. 31b). In dogs treated with L-dopa, D cells fail to develop dopamine-related fluorescence or the argentaffin reaction (Fig. 15b).

D cells are scattered in the stomach and upper small intestine, especially in the pyloric glands and the crypts of the proximal duodenal tract. They are consistently represented in the fundic gastric mucosa, although they are relatively few in number; also, few D cells are present in the Brünner glands, as well as in the jejunal crypts of dog and man.

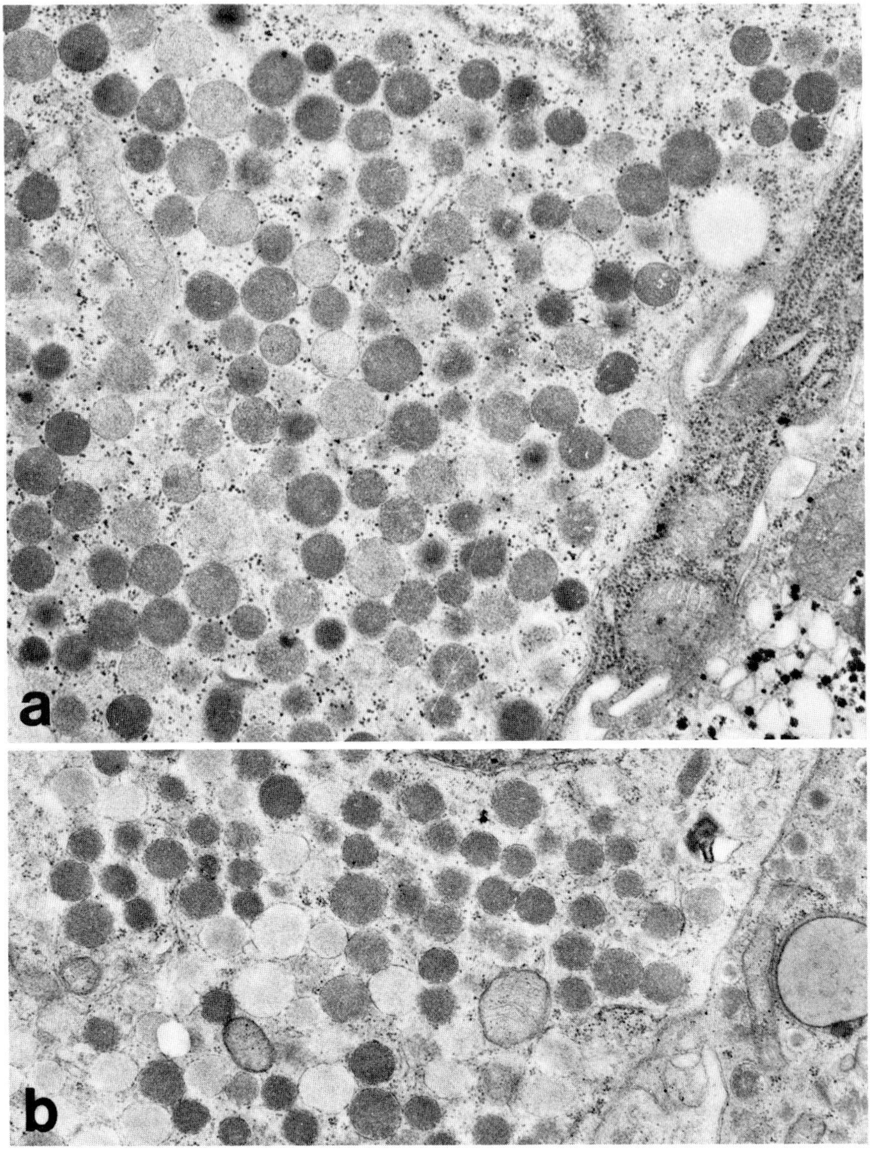

FIG. 30. (a) D cell of dog fundic mucosa. ×20,500. (b) D-cell granules in human pancreatic islet. ×17,000.

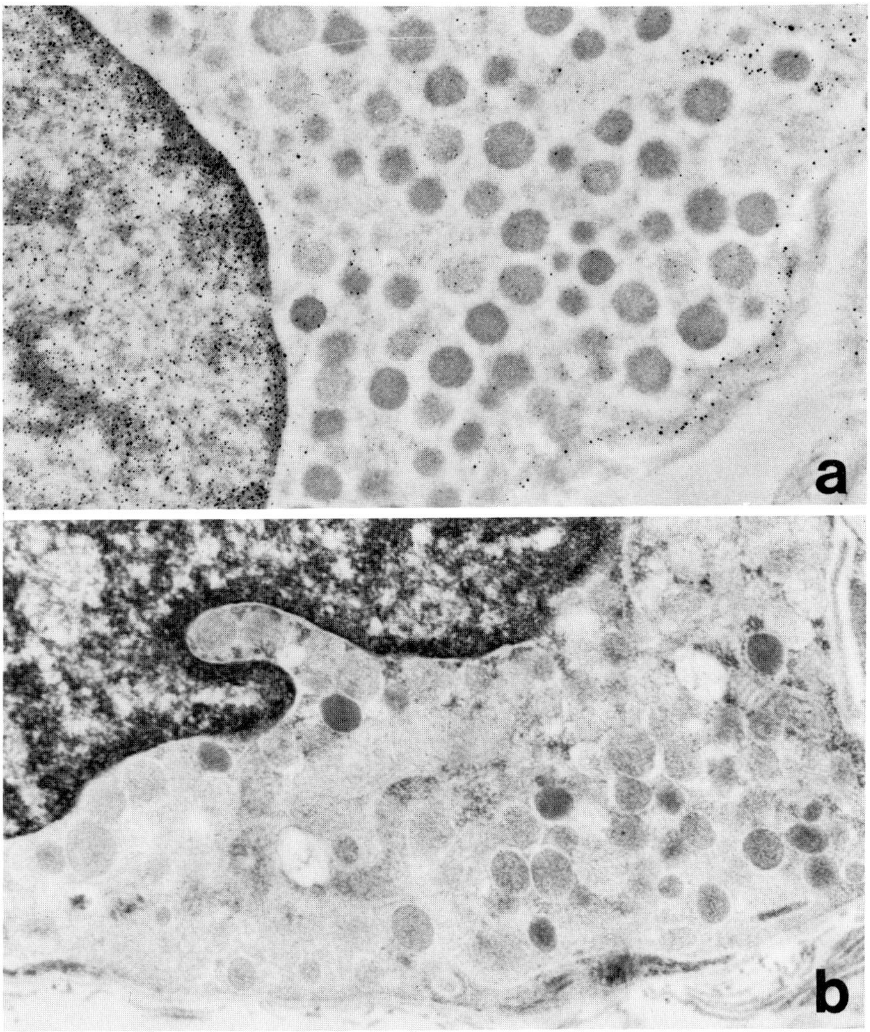

FIG. 31. Dog fundic mucosa fixed in formaldehyde–glutaraldehyde. (a) D-cell granules unreactive with Grimelius' silver, stained with uranyl. ×21,000. (b) 2% aqueous phosphotungstic acid (pH 2); poor staining of D-cell granules. ×21,000.

The function of D cells remains obscure. The reactivity of pancreatic D cells with some antigastrin antisera (Lomsky *et al.*, 1969; Greider and McGuigan, 1971; Polak *et al.*, 1972) has not been reproduced in other studies (Creutzfeldt *et al.*, 1971; Lotstra *et al.*, 1974; Solcia *et al.*, 1974d). The possibility that a gastrin cross-reactant, rather than gastrin itself, is involved in this reactivity should be considered, given the different technical requirements which proved optimal for pancreatic D cells as compared to pyloric G cells (Greider and McGuigan, 1971; J. M. Polak, personal communication), as well as the controversial results of various attempts to extract and purify gastrin from pancreatic tissue (Hallenbeck *et al.*, 1963; Nilsson *et al.*, 1970; Creutzfeldt *et al.*, 1971; Greider and McGuigan, 1971; Omole *et al.*, 1972; Rehfeld and Iversen, 1973; Lotstra *et al.*, 1974). Presently available data do not support the reactivity of gastrointestinal D cells with antigastrin antisera (McGuigan and Greider, 1971; Bussolati and Canese, 1972; Greider *et al.*, 1972; Canese and Bussolati, 1974). In dog pyloric mucosa, D cells were shown to release their granules after hydrochloric acid had been instilled into the stomach (Fujita and Kobayashi, 1971); this is in sharp contrast with the well-known inhibition of gastrin release from the pyloric mucosa under the same experimental conditions. The results of Fujita and Kobayashi may support the hypothesis that D cells produce a hydrochloric acid–releasable inhibitor of gastric secretion, such as antral chalone (Thompson, 1966) or bulbogastrone (Andersson *et al.*, 1965). Bulbogastrone, a hydrochloric acid–releasable selective (competitive?) inhibitor of gastrin, seems to be particularly considered in this respect.

A duodenal tumor partly made up of well-differentiated D cells has been described by Weichert *et al.* (1971). No D cell tumor has been found so far in the stomach. D cells occur only seldom in gastrin-producing pancreatic tumors, and seem unrelated to the pathogenesis of the Zollinger-Ellison syndrome (Creutzfeldt *et al.*, 1971; Solcia *et al.*, 1974d).

F. D_1 Cells

Small cells with small (about 150-nm), round granules showing a moderately osmiophilic compact core and a very thin clear halo (Figs. 32 and 33) have been observed repeatedly in the gastric mucosa (Solcia *et al.*, 1967; Orci *et al.*, 1968; Forssmann *et al.*, 1969; Forssmann, 1970; Sasagawa *et al.*, 1970; Kobayashi *et al.*, 1971; Rubin *et al.*, 1971; Ericson *et al.*, 1972), as well as in the duodenal mucosa (Kobayashi *et al.*, 1970; Pearse *et al.*, 1970). They have been reported under various names and more frequently as a D-cell variant because of their similarity to some pancreatic cells reputed to belong to the D-cell type. The observation that, in both the gastrointestinal mucosa and the pancreas, such cells with small granules react consistently with Grimelius' silver at the ultrastructural level

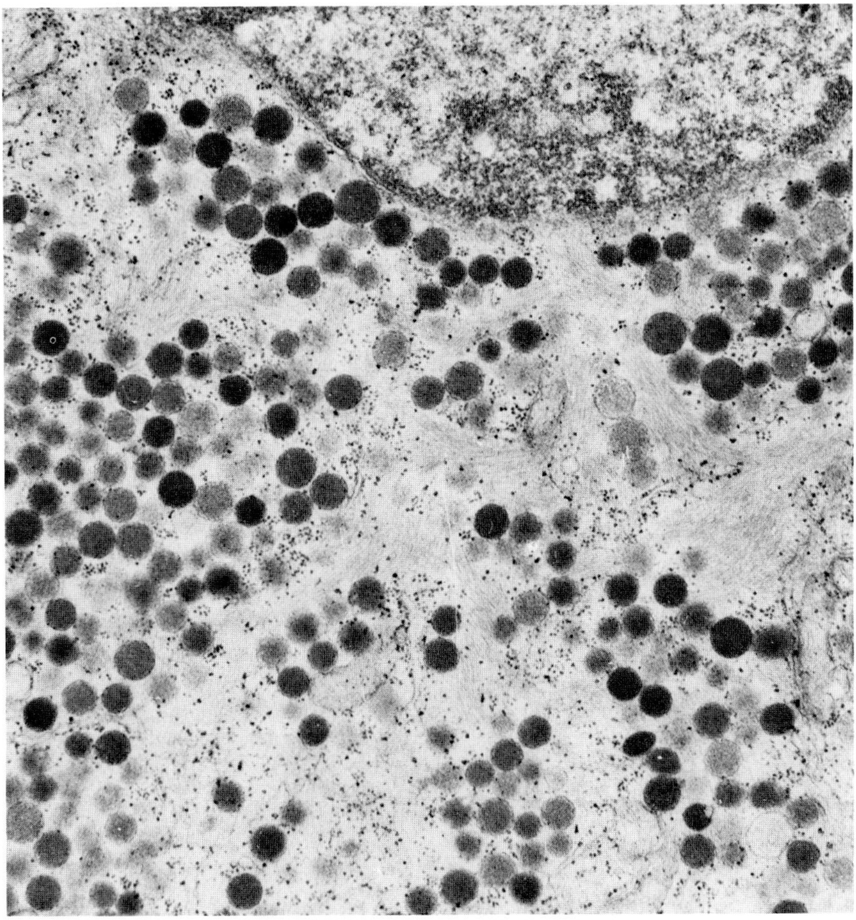

FIG. 32. D_1 cell of pig pyloric mucosa. ×25,000.

(Fig. 34a), while D cell granules remain mostly unreactive, supports the opinion that the small granule cells, under the tentative name of D_1 cells, be distinguished from D cells (Capella *et al.*, 1971; Vassallo *et al.*, 1971b, 1972; Capella and Solcia, 1972). Unlike granules of D cells, granules of D_1 cells are well preserved with simple fixation in osmium tetroxide; both types of granules are well preserved with aldehyde–osmium fixation.

As first noted by Sasagawa *et al.* (1970: see type III cells), cytoplasmic filaments are abundant in D_1 cells, particularly around the nucleus (Fig. 32). Besides staining with Grimelius' silver, D_1 granules are also stained with Sevier-Munger's silver (Fig. 34b),

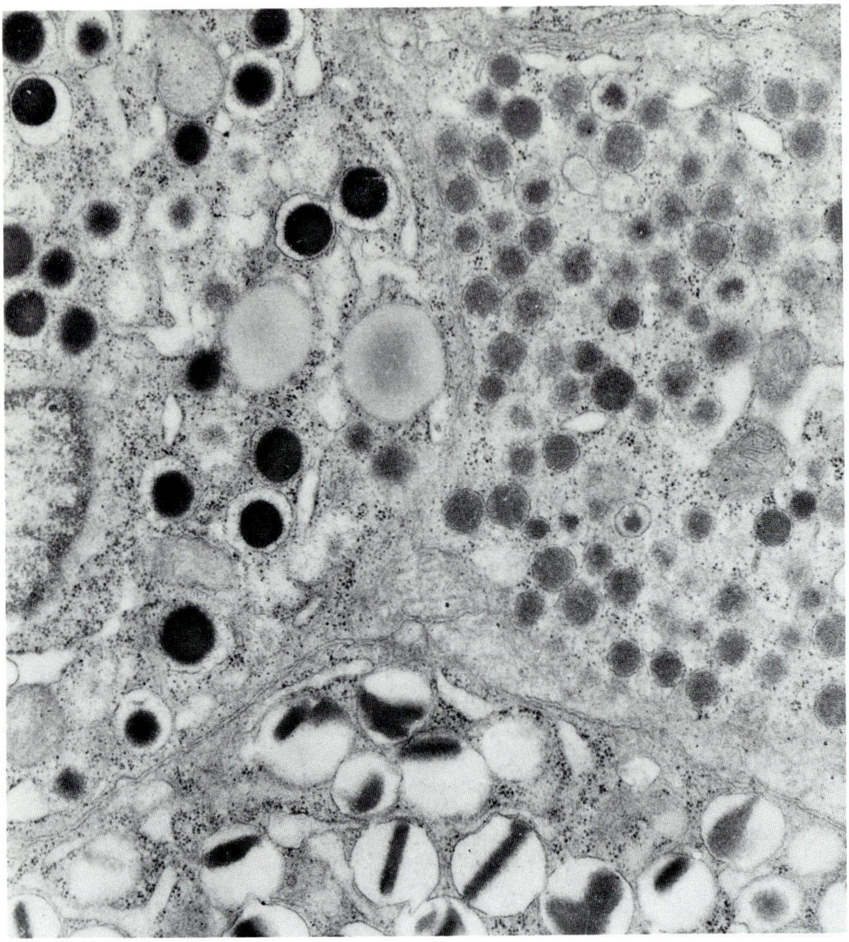

FIG. 33. Dog pancreatic islet showing granules of A (top left), B (bottom), and D_1 (top right) cells. ×24,000.

phosphotungstic acid (Fig. 34c), lead–hematoxylin, the hydrochloric acid–basic dye technique, and sometimes even with Davenport's silver. With silver techniques at the ultrastructural level, especially with Sevier-Munger's silver, some preferential reactivity at the periphery of the granules has been noted.

In both the pyloric and fundic mucosa, D_1 cells are mostly found in the deep part of the glands. They are represented throughout the small and large bowel, including the Brünner glands, although they do not occur in significant concentrations at any site. In the pancreas, D_1 cells are the least frequent of the four cell types commonly

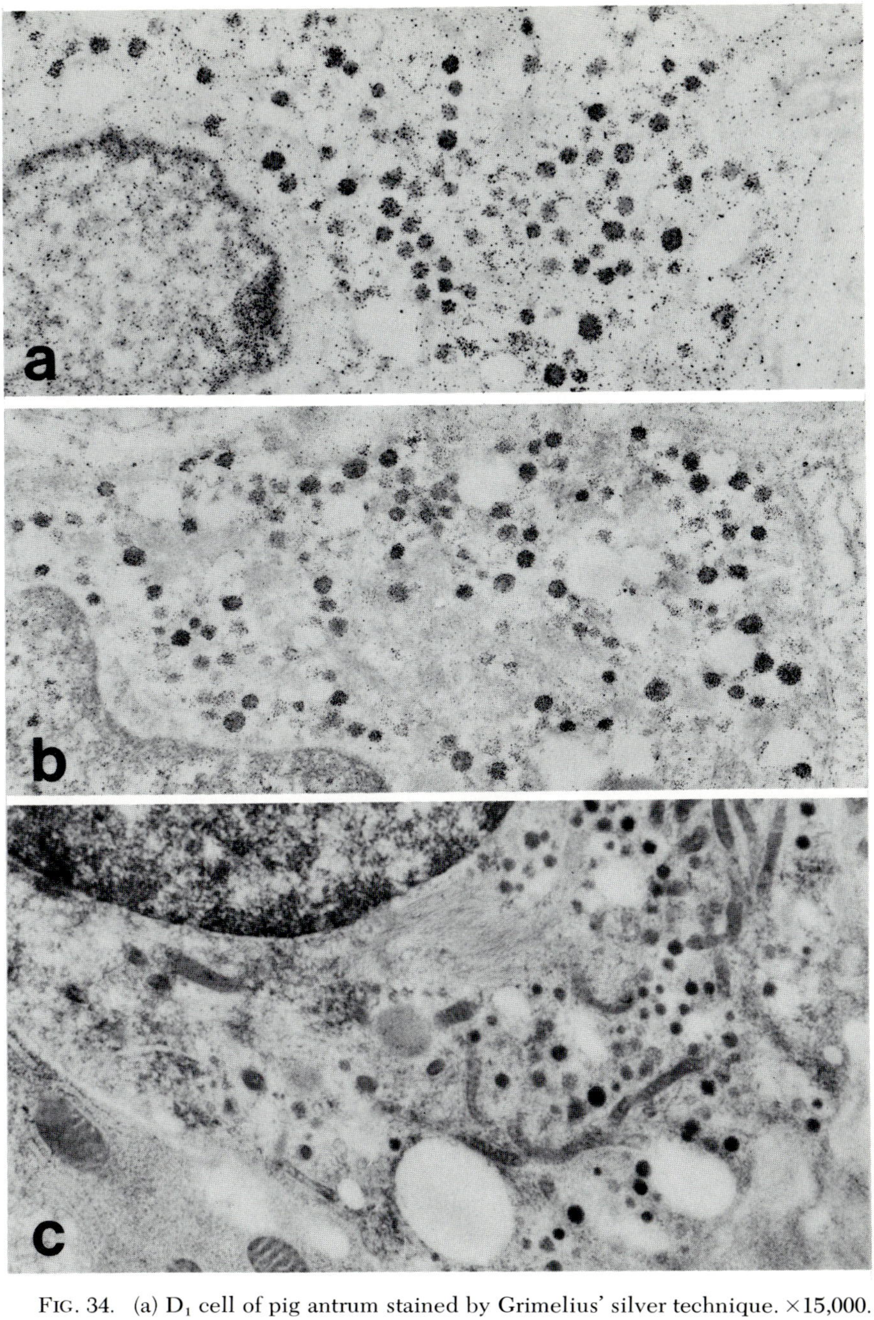

FIG. 34. (a) D_1 cell of pig antrum stained by Grimelius' silver technique. ×15,000. (From Capella and Solcia 1972.) (b) D_1 cell of human fundic mucosa; Sevier-Munger silver and uranyl. Note argyrophilia of the granules, often more evident at their periphery. ×18,000. (c) D_1 cell of human fundic mucosa; phosphotungstic acid. Note staining of the granules. ×18,000.

represented in the islets, but are, together with EC and F cells, among the most frequent endocrine cells scattered in the exocrine parenchyma. They are well represented in the small islets of the dog uncinate process, devoid of both A and D cells, and are absent in the "dark" islets of the duck, where only A and D cells have been found.

The exact relationship between D_1 cells and the enterocatecholamine cells first described by Forssmann *et al.* (1969) in the rat gastric mucosa remains to be ascertained. In addition to in the rat we found similar cells also in the rabbit, cat, monkey, and human stomach. They showed vesicles storing irregular, highly osmiophilic cores eccentrically located in a wide space delimited by a clear-cut membrane. These granules have been compared with those of some catecholamine-storing cells (Forssmann *et al.*, 1969). In fact, a few dopamine-storing cells have been found in rabbit and cat stomach (Håkanson *et al.*, 1970). However, in our experience many of the cells with enterocatecholamine granules also showed round, regular granules resembling those of D_1 cells, so that distinction between the two cell types was sometimes difficult (Fig. 35).

The function of D_1 cells is unknown. Gastric endocrine cells undergo diffuse micronodular proliferation in chronic atrophic gastritis, either with or without pernicious anemia (Rubin, 1969, 1972a; Solcia

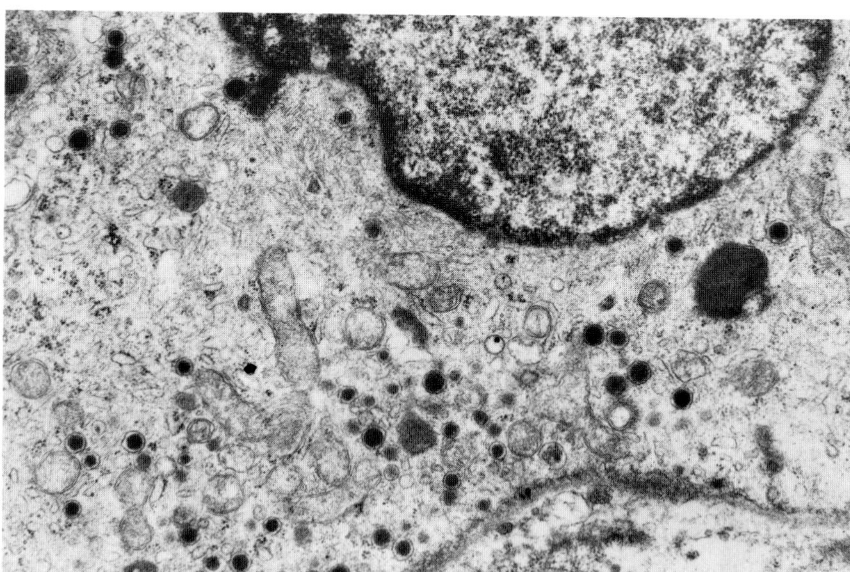

FIG. 35. Endocrine cell of rabbit pyloric mucosa, resembling D_1 cells as well as enterocatecholamine cells. ×15,000.

et al., 1970a). Although the number of ECL and EC cells may also increase, the cells composing such micronodular growths in gastric-type mucosa (not affected by intestinal metaplasia) are mostly to be interpreted as D_1 cells, owing to their small, round granules with peripheral argyrophilia and to their abundant cytoplasmic filaments (Fig. 36). Sometimes gastric argyrophilic nonargentaffin carcinoids

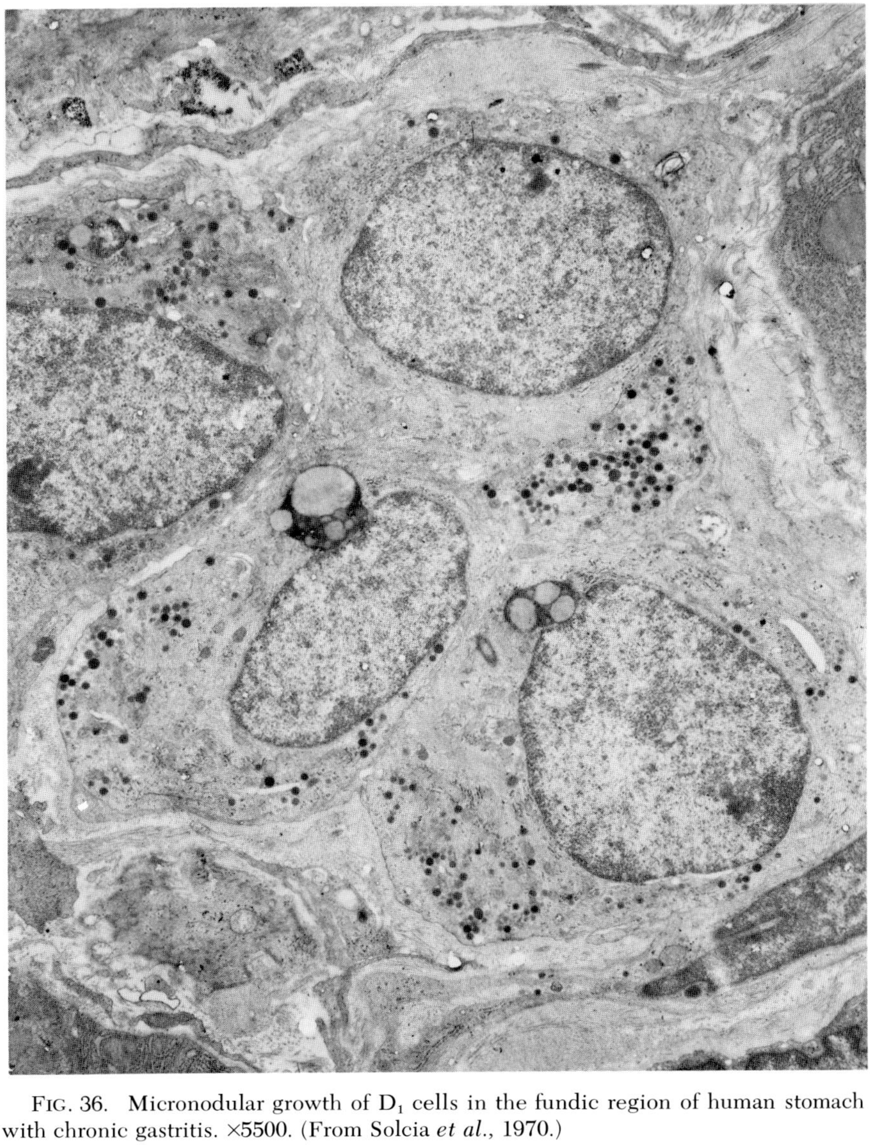

FIG. 36. Micronodular growth of D_1 cells in the fundic region of human stomach with chronic gastritis. ×5500. (From Solcia *et al.*, 1970.)

arise in patients with chronic gastritis with or without anemia (Lattes and Grossi, 1956; Pestana *et al.*, 1963; one of our unpublished cases). Gastric hypersecretion or histamine overproduction has not been found in these patients. The contribution of D_1 cells to these tumors, and probably also to some nonargentaffin gastric carcinoids unrelated to chronic gastritis, seems likely. Such tumors show histological and biochemical similarities with many bronchial nonargentaffin carcinoid tumors (Williams and Sandler, 1963; Bensch *et al.*, 1965; Hage, 1973b); moreover, many of the endocrine P cells (Feyrter, 1953) scattered in the bronchial mucosa show some ultrastructural and staining similarities with gut D_1 cells (Figs. 37 and 38). In particular, type 2 cells described by Hage (1973a) in the human bronchial mucosa resemble D_1 cells of the human gut, while Hage's type 1 cells resemble gut enterocatecholamine cells.

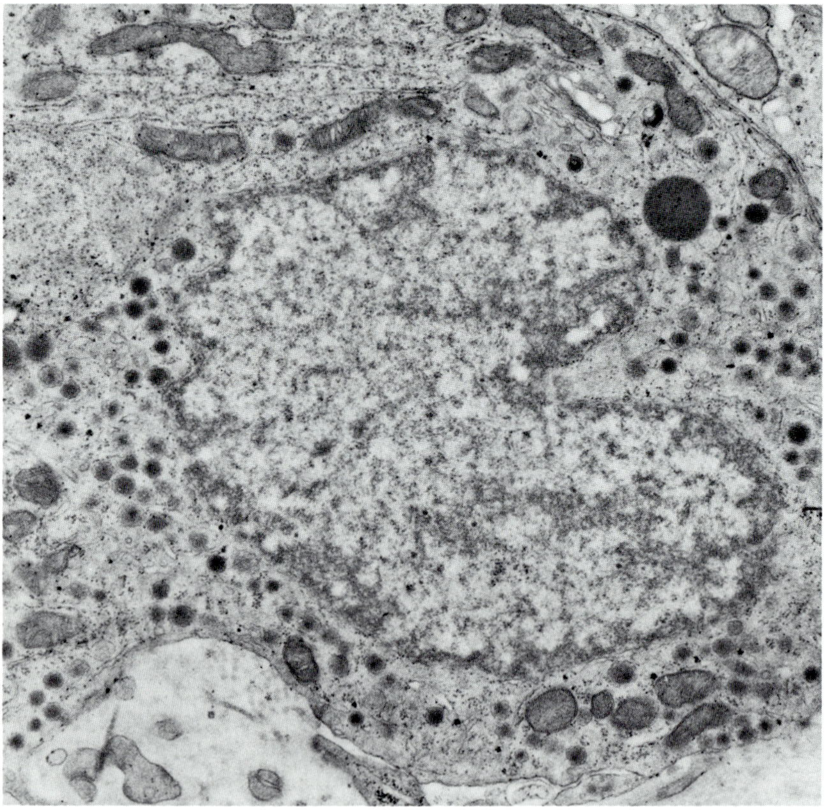

FIG. 37. Endocrine cell of human bronchial mucosa showing small granules with a thin halo, somewhat resembling those of gut D_1 cells. ×18,000.

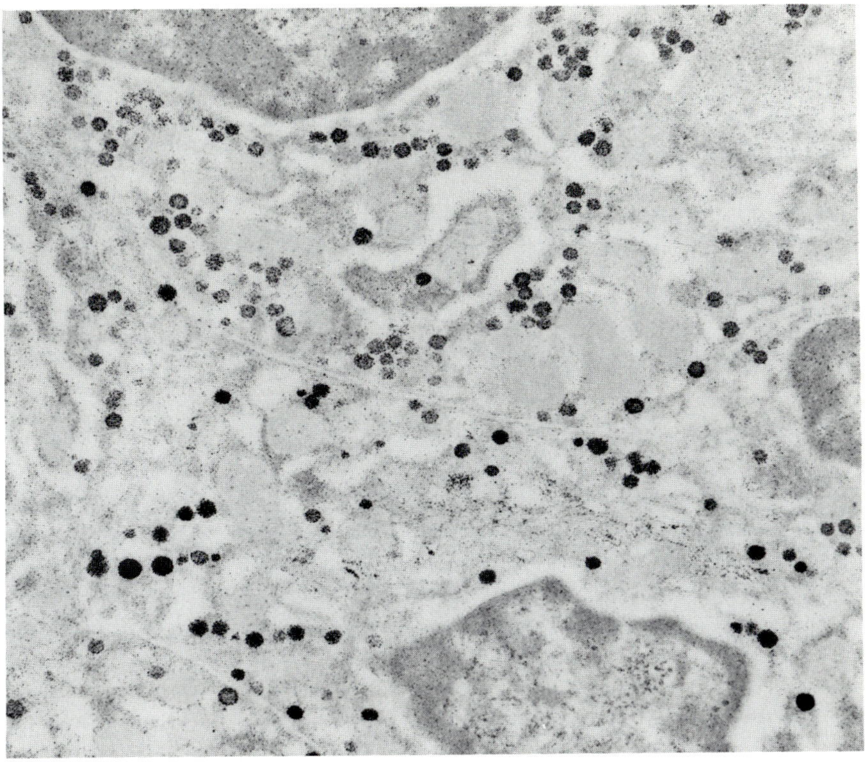

FIG. 38. Bronchial carcinoid tumor stained by Sevier-Munger's silver techique. Note intense reactivity of secretory granules, especially their halos. ×18,000.

For unknown reasons infants with idiopathic hypoglycemia show increased numbers of D_1 cells in their pancreas (Misugi *et al.*, 1970). In our experience well-differentiated D_1 cells also occur in some insulinomas and gastrinomas, although they fail to react with antigastrin or antiinsulin antibodies in immunofluorescence tests. Tumor D_1 cells are often difficult to distinguish from the undifferentiated tumor cells (stem cells?) showing small, poorly diagnostic granules which are usually found in atypical pancreatic tumors (mostly carcinomas) producing gastrin, insulin, or/and glucagon.

G. A CELLS

In 1948, Sutherland and De Duve extracted from the fundic mucosa of dog and rabbit a hyperglycemic-glycogenolytic factor with the same chemical and biological properties of pancreatic glucagon (Sutherland *et al.*, 1949). Later, a peptide material reacting with antiglucagon antisera was extracted by Unger *et al.* (1966) from the gastrointestinal mucosa. This glucagonlike immunoreactive (GLI) material proved to be chemically heterogeneous (Valverde *et al.*, 1968; Moody *et al.*, 1970). In dog and pig intestine, the main immunoreactive component was identified as a peptide of molecular weight much larger than that of pancreatic glucagon and devoid of any hyperglycemic-glycogenolytic activity (Valverde *et al.*, 1970; Murphy *et al.*, 1973). Using antiglucagon antisera on carbodiimide-fixed tissue, Polak *et al.* (1971a) selectively stained numerous immunofluorescent endocrine cells in the jejunum, ileum, and colon; morphological features of these cells were found to correspond with those of ultrastructurally identified L cells (Fig. 43), regular components of mammalian intestine (Solcia *et al.*, 1970b; Bussolati *et al.*, 1971; Capella *et al.*, 1972).

In dog fundic mucosa cells reacting with antiglucagon antibodies have been described (Polak *et al.*, 1971a) despite the absence of L cells, while cells reproducing several staining and ultrastructural aspects of pancreatic A cells have been found in the deep half of the glands (Solcia and Sampietro, 1965a; Forssmann, 1970; Solcia *et al.*, 1970b; Vassallo *et al.*, 1971a). It seems noteworthy that in human and pig stomach, from which very little GLI material has been extracted (Unger and Eisentraut, 1967; Murphy *et al.*, 1973), neither glucagon immunofluorescent cells (Polak *et al.*, 1971b; J. M. Polak, personal communication) nor A-like cells have been detected so far (Vassallo *et al.*, 1971b; Capella and Solcia, 1972). In rat stomach results are quite contradictory. While some investigators found very little GLI material (Unger and Eisentraut, 1967; Gutman *et al.*, 1973), Assan *et al.* (1969) extracted it in large amounts. A-like cells have been described by electron microscopists (Orci *et al.*, 1968; Forssmann *et al.*, 1969), while glucagon immunofluorescent cells have never been detected.

In addition, A cells have also been found in cat fundic mucosa (Forssmann *et al.*, 1968; Vassallo *et al.*, 1969). Fundic A cells are characterized by round granules with a dense, round, phosphotungstic acid–reactive core and a regular clear halo (Fig. 39). As in the pancreas, the halo is filled with Grimelius-reactive material (Fig. 40) to which the dark-field luminosity of the granules is probably related (Solcia *et al.*, 1974c).

The presence in the dog fundic mucosa of a peptide chemically and biologically indistinguishable from pancreatic glucagon (gastroglucagon), but clearly different from intestinal GLI material, has been confirmed recently (Unger, 1974). Thus, morphological find-

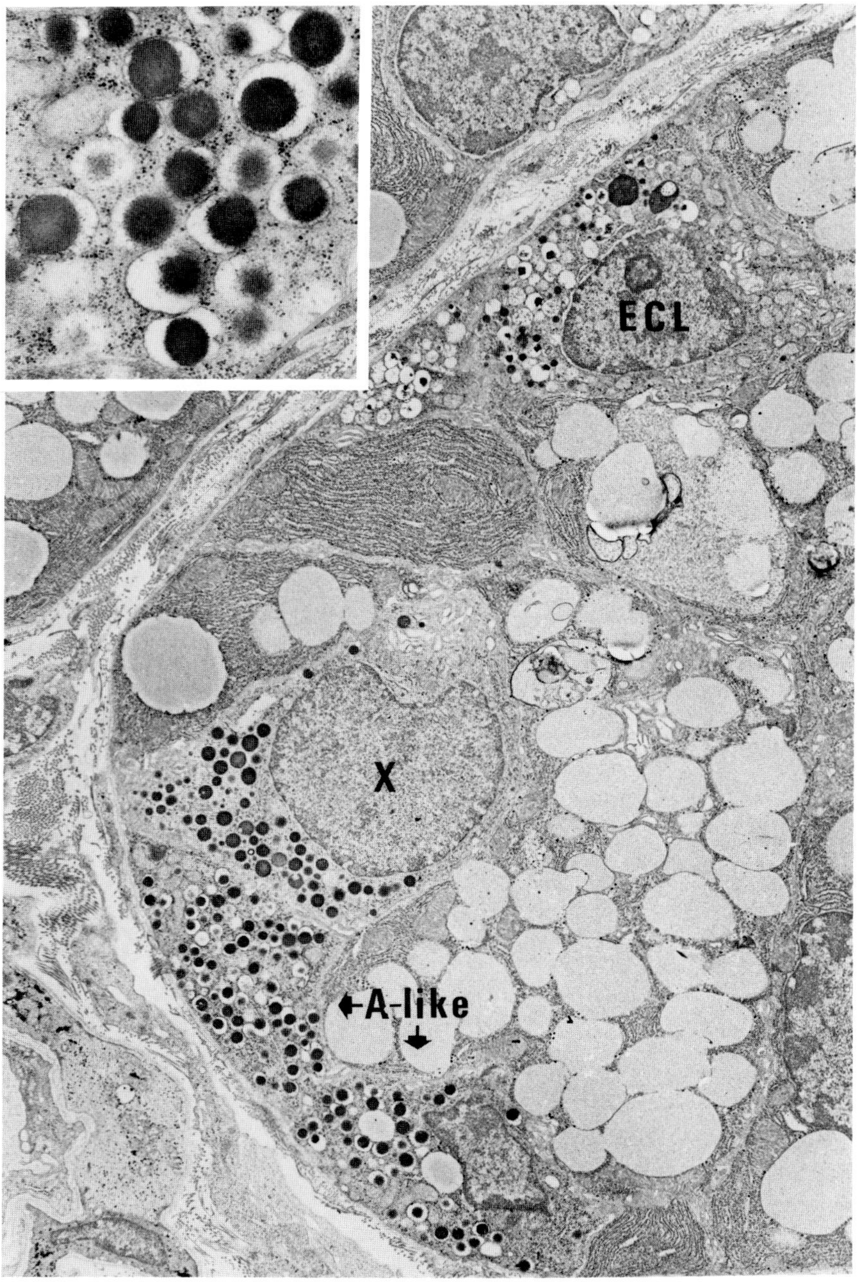

FIG. 39. Dog fundic gland showing two A cells, X cell, and an ECL cell. ×6300. Inset: Granules of an A cell in dog fundic mucosa; to be compared with granules of pancreatic A cell in Fig. 33. ×22,000.

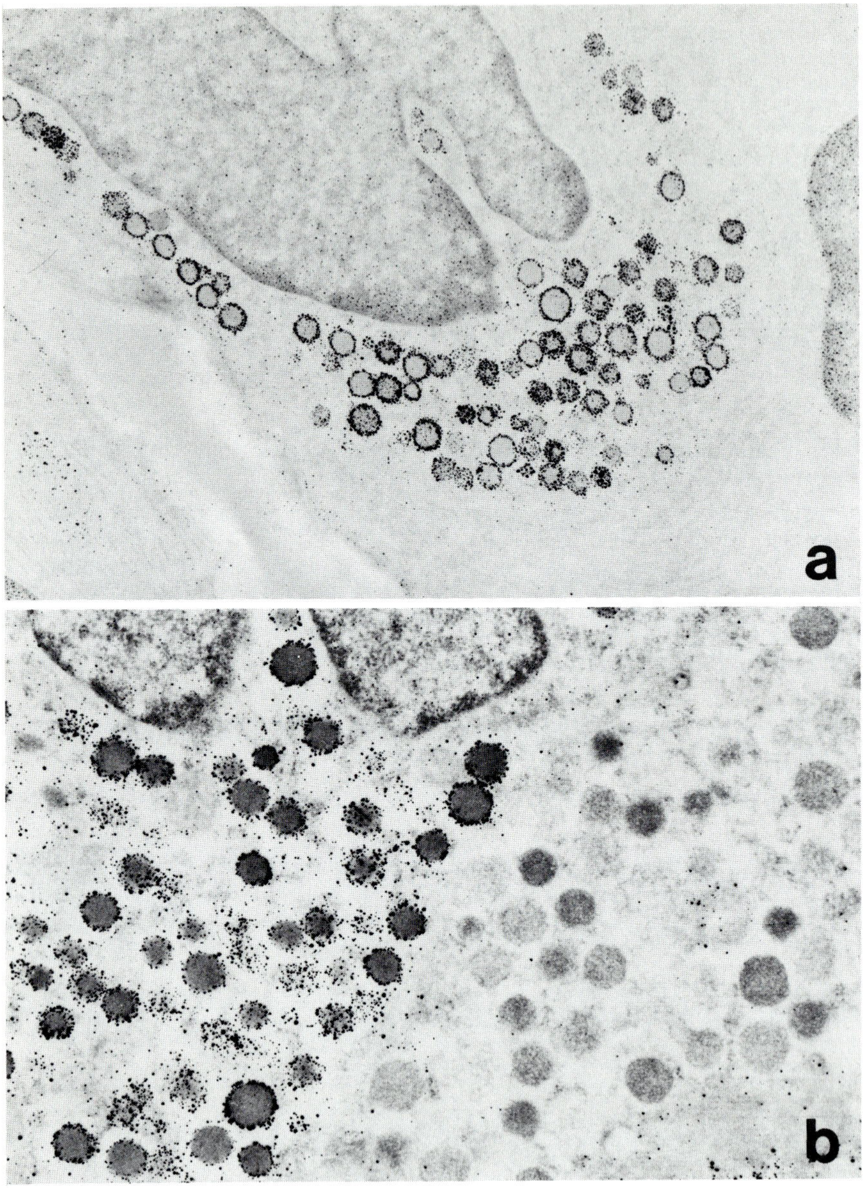

FIG. 40. (a) A cell of dog fundic mucosa showing granules with intense Grimelius reactivity restricted to the halo. ×13,000. (b) Dog pancreatic islet; Grimelius-reactive halo of A-cell granules (left) and Grimelius-negative granules of D cell (right). ×17,000.

ings are now well substantiated by biochemical researches. Gastroglucagon may well explain the elevated plasma glucagon levels found by Pek and co-workers (1974) in depancreatized dogs.

H. X Cells

In 1954, Davis found in the fundic mucosa of rabbit medium-sized cells with acidophilic granules intensely reacting with phosphotungstic hematoxylin and resembling only in part pancreatic A cells; they were called X cells and were reputed to be somewhat related to oxyntic cells. Later, Capella *et al.* (1969) recognized in light and electron microscopy investigations the unequivocal endocrine morphology of such cells. Davis' cells probably included EC cells, which are also stained with acid dyes. However, when Capella *et al.* (1969) stained the rabbit stomach first with Masson's argentaffin reaction and then with phosphotungstic hematoxylin, in addition to brown-stained EC cells, they observed numerous blue-stained round or ovoid *X cells.* Ultrastructurally, X cells show numerous moderately osmiophilic granules of round or fairly irregular shape (Figs. 39 and 41). Unlike in previous articles (Solcia *et al.*, 1970b), X cells are now clearly distinguished from D cells, whose granules are less osmiophilic and more round and uniform. X-cell granules also differ from those of D cells in being quite reactive with phosphotungstic acid (Fig. 41) and Grimelius' silver (Fig. 42), poorly reactive with Davenport's silver, and red-stained with the Mallory-Heidenhain azan technique. Moreover, D cells do not stain blue with phosphotungstic hematoxylin, react poorly with phosphotungstic acid at the ultrastructural level and are better represented in the pyloric than in the fundic mucosa. X cells are heavily stained with the hydrochloric acid–basic dye technique (red-violet with hydrochloric acid–toluidine blue) and lead–hematoxylin.

In several species X cells are also present in the pyloric mucosa—and perhaps also in the duodenal mucosa—although generally less numerous than in the fundic mucosa; in dog, and probably also in rat and mouse, they seem to be exclusive to the fundic mucosa. Most of the gastric cells that have been interpreted as L cells in previous studies (Solcia *et al.*, 1970b; Kobayashi *et al.*, 1971; Sasagawa *et al.*, 1973) are likely to be reinterpreted as X cells. At least in dog, the L cells of the jejunum, ileum, and colon differ clearly from X cells because of their larger granules which react strongly with the Gram-Weighert staining technique and show a Grimelius-negative core surrounded by a very thin peripheral rim of Grimelius-

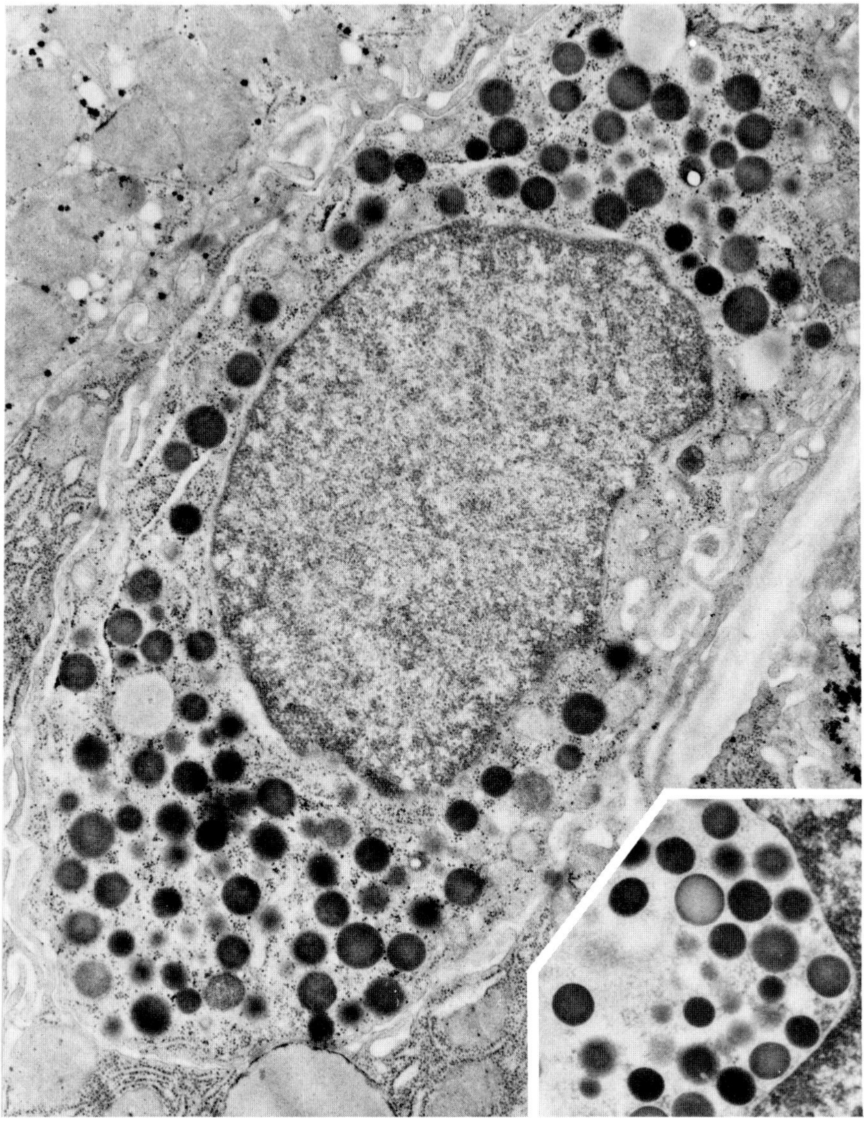

FIG. 41. X cell of dog fundic mucosa. ×14,000. Inset: Phosphotungstic acid–reactive granules of X cell of dog fundic mucosa. ×14,000.

reactive material. E cells observed by Thomas (1937) in the pancreas of opossum, and studied ultrastructurally by Munger *et al.* (1965), show secretory granules resembling those of gut X cells in at least two aspects, namely, their intense red staining with the Mallory-Heidenhain technique and their ultrastructural features. Based on

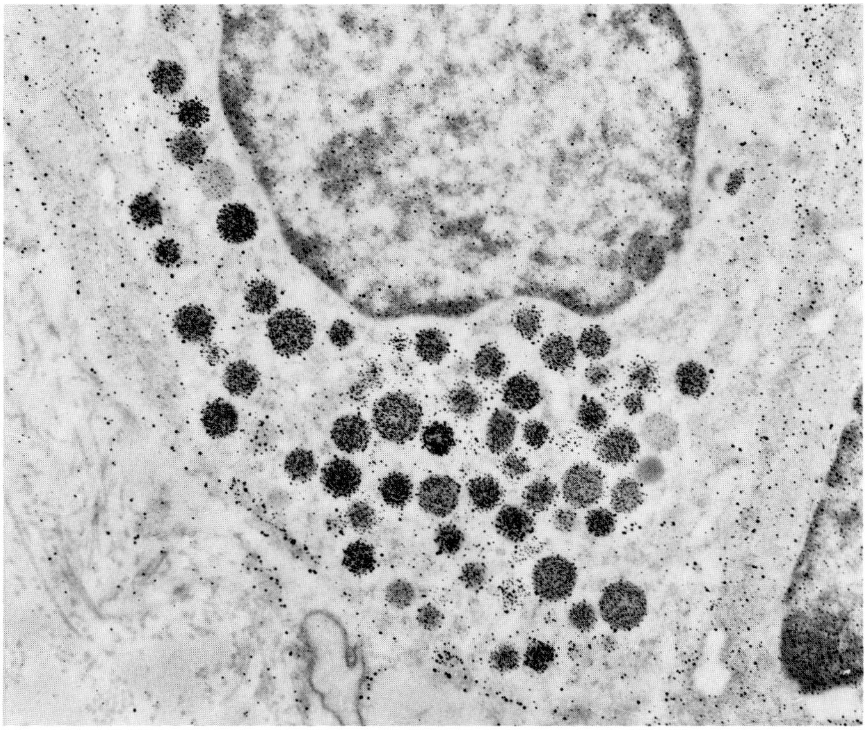

FIG. 42. Dog fundic mucosa (same specimen as in Fig. 31a). X cell granules with Grimelius-reactive core. ×14,000.

their staining and ultrastructural patterns, gut X cells, as well as E cells of opossum pancreas, must be distinguished from the cells of the dog and cat uncinate process described by Bencosme and Liepa (1955) under the name of X cells, and relabeled *F cells* by Munger *et al.* (1965). There is now growing evidence that F cells are also present in the pancreas of many other species, including mammals, as man, rat, mouse, pig, and other classes, such as birds, reptiles, and amphibia. A few F-like cells have been observed in the gastrointestinal mucosa of several species (Forssmann, 1970; Solcia *et al* , 1974b).

The function of X cells remains obscure. Their possible relationship with the type-3 endocrine cells found by Hage (1973a,b) in the human bronchial mucosa and in some bronchial carcinoids should be investigated. At least part of the round granule cells occurring in gastric argyrophilic carcinoids of *M. natalensis* (Capella *et al.*, 1973), must be interpreted as X cells.

III. Intestinal Endocrine Cells

Among intestinal endocrine cells are some of the above-reported cells occurring in the gastric mucosa, such as EC, G, D, D_1, or X cells, as well as some cells which are peculiar to the intestinal mucosa, such as S, I, K, and L cells (Capella *et al.*, 1969; Cavallero *et al.*, 1969, 1972; Forssmann, 1970; Kobayashi *et al.*, 1970; Solcia *et al.*, 1970b; Bussolati *et al.*, 1971; Ferreira, 1971; Polak *et al.*, 1971a,b,d; Frexinos *et al.*, 1973). In particular, S, I, and K cells seem to be restricted to the small intestine while, as reported above (see Section II,G), L cells are widely represented in both the small and large intestine (Fig. 43).

S cells (Figs. 44 and 45) are mostly found in the duodenum and upper jejunum; they are reputed to produce secretin (Capella *et al.*, 1969; Vassallo *et al.*, 1969; Solcia *et al.*, 1970b, 1972; Bussolati *et al.*, 1971; Polak *et al.*, 1971b,d). Their osmiophilic Grimelius-reactive granules, which are relatively small (about 200 nm), fairly irregular, and have a thin, clear halo, must be carefully distinguished from the granules of D_1 cells, which are slightly smaller (about 150 nm), less osmiophilic, more regular, and have a less evident halo filled with weakly osmiophilic material.

I cells (Fig. 46) differ from S cells in showing round, compact, uniform granules which at the ultrastructural level lack argyrophilia. A possible connection of these cells with the hormone cholecystokinin-pancreozymin remains to be investigated (Bussolati *et al.*, 1971; Capella *et al.*, 1972; Capella and Solcia, 1972).

K cells (Fig. 47) have been recognized only recently as an independent cell type (Solcia *et al.*, 1974c). Very likely, the so-called A-like cells found by Osaka *et al.* (1973b) in the human small intestine correspond to this type of cell. Granules of K cells are medium-sized (about 300 nm, with marked differences even in the same cell), haloed, and react strongly with both Grimelius' and Sevier-Munger's silver. Their wide distribution in the duodenum, jejunum, and ileum corresponds well with the distribution of "gastric inhibitory peptide" (GIP) (Brown and Dryburg, 1971; Bloom *et al.*, 1974), as well as of cells reacting with anti-GIP antibodies in immunofluorescence tests (Polak *et al.*, 1973). A correlative immunohistochemical staining and ultrastructural study we have in progress seems to confirm the presence of GIP in K cells. On restaining immunofluorescent sections, GIP cells proved to react strongly with Sevier-Munger silver technique (Solcia *et al.*, 1974e).

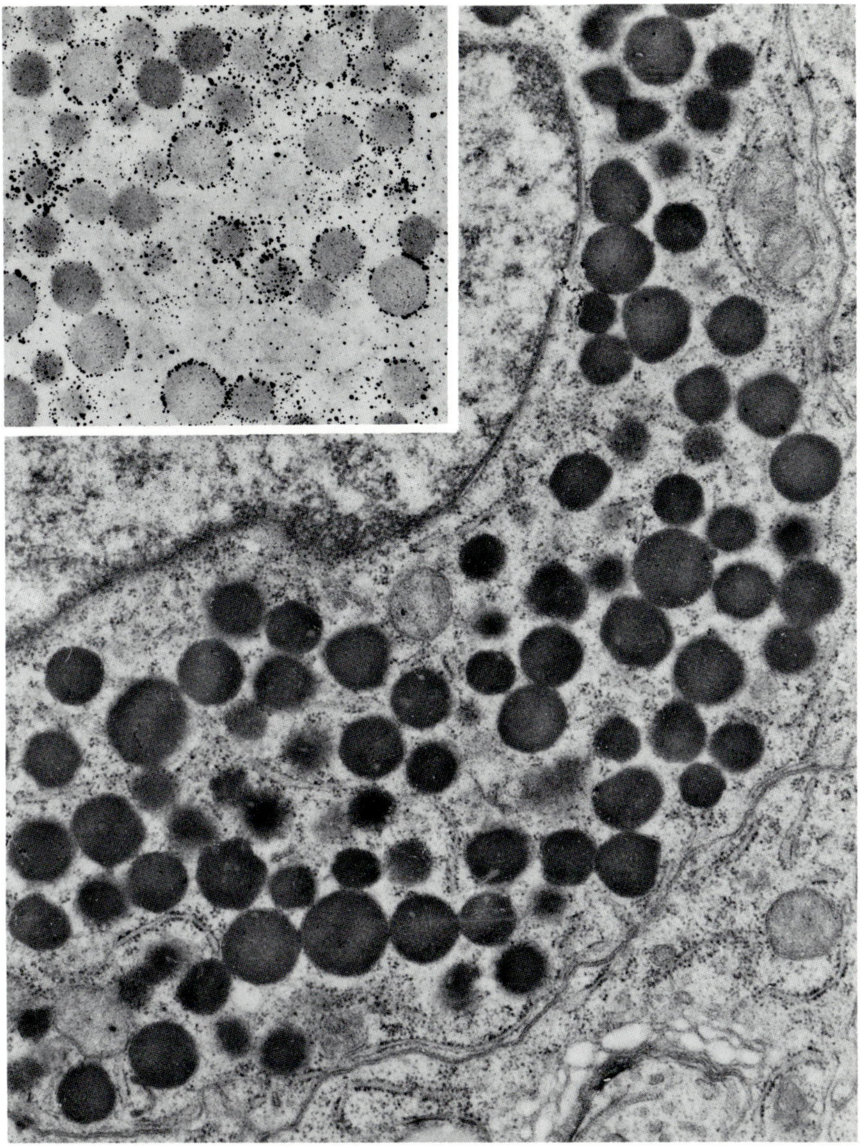

FIG. 43. L cell of dog colon, reputed to store enteroglucagon. ×19,000. Inset: Granules with a thin peripheral rim of Grimelius reactivity in another L cell of dog colon. ×19,000. Compare L-cell granules with those of pancreatic A cells (Figs. 33 and 40b), gastric A cells (Figs. 39 and 40a), and X cells (Figs. 41 and 42).

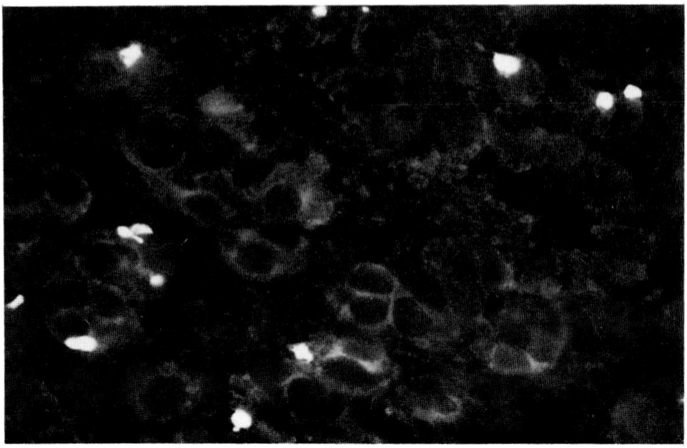

FIG. 44. Secretin cells of pig duodenum fixed in carbodiimide and stained with immunofluorescence using antibodies against synthetic porcine secretin. ×400. (From Solcia *et al.*, 1972.)

V. Concluding Remarks

Up to seven types of endocrinelike cells have been identified ultrastructurally in the gastric mucosa of some mammals, for instance, the dog. In man and some other mammals, six types of cells have been found. Staining patterns in light and electron microscopy,

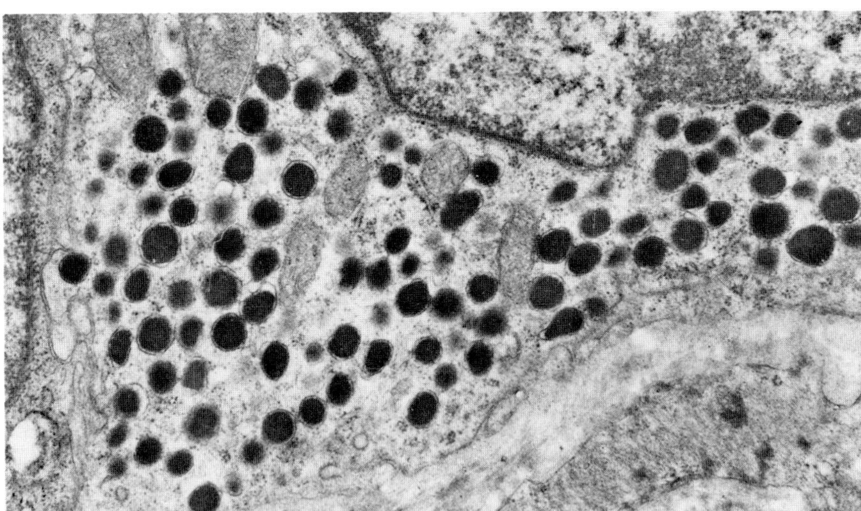

FIG. 45. S cell of dog duodenum. ×19,000.

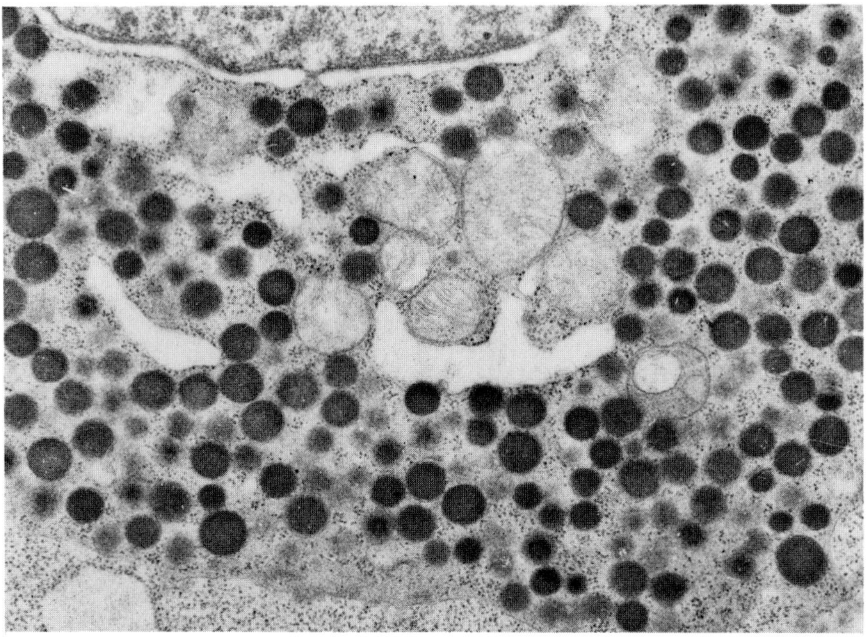

FIG. 46. I cell of dog duodenum. ×19,000.

amine histochemistry, and/or immunohistochemistry confirm the existence of this manifold population of endocrine cells. So far, biochemical and functional studies support the existence of four endocrine products, namely, 5HT, gastrin, histamine, and gastroglucagon. Thus, based on morphological evidence, more hormones than those presently reported are to be expected from the gastric mucosa.

Clarification of the morphological aspects of gut endocrine functions should contribute a new basis for overcoming some controversial problems; it should also open new fields of investigation to experimental morphologists and pathologists. Among topics deserving further study, the following may be noted.

1. The interaction between endocrine cells and stimuli coming from the lumen or from blood and the nervous system; both the hypothesis of a direct interaction of luminal stimuli with the endocrine cell (Solcia *et al.*, 1967, 1970b; Fujita and Kobayashi, 1973) and a more complex mechanism involving specialized receptors and local or long central reflexes (Emås *et al.*, 1967; Schofield *et al.*, 1967; Debas *et al.*, 1974) must be considered. Possible reception of

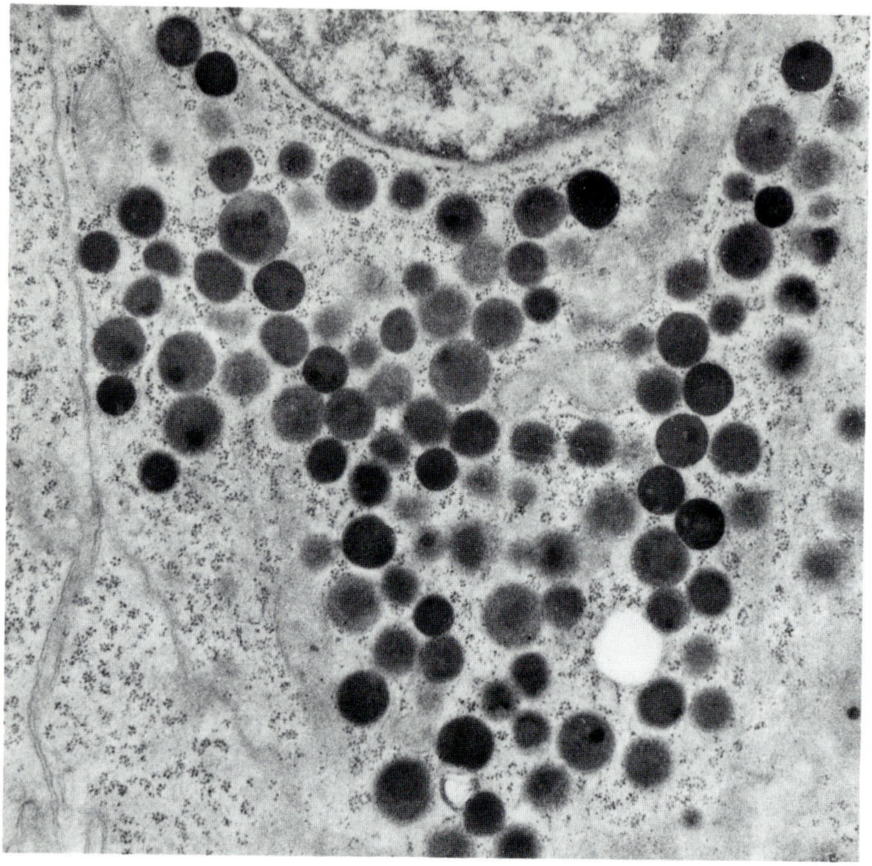

FIG. 47. K cell of dog duodenum. ×19,000.

luminal stimuli by the "fibrillovesicular" or "tuft" cells scattered in the gastrointestinal mucosa, especially in the surface epithelium (Silva, 1966; Hammond and La Deur, 1968; Isomäki, 1973), should also be investigated (Fig. 48).

2. The fate of secretory products released by the endocrine cells. Besides acting through vessels of the portal system and general circulation, hormonelike substances could act locally on neighboring secretory cells of the mucosa (as histamine released by ECL cells seems to do) independently of blood supply; or they could reach their target cells through local anastomoses between the circulation of various parts of the gut and pancreas. Direct vascular connections of a "portal" type have been demonstrated between pancreatic islets and exocrine parenchyma of the pancreas (Fujita, 1973; Fujita and Murakami, 1973).

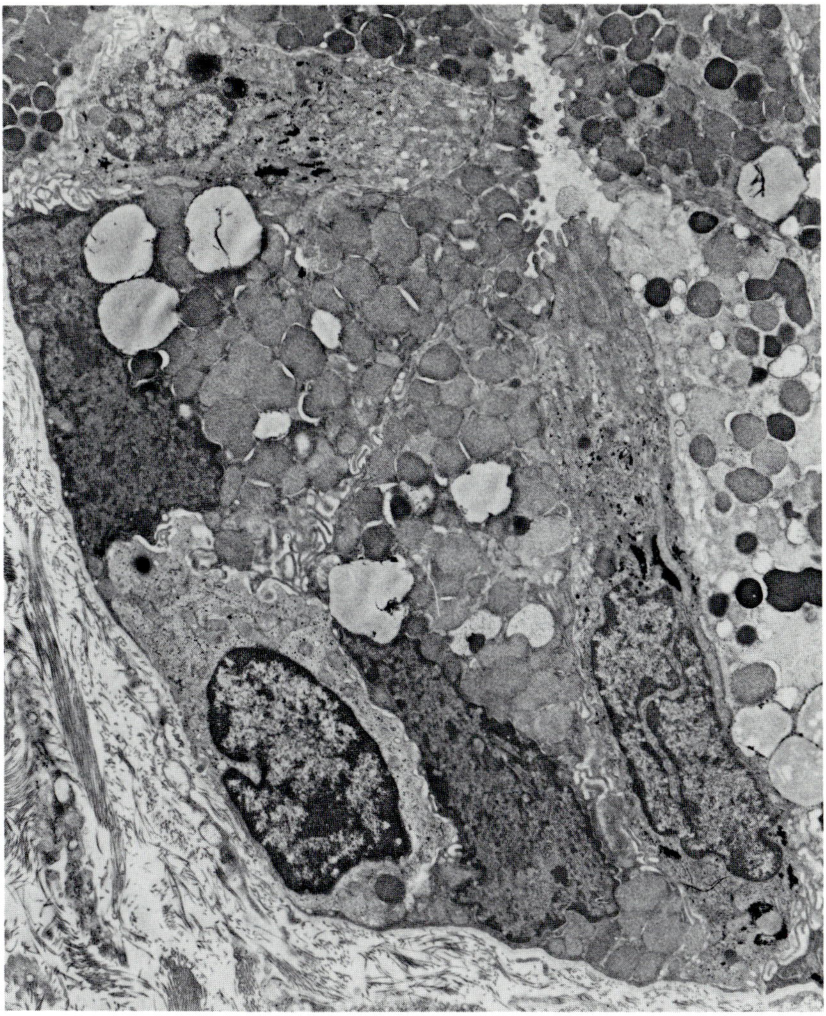

FIG. 48. Two tuft cells of dog pyloric mucosa. ×6000.

3. The interrelationship between gastrointestinal and pancreatic endocrine cells. Several types of endocrine cells found in the gastrointestinal mucosa for instance EC, A, D, and D_1 cells, also seem to occur in the pancreas. A common origin of such cells, either from the endoderm or from neural crest derivatives (Monesi, 1960; Andrew, 1963; Pearse and Polak, 1971; Pictet and Rutter, 1972; Le Douarin and Teillet, 1973; Pearse *et al.*, 1973), may explain their wide distribution in different tissues. Although other cells are restricted to

special areas of the gut (ECL, G, and S cells) or to the pancreas (B cells), a close functional relationship has been demonstrated among the endocrine functions of the gut and pancreas. In particular, the so-called enteroinsular axis, involving modulation of insulin release by intestinal hormones, is under extensive investigation by physiologists (Rabinowitch and Dupré, 1974). The morphological equivalents of these functional connections remain to be explored.

4. Based on the assumption that each cell type occurring in normal tissues may undergo abnormal proliferation or represent a potential pattern of stem cell differentiation in tumor growth, at least as many types of endocrine tumors as the above reported endocrine cells may be expected to arise in the gastric mucosa. Preliminary observations reported in this review seem to support this expectation in showing the heterogeneity of gut endocrine growths and their obvious or possible relationship with normal endocrine cells. However, much work remains to be done in this respect.

5. Given the close dependence of most digestive functions on gastrointestinal hormones, the behavior of gastrointestinal endocrine cells in some digestive diseases associated with severe functional derangements—such as peptic ulcer, malabsorption, pancreatitis, and the sequelae of surgical procedures affecting the alimentary canal—should be extensively investigated.

References

Andersson, S., and Elwin, C. E. (1972). *In* "Gastrointestinal Hormones" (L. Demling, ed.), p. 1. Thieme, Stuttgart.

Andersson, S., Nilsson, G., and Uvnäs, B. (1965). *Acta Physiol. Scand.* **65**, 191.

Andrew, A. (1963). *J. Embryol. Exp. Morphol.* **11**, 307.

Assan, R., Tchobroutsky, G., and Rosselin, G. (1969). *Pathol. Biol.* **17**, 747.

Aures, D., Håkanson, R., and Shauer, A. (1968). *Eur. J. Pharmacol.* **3**, 217.

Aures, D., Johnson, L. R., and Way, L. W. (1970). *Amer. J. Physiol.* **219**, 214.

Baugh, C. M., Bravo, J. L., Barcena, J., and Dragstedt, L. R. (1958). *Arch. Surg. (Chicago)* **76**, 441.

Becker, V. (1969). *In* "Non-Insulin-Producing Tumors of the Pancreas" (L. Demling and R. Ottenjann, eds.), p. 49. Thieme, Stuttgart.

Beltrami, C. A., Fabris, G., Marzola, A., Nenci, I., Lanza, G., and Vezzadini, P. (1975). *Histochem J.* **7**, 95.

Bencosme, S. A., and Lechago, J. (1973). *J. Clin. Endocrinol. Metab.* **26**, 427.

Bencosme, S. A., and Liepa, E. (1955). *Endocrinology* **57**, 588.

Bensch, K. G., Gordon, G. B. and Miller, L. R. (1965). *Cancer* **18**, 592.

Berson, S. A., and Jalow, R. S. (1971). *Gastroenterology* **60**, 215.

Block, M. A., Kelly, A. R., and Horn, C. R. (1969). *Arch. Surg. (Chicago)* **98**, 734.

Bloom, S. R., Polak, J. M., and Pearse, A. G. E. (1974). *In* "Endocrinology 1973" (in press).

Brodie, D. A. (1966). *Amer. J. Dig. Dis.* **11**, 231.

Bromé, A., Fyrö, B., and Olbe, L. (1968). *Acta Physiol. Scand.* **74**, 331.

Brown, J. C., and Dryburgh, J. R. (1971). *Can. J. Biochem.* **49**, 867.

Brown, J. C., Cook, M. A., and Dryburgh, J. R. (1973). *Can. J. Biochem.* **51**, 533.

Bülbring, E., and Crema, A. (1959). *J. Physiol. (London)* **146**, 18.

Bussolati, G., and Canese, M. G. (1972). *Histochemie* **29**, 198.

Bussolati, G., and Monga, G. (1973). *Virchows Arch., B* **13**, 55.

Bussolati, G., and Pearse, A. G. E. (1970). *Histochemie* **21**, 1.

Bussolati, G., Capella, C., Solcia, E., Vassallo, G., and Vezzadini, P. (1971). *Histochemie* **26**, 218.

Bussolati, G., Polak, J. M., and Coulling I. (1972). *In* "Endocrinology 1971," (S. Taylor, ed.), p. 164. Heinemann, London.

Campbell, A. C. P., Gowenlock, A. H., Platt, D. S., and Snow, P. J. D. (1963). *Gut* **4**, 61.

Canese, M. G., and Bussolati, G. (1974). *Rend. Gastroenterol.* **6**, 12.

Capella, C., and Solcia, E. (1972). *Arch. Histol. Jap.* **35**, 1.

Capella, C., Solcia, E., and Vassallo, G. (1969). *Arch. Histol. Jap.* **30**, 479.

Capella, C., Vassallo, G., and Solcia, E. (1971). *Z. Zellforsch. Mikrosk. Anat.* **118**, 68.

Capella, C., Solcia, E., and Vassallo, G. (1972). *In* "Endocrinology 1971" (S. Taylor, ed.), p. 282. Heinemann, London.

Capella, C., Solcia, E., and Snell, K. C. (1973). *J. Nat. Cancer Inst.* **50**, 1471.

Carlsson, A. (1966). *In* "Handbuch der experimentellen Pharmalkologie" (V. Erspamer, ed.), Vol. 19, p. 529. Springer-Verlag, Berlin and New York.

Carvalheira, A. F., Welsch, U., and Pearse, A. G. E. (1968). *Histochemie* **14**, 33.

Cavallero, C., Solcia, E., Vassallo, G., and Capella, C. (1969). *Rend. R. R. Gastroenterol.* **1**, 51.

Cavallero, C., Vassallo, G., Capella, C., and Solcia, E. (1972). *In* "Advances in Gastrointestinal Endoscopy," *Proc. 2nd World Congr. Gastroint. Endosc.*, p. 449. Piccin, Padova.

Cegrell, L., Falck, B., and Rosengren, E. (1968). *Acta Physiol. Scand., Suppl.* **314**, 8.

Ciaccio, C. (1907). *Arch. Ital. Anat. Embriol.* **6**, 482.

Clara, M. (1957). *Acta Neuroveg.* **16**, 294.

Cordier, R., and Lison, L. (1930). *Bull. Histol. Appl.* **7**, 140.

Cowley, D. J., Dymock, I. W., Boyes, B. E., Wilson, R Y., Stagg, B. H., Lewin, M. R., Polak, J. M., and Pearse, A. G. E. (1973). *Gut* **14**, 25.

Creutzfeldt, W., Arnold, R., Creutzfeldt, C., Feurle, G., and Ketterer, H. (1971). *Eur. J. Clin. Invest.* **1**, 461.

Davis, J. C. (1954). *J. Pathol. Bacteriol.* **67**, 237.

Dawson, A. B. (1948). *Anat. Rec.* **100**, 319.

Debas, H. T., Walsh, J. H., and Grossman, M. I. (1974). *Abstr., Annu. Meet. Amer. Gastroenterol. Ass.* p. A-213.

Deconinck, J. F., Van Assche, F. A., Potvliege, P. R., and Gepts, W. (1972). *Diabetologia* **8**, 326.

Dotevall, G., and Walan, A. (1970). *In* "Origin, Chemistry, Physiology and Pathophysiology of the Gastrointestinal Hormones" (W. Creutzfeldt, ed.), p. 227. Schattauer, Stuttgart.

Drapanas, F., McDonald, S. C., and Stewart, J. I. (1962). *Ann. Surg.* **156**, 528.

Emås, S., and Fyrö, B. (1968). *Acta Physiol. Scand.* **74**, 359.

Emås, S., Fyrö, B., and Uvnäs, B. (1967). *In* "Gastrin, Secretion, Mechanism and Control" (T. K. Schnitka, J. A. L. Gilbert, and R. C. Harrison, eds.), p. 111. Pergamon, Oxford.

Emås, S., Borg, I., and Fyrö, B. (1971). *Scand. J. Gastroenterol.* **6,** 39.

Enerback, L. (1966). *Acta Pathol. Microbiol. Scand.* **67,** 365.

Ericson, L. E., Ekholm, R., and Lager, I. (1972). *In* "Endocrinology 1971" (S. Taylor, ed.), p. 299. Heinemann, London.

Erspamer, V. (1939). *Z. Anat. Entwicklungsgesch.* **109,** 586.

Erspamer, V., and Asero, B. (1952). *Nature (London)* **169,** 800.

Falck, B., Larson, B., Mecklenburg, C. V., Rosenberg, E., and Svenaeus, K. (1964). *Acta Physiol. Scand.* **62,** 491.

Ferreira, M. N. (1971). *Amer. J. Anat.* **141,** 315.

Feyrter, F. (1953). "Über die peripheren endokrinen (parakrinen) Drüsen des Menshen." W. Maudrich, Vienna.

Forssmann, W. G. (1970). *In* "Origin, Chemistry, Physiology and Pathophysiology of the Gastrointestinal Hormones" (W. Creutzfeldt, ed.), p. 31. Schattauer, Stuttgart.

Forssmann, W. G., and Orci, L. (1969). *Z. Zellforsch. Mikrosk. Anat.* **101,** 419.

Forssmann, W. G., Orci, L., and Rouiller, C. (1968). *Symp. Deut. Ges. Endokrinol.* **14,** 251.

Forssmann, W. G., Orci, L., Pictet, R., Renold, A. E., and Rouiller, C. (1969). *J. Cell Biol.* **40,** 692.

Frexinos, J., Vigoni, F., and Ribet, A. (1972). *Biol. Gastro-enterol.* **5,** 21.

Frexinos, J., Balas, D., and Ribet, A. (1973). *Biol. Gastro-enterol.* **6,** 105.

Fujita, T. (1973). *Arch. Histol. Jap.* **35,** 161.

Fujita, T., and Kobayashi, S. (1971). *Z. Zellforsch. Mikrosk. Anat.* **116,** 52.

Fujita, T., and Kobayshi, S. (1973). *In* "Gastro-entero-pancreatic Endocrine System. A Cell-biological Approach" (T. Fujita, ed.), p. 1. Igaku Shoin, Tokyo.

Fujita, T., and Murakami, T. (1973). *Arch. Histol. Jap.* **35,** 255.

Ganguli, P. C., Polak, J. M., Pearse, A. G. E., Elder, J. B., and Hegarty, M. (1973). *Gut* **14,** 822.

Geyer, G. (1968a). *Acta Histochem.* **31,** 141.

Geyer, G. (1968b). *Acta Histochem.* **31,** 219.

Granérus, G., Lindell, S. E., Waldenström, J., Westling, H., and White, T. (1966). *Lancet* **1,** 1267.

Gregory, R. A. (1962). "Secretory Mechanisms of the Gastrointestinal Tract." Arnold, London.

Greider, M. H., and McGuigan, J. E. (1971). *Diabetes* **20,** 389.

Greider, M. H., Bencosme, S. A., and Lechago, J. (1970). *Lab. Invest.* **22,** 344.

Greider, M. H., Steinberg, V., and McGuigan, J. E. (1972). *Gastroenterology* **63,** 572.

Greider, M. H., Rosai, J., and McGuigan, J. E. (1974). *Cancer* **33,** 1423.

Grimelius, L. (1968). *Acta Soc. Med. Upsal.* **73,** 243.

Grützner, P., and Menzel, H. (1879). *Arch. Gesamte Physiol. Menschen Tiere* **20,** 395.

Gutman, R. A., Fink, G., Voyles, N., Selawry, H., Penhos, J. C., Lepp, A., and Recant, L. (1973). *J. Clin. Invest.* **52,** 1165.

Hage, E. (1973a). *Z. Zellforsch. Mikrosk. Anat.* **141,** 401.

Hage, E. (1973b). *Virchows Arch., A* **361,** 121.

Håkanson, R., and Owman, C. (1967). *Life Sci.* **6,** 759.

Håkanson, R., and Owman, C. (1969). *Experientia* **25,** 625.

Håkanson, R., Owman, C., Sjöberg, N.-O., and Sporrong, B. (1970). *Histochemie* **21,** 189.

Håkanson, R., Owman, C., Sporrong, B., and Sundler, F. (1971). *Z. Zellforsch. Mikrosk. Anat.* **122,** 460.

Håkanson, R., Lindstrand, K., Owman, C., and Sundler, F. (1972). *Biochem. Pharmacol.* **21,** 1703.

Håkanson, R., Larsson, L.-I., Nishizaki, H., Owman, C., and Sundler, F. (1973). *Histochemie* **34,** 1.
Hallenbeck, G. A., Code, F. C., and McIllrath, D. C. (1963). *Gastroenterology* **44,** 627.
Hamberger, C. A., Ritzen, M., and Wersäll, J. (1966). *J. Pharmacol. Exp. Ther.* **152,** 197.
Hammond, J. B., and La Deur, L. (1968). *Anat. Rec.* **161,** 393.
Hamperl, H. (1952). *Virchows Arch. Pathol. Anat. Physiol.* **321,** 482.
Haverback, B. J., and Davidson, J. D. (1958). *Gastroenterology* **35,** 570.
Heidenhain, R. (1870). *Arch. Mikrosk. Anat.* **6,** 368.
Hosoda, S., Nakamura, W., Snell, K. C., and Stewart, H. L. (1970). *Science* **170,** 454.
Isomäki, A. M. (1973). *Acta Pathol. Microbiol. Scand., Sec. A., Suppl* **240,** 1.
Jalow, R., and Berson, S. A. (1972). *Biochem. Biophys. Res. Commun.* **48,** 391.
Kaltiala, E. H. (1971). *Acta Chir. Scand.* **137,** 563.
Kobayashi, S., and Fujita, T. (1973). *In* "Gastro-entero-pancreatic Endocrine System. A Cell-biological Approach" (T. Fujita, ed.), p. 49. Igaku Shoin, Tokyo.
Kobayashi, S., Fujita, T., and Sasagawa, T. (1970). *Arch. Histol. Jap.* **31,** 477.
Kobayashi, S., Fujita, T., and Sasagawa, T. (1971). *Arch. Histol. Jap.* **32,** 429.
Kow, B. K., and Weissman, A. (1966). *J. Pharmacol. Exp. Ther.* **154,** 499.
Kull, H. (1913). *Arch. Mikrosk. Anat.* **81,** 185.
Kultschitzky, N. (1897). *Arch. Mikrosk. Anat.* **49,** 7.
Larsson, L. I., Ljungberg, O., Sundler, F., Håkanson, R., Svensson, S. O., Rehfeld, J., Stadil, F., and Holst, J. (1973a). *Virchows Arch., A* **360,** 305.
Larsson, L. I., Sundler, F., Håkanson, R., Reheld, J. F., and Stadil, F. (1973b). *Histochemie* **37,** 81.
Lattes, R., and Grossi, C. (1956). *Cancer* **9,** 698.
Law, D. H., Liddle, G. W., Scott, H. W., and Tauber, S. D. (1965). *N. Engl. J. Med.* **273,** 292.
Lechago, J., and Bencosme, S. A. (1973). *Int. Rev. Exp. Pathol.* **12,** 119.
Le Douarin, N. M., and Teillet, M.-A. (1973). *J. Embryol. Morphol.* **30,** 31.
Lefranc, G., and Pradal, G. (1971). *Bull. Ass. Anat., Paris* **56,** No. 150, 1.
Like, A. A. (1967). *Lab. Invest.* **16,** 937.
Like, A. A., and Orci, L. (1972). *Diabetes* **21,** 511.
Lillie, R. D., Pizzolato, P., Vacca, L. L., Catalano, R. A., and Donaldson, P. T. (1973). *J. Histochem. Cytochem.* **21,** 455.
Lomsky, R., Langr, F., and Vortel, V. (1969). *Nature (London)* **223,** 618.
Lotstra, F., Van Der Loo, W., and Gepts, W. (1974). *Diabetologia* **10,** 291.
McGuigan, J. E. (1968). *Gastroenterology* **33,** 315.
McGuigan, J. E., and Greider, M. H. (1971). *Gastroenterology* **60,** 223.
Masson, P. (1914). *C. R. Acad. Sci.* **158,** 59.
Misugi, K., Misugi, N., Sotos, J., and Smith, B. (1970). *Arch. Pathol.* **89,** 208.
Mitschke, H., and Becker, M. (1973). *Virhows Arch., A* **358,** 81.
Monesi, V. (1960). *Acta Anat.* **41,** 97.
Monga, G., Bonfanti, S., and Bussolati, G. (1974). *Histochem. J.* **6,** 559.
Moody, A. J., Markussen, J., Schaich Fries, A., Steenstrup, C., and Sundby, F. (1970). *Diabetologia* **6,** 135.
Munger, B. L. (1972). *In* "Handbook of Physiology" (Amer. Physiol. Soc., J. Field, ed.), Sect. 7, Vol. I, p. 305. Williams & Wilkins, Baltimore, Maryland.
Munger, B. L., Caramia, F., and Lacy, P. E. (1965). *Z. Zellforsch. Mikrosk. Anat.* **67,** 776.
Murphy, R. F., Buchanan, K. D., and Elmore, D. T. (1973). *Biochim. Biophys. Acta* **303,** 118.

Mutt, V., and Jorpes, J. E. (1967). *Biochem. Biophys. Res. Commun.* **26**, 392.
Nicolas, A. (1891). *Int. Monatsschr. Anat. Physiol.* **8**, 1.
Nilsson, G., Yalow, R. S., and Berson, S. A. (1970). *In* "Frontiers in Gastrointestinal Hormone Research" Nobel Symp. XVI (S. Andersson, ed.). Almqvist & Wiksell, Stockholm (in press).
Noll, W. W., and Levine, R. J. (1970). *Biochem. Pharmacol.* **19**, 1043.
Nussbaum, M. (1879). *Arch. Mikrosk. Anat.* **16**, 532.
Oates, J. A., and Sjoerdsma, A. (1962). *Amer. J. Med.* **32**, 333.
Oates, J. A., Melmon, K., Sjoerdsma, A., Gillespie, L., and Mason, A. S. (1964). *Lancet* **1**, 514.
Oberhelman, H. A., Nelsen, T. S., Johnson, A. N., and Dragstedt, L. R. (1961). *Ann. Surg.* **153**, 214.
Omole, A. A., Amure, B. O., Ezimokhai, M., and Ayandipo, A. (1972). *Nature (London)* **237**, 183.
O'Neal, L. W., Kipnis, D. M., Luse, S. A., Lacy, P. E., and Jarrett, L. (1968). *Cancer* **21**, 1219.
Orci, L., Pictet, R., Forssmann, W. G., Renold, A. E., and Rouiller, C. (1968). *Diabetologia* **4**, 56.
Osaka, M., Sasagawa, T., and Fujita, T. (1973a). *In* "Gastro-entero-pancreatic Endocrine System. A Cell-biologic Approach" (T. Fujita, ed.), p. 59. Igaku Shoin, Tokyo.
Osaka, M., Sasagawa, T., and Fujita, T. (1973b). *Arch. Histol. Jap.* **35**, 235.
Pearse, A. G. E. (1969). *J. Histochem. Cytochem.* **17**, 303.
Pearse, A. G. E., and Bussolati, G. (1972). *Virchows Arch., A* **433**, 99.
Pearse, A. G. E., and Polak, J. M. (1971). *Gut* **12**, 783.
Pearse, A. G. E., Coulling, I., Weavers, B., and Friesen, S. (1970). *Gut* **11**, 64.
Pearse, A. G. E., Polak, J. M., and Heath, C. M. (1973). *Diabetologia* **9**, 120.
Pearse, A. G. E., Polak, J. M., Bloom, S. R., Adams, C., Dryburgh, J. R., and Brown, J. C. (1974). *Virchows Arch., B* **16**, 111.
Pek, S., Vranic, M., and Kawamori, R. (1974). Proceedings of Endocrine Society and ADA Meetings. Atlanta, June 1974.
Pestana, C., Beahrs, O. H., and Woolner, L. B. (1963). *Mayo Clin. Proc.* **38**, 453.
Pictet, R., and Rutter, W. J. (1972). *In* "Handbook of Physiology" (Amer. Physiol. Soc., Wash., J. Field, ed.), Sect. 7, Vol. I, p. 25. Williams & Wilkins, Baltimore, Maryland.
Polak, J. M., Bloom, S., Coulling, I., and Pearse, A. G. E. (1971a). *Gut* **12**, 31.
Polak, J. M., Bloom, S., Coulling, I., and Pearse, A. G. E. (1971b). *Gut* **12**, 605.
Polak, J. M., Bussolati, G., and Pearse, A. G. E. (1971c). *Virchows Arch., B* **9**, 187.
Polak, J. M., Coulling, I., Bloom, S., and Pearse, A. G. E. (1971d). *Scand. J. Gastroenterol.* **6**, 739.
Polak, J. M., Stagg, B., and Pearse, A. G. E. (1972). *Gut* **13**, 501.
Polak, J. M., Bloom, S., Kuzio, M., Brown, J. C., and Pearse, A. G. E. (1973). *Gut* **14**, 284.
Rabinowitch, A., and Dupré, J. (1974). *Endocrinology* **94**, 1139.
Rehfeld, J. F., and Iversen, J. (1973). *Abstr. Congr. Int. Diabetes Fed. 8th*, p. 52. Excerpta Medica, Amsterdam.
Reichle, F. A., Brigham, M. P., and Rosemond, G. P. (1967). *J. Amer. Med. Ass.* **199**, 914.
Royston, C. M. S., Brew, D. St. J., Garnham, J. R., Stagg, B. H., and Polak, J. (1972). *Gut* **13**, 638.

Rubin, W. (1969). *Gastroenterology* **57**, 641.
Rubin, W. (1972a). *J. Cell Biol.* **52**, 219.
Rubin, W. (1972b). *Gastroenterology* **62**, 784.
Rubin, W., Gershon, M. D., and Ross, L. L. (1971). *J. Cell Biol.* **50**, 399.
Sandler, M., and Snow, P. D. J. (1958). *Lancet* **1**, 137.
Sandler, M. S., Karim, M. M., and Willimas, E. D. (1968). *Lancet* **2**, 1053.
Sasagawa, T., Kobayashi, S., and Fujita, I. (1970). *Arch. Histol. Jap.* **32**, 275.
Sasagawa, T., Kobayashi, S., and Fujita, I. (1973). *In* "Gastro-entero-pancreatic Endocrine System. A Cell-biological Approach" (T. Fujita, ed.), p. 17. Igaku Shoin, Tokyo.
Schmidt, J. E. (1905). *Arch. Mikrosk. Anat.* **66**, 12.
Schofield, B., Redford, M., Grabham, A. H., and Nuaimi, K. (1967). *In* "Gastrin: Secretion, Mechanism and Control" (T. K. Schnitka, J. A. L. Gilbert, and R. C. Harrison, eds.), p. 91. Pergamon, Oxford.
Shibasaki, S., and Ito, T. (1969). *Arch. Histol. Jap.* **31**, 119.
Silva, D. G. (1966). *J. Ultrastruct. Res.* **16**, 693.
Sircus, W. (1969). *Gut* **10**, 506.
Solcia, E., and Buffa, R. (1974). *J. Histochem. Cytochem.* **22**, 127.
Solcia, E., and Sampietro, R. (1965a). *Z. Zellforsch. Mikrosk. Anat.* **68**, 689.
Solcia, E., and Sampietro, R. (1965b). *Riv. Istochim. Norm. Patol.* **11**, 265.
Solcia, E., and Sampietro, R. (1966). *Riv. Istochim. Norm. Patol.* **12**, 122.
Solcia, E., Sampietro, R., and Vassallo, G. (1966). *J. Histochem. Cytochem.* **14**, 691.
Solcia, E., Vassallo, G., and Sampietro, R. (1967). *Z. Zellforsch. Mikrosk. Anat.* **81**, 474.
Solcia, E., Vassallo, G., and Capella, C. (1968). *Stain Technol.* **43**, 257.
Solcia, E., Sampietro, R., and Capella, C. (1969a). *Histochemie* **17**, 273.
Solcia, E., Capella, C., and Vassallo, G. (1969b). *Histochemie* **20**, 116.
Solcia, E., Vassallo, G., and Capella, C. (1969c). *Gut* **10**, 379.
Solcia, E., Capella, C., and Vassallo, G. (1970a). *Rend. R. Gastroenterol.* **2**, 147.
Solcia, E., Vassallo, G., and Capella, C. (1970b). *In* "Origin, Chemistry Physiology and Pathophysiology of the Gastrointestinal Hormones" (W. Creutzfeldt, ed.), p. 3. Schattauer, Stuttgart.
Solcia, E., Capella, C., Vezzadini, P., Barbara, L., and Bussolati, G. (1972). *Experientia* **28**, 549.
Solcia, E., Pearse, A. G. E., Grübe, D., Kobayashi, S., Bussolati, G., Creutzfeldt, W., and Gepts, W. (1973). *Rend. Gastroenterol.* **5**, 13.
Solcia, E., Capella, C., and Buffa, R. (1974a). *In* "Peptic Ulcer Disease," Proceedings of Postgraduate Course, San Francisco, p. 23-1. Amer. Gastroenterol. Ass.
Solcia, E., Capella, C., Buffa, R., and Bettini, R. (1974b). *In* "Endocrinology 1973" (in press).
Solcia, E., Capella, C., Buffa, R., Bettini, R., and Vassallo, G. (1974c). *Riv. Istochim. Norm. Patol.* (in press).
Solcia, E., Capella, C., Vassallo, G., Buffa, R., Passaro, E., and Basso, N. (1974d). *Gastroenterology* (submitted for publication).
Solcia, E., Polak, J. M., Buffa, R., Capella, C., and Pearse, A. G. E. (1974e). *In* "Gastrointestinal Hormones" (J. C. Thompson, ed.). Univ. of Texas Press, Austin (in press).
Stern, D. H., and Walsh, J. H. (1973). *Gastroenterology* **64**, 363.
Sutherland, E. W., and De Duve, C. (1948). *J. Biol. Chem.* **175**, 663.
Sutherland, E. W., Cori, C. F., Haynes, R., and Olsen, N. S. (1949). *J. Biol. Chem.* **180**, 825.
Thomas, T. B. (1937). *Amer. J. Anat.* **62**, 31.

Thompson, J. C. (1966). *In* "Gastrin" (M. I. Grossman, ed.), p. 193. Univ. of California Press, Berkeley.

Tjälve, H. (1971). *Experientia* **27**, 307.

Toldt, C. (1880). *Sitzungsber. Akad. Wiss. Wien, Math.-Naturwiss. KI., Abt. 3* **82,** 57.

Unger, R. H. (1974). *In* "Gastrointestinal Hormones" (J. C. Thompson, ed.). Univ. of Texas Press, Austin (in press).

Unger, R. H., and Eisentraut, A. M. (1967). *In* "Hormones in Blood" (C. H. Gray and A. L. Bacharach, eds.), Vol. 1, p. 83. Academic Press, New York.

Valverde, I., Ringopoulou, D., Exton, J., Ohneda, A., Eisentraut, A. M., and Unger, R. H. (1968). *Amer. J. Med. Sci.* **255,** 415.

Valverde, I., Ringopoulou, D., Marco, J., Faloona, G. R., and Unger, R. H. (1970). *Diabetes* **19,** 135.

Vassallo, G., Solcia, E., and Capella, C. (1969). *Z. Zellforsch. Mikrosk. Anat.* **98,** 333.

Vassallo, G., Capella, C., and Solcia, E. (1971a). *Stain Technol.* **46,** 7.

Vassallo, G., Capella, C., and Solcia, E. (1971b). *Z. Zellforsch. Mikrosk. Anat.* **118,** 49.

Vassallo, G., Solcia, E., Bussolati, G., Polak, J. M., and Pearse, A. G. E. (1972). *Virchows Arch., B* **11,** 66.

Vialli, M., and Erspamer, V. (1937a). *Boll. Soc. Med.-Chir. Pavia* **51,** 111.

Vialli, M., and Erspamer, V. (1937b). *Boll. Soc. Med.-Chir. Pavia* **51,** 357.

Vialli, M., and Erspamer, V. (1937c). *Boll. Soc. Med.-Chir. Pavia* **51,** 443.

Vialli, M., and Erspamer, V. (1942). *Arch. Sci. Biol. (Bologna)* **28,** 101.

Weichert, R. F., Reed, and Creech, O. (1967). *Ann. Surg.* **165,** 660.

Weichert, R. F., Roth, L. M., and Harkin, J. C. (1971). *Cancer* **27,** 910.

Wetzstein, R., Doerfler, W., and Schwink, A. (1964). *Ger. Med. Res.* **22,** 2.

White, T. T., and Magee, D. F. (1958). *Gastroenterology* **35,** 289.

Williams, E. D., and Sandler, M. (1963). *Lancet* **1,** 238.

Willie, J. H., Hesselbo, T., and Black, J. W. (1972). *Lancet* **2,** 1117.

Zeitlin, I. J., and Smith, A. N. (1970). *Rend. R. Gastroenterol.* **2,** 176.

Zollinger, R. M., and Ellison, E. H. (1955). *Ann. Surg.* **142,** 709.

Membrane Transport of Purine and Pyrimidine Bases and Nucleosides in Animal Cells

RICHARD D. BERLIN AND JANET M. OLIVER

Department of Physiology, School of Medicine,
The University of Connecticut Health Center, Farmington, Connecticut

I. Introduction

The study of transport of purine and pyrimidine bases and nucleosides has acquired new significance with the discovery that specific membrane "carrier" systems exist for these compounds and that transport capacity may be closely coupled with physiological events such as cell growth and malignant transformation. In addition, it has become clear that interpretation of the enormous literature on nucleic acid synthesis based on labeled precursor incorporation depends on a clear understanding of transport mechanisms and of the relationship between transport and metabolism.

In this article we establish some general principles about transport processes in animal cells, and then examine: the evidence for specific transport systems for bases and nucleosides, attempting to differentiate between membrane and metabolic events; the detailed characteristics of the systems, with particular attention to implications for carrier structure based on substrate specificity and sensitivity to inhibition by chemical agents; the physiological control of transport; and the evidence that carriers are mobile components of biological membranes and as such may be functionally modified by interactions with other membrane components.

II. General Principles

A. Mechanisms for Transport across Biological Membranes

Low-molecular-weight compounds may cross biological membranes by one of several mechanisms. Hydrophobic molecules that are readily soluble in the lipid of the plasma membrane may permeate by passive diffusion. The rate of permeation is usually a direct function of the oil–water partition coefficient of the compound. Experimentally, this mechanism is recognized by independence from metabolic energy sources, low temperature dependence, lack of competition between related compounds for uptake, and the absence of saturability. However, hydrophilic compounds unable to diffuse freely through a lipid bilayer and large enough (>5Å) to be excluded from aqueous pores or channels are instead often transported by specific carriers, presumably proteins, which exist in the membrane. Carrier-mediated transport is characterized by Michaelis–Menten kinetics (saturability and competition between related compounds for uptake) and high temperature dependence. The term *active transport* is reserved for carrier systems that depend on a direct (ATP) or indirect (proton and ion gradients) input of energy and lead to accumulation of a metabolically unaltered substrate against an electrochemical gradient. Facilitated diffusion is carrier-mediated transport driven by the electrochemical concentration gradient of an unchanged permeant. "Uphill" movement of a substrate of a facilitated diffusion system never occurs except when driven by a concentration gradient of a second substrate of the same carrier system. This latter phenomenon is known as counterflow, and provides a useful tool for analysis of the substrate specificity of the facilitated diffusion system.

These criteria for distinguishing transport mechanisms have been lucidly described by Stein (1967) and are not elaborated on here. It is, however, important for us to stress the difference between measurement of uptake and transport, since abuse of these terms has introduced great difficulties in the literature on permeation of purine and pyrimidine compounds.

B. On Transport and Uptake

By *transport* we mean the events essential to translocation of the substrate across the cytoplasmic membrane. As a consequence of this translocation, substrates may become available to intracellular enzymes and metabolized to chemical forms that are not substrates for the transport systems. The accumulation of such metabolites (as well as of the chemically unaltered substrate) is properly referred to as *uptake*. Clearly, uptake is several steps removed from transport. The availability of energy, which may determine the concentration of substrates against electrochemical gradients and the activities of intracellular enzymes which affect their metabolism, may limit uptake but are not directly related to the transport event.

The most serious problem in studying the process of transport as distinct from uptake is the failure to determine rates at sufficiently early times. As with enzyme reactions, it is essential to measure initial rates in order to determine unidirectional flux. At later times there is an apparent reduction in forward rate due to the superimposition of the reverse reaction (i.e., backflux). In our experience the period in which initial rates of transport can be measured is of the order of a minute (with some exceptions). This period is clearly incompatible with prolonged washings or centrifugations, and rapid sampling methods must be employed. We have used a monolayer technique—described in Section III,A—which facilitates rapid measurement in white cells, and a procedure for erythrocytes based on rapid separation of medium from cells by centrifugation through an inert oil (Section IV,A). It should be emphasized that, when radioactive substrates are used (as is nearly always the case), the metabolism of substrate often serves to prolong the period over which initial rates can be determined. Nonmetabolizable substrates facilitate analysis of transport mechanisms only if a concentration gradient can be maintained sufficiently long to permit sampling. Thus, for very rapid transport systems, it is often useful to employ permeants that are rapidly metabolized inside the cell to impermeant compounds so that backflux is reduced.

C. Measurement of Transport Rates

As indicated above, the experimenter usually seeks a unidirectional flux measurement, that is, one uncomplicated by backflux. In practice this requires rapid measurements. How does one know if one has assayed with sufficient rapidity? (1) The uptake should be linear with time over the measurement period. Obviously, it is essential that the method employed be rapid compared to the period of linearity. (2) The plot of uptake versus time should extrapolate back through zero after corrections have been made for mechanically trapped media and simple diffusional processes. A common error in this connection is to take, for example, 5 minutes as the first point. Of course, the line is then governed by this and later points. However, if the initial rate (unidirectional) period were only 1 minute, as is frequently the case, the initial rate value will make an insignificant contribution to the apparent shape of the uptake curve compared to the higher values obtained at later times. Thus the line may appear to extrapolate through zero, even though its slope does not correspond to an initial rate. Unfortunately, no matter how rapid the method, there may yet be a more rapid translocational membrane event. (3) As noted, metabolism of a substrate to a form that is trapped within the cell prolongs the period of linearity. In this case the absence of nonmetabolized substrate within the cell suggests that trapping has so reduced the intracellular pool of transportable substrate that backflux cannot occur (but see the next paragraph).

If substrate metabolism may figure so importantly in transport measurements, how can one be sure that the measured rate is not limited by enzyme activity? The use of nonmetabolizable substrates, or cells totally defective in the appropriate enzyme, is one relatively incontrovertible approach that has been useful in studies of nucleoside transport in red cells and in certain mutant strains of tissue culture cells. When such permeants are available, counterflow can be used to expand the utilizable period and to facilitate analysis of substrate specificity for the membrane carrier. With these substrates it is possible to study the kinetics of efflux as well as influx, if a rapid sampling procedure is available, and also to measure rates of exchange of permeants at equilibrium. Where metabolism occurs, other tests can be employed. (1) A sufficiently low level of intracellular nonmetabolized substrate indicates that the rate of metabolism is not limiting for uptake. One must caution here that for facilitated diffusion only equilibration can occur, and so the concentration of nonmetabolized substrate cannot be expected to exceed that of the medium. Since this concentration is small, once cells are removed from a medium containing substrate, any lag in the quenching of enzyme activity permits metabolism and spurious lowering of substrate concentrations. Lowering the temperature of wash fluids is the almost universal method for removing medium while inhibiting enzyme. Significant enzyme activity relative to the small amounts of substrate present may well persist at reduced temperatures. (2) Demonstration that the rate of uptake in whole cells is lower than enzyme activity measured in cell extracts is also evidence that intracellular metabolism is not rate-limiting for transport. However, inside the cell enzyme activity is usually reduced by inhibitors such as nucleotides, so that values obtained in extracts may be an unreliable guide to the available activity *in vivo*.

D. Properties of Purine and Pyrimidine Compounds

With these points in mind, it is useful to review some important properties of purine and pyrimidine compounds very briefly.

The solubility properties of bases are to a large degree determined by the nature of the substituents on the heterocyclic ring. For example, the parent compound purine shows appreciable solubility in organic solvents, whereas a single substitution of a hydroxyl (hypoxanthine) or amino (adenine) group on the pyrimidine ring renders the molecule quite insoluble in nonpolar solvents. Even here, adenine is rather soluble in *n*-butanol, a property that has been used to effect its separation (Hori and Henderson, 1966). Hydrogen bonding by bases, so familiar to molecular biologists as a mechanism for stabilizing nucleic acid structure, also determines their solubility characteristics. Although xanthine is quite insoluble in organic solvents, methylation of ring nitrogens, which prevents tautomerization of the hydroxyl groups, hence the formation of hydrogen bonds, results in derivatives such as caffeine and theophylline with appreciable lipid solubility. Analogous arguments apply to pyrimidine bases.

Thus the biologically important purine and pyrimidine bases are generally hydrophilic molecules and, a priori, it may be assumed that they require some special mechanism of membrane transport even though certain modifications of their structures can lead to appreciable lipid solubility, permitting passive diffusion.

The formation of nucleosides, which may be viewed here as the addition of a bulky hydrophilic group, of course reduces lipid solubility greatly, and we may assume that with certain rare exceptions nucleoside uptake signifies the existence of a specific membrane transport system.

III. Transport of Purine and Pyrimidine Bases

A. Mechanism

Purine and pyrimidine bases were the first nucleic acid precursors whose uptake was studied in mammalian cells. Whittam (1960) observed that human red cells were freely permeable to adenine and hypoxanthine, and Lassen (1961, 1962) and Lassen and Overgaard-Hansen (1962a,b) followed this report by kinetic analyses of uric acid and hypoxanthine uptake in erythrocytes. Their studies showed that urate transport is a temperature-dependent, saturable process which leads to equilibration of the permeant between the medium and the cytoplasm. Hypoxanthine and other purine and 8-azapurine derivatives competitively inhibited the flux of uric acid. This was the first indication that a carrier system obeying Michaelis–Menten kinetics, rather than a simple diffusional process, is responsible for purine uptake in mammalian cells. Uptake of hypoxanthine occurred much more rapidly than uptake of urate, so that in these early experiments satisfactory initial rates could not be obtained. It was only possible to establish that hypoxanthine, like urate, was also transported by a saturable process. A technique for rapid sampling by filtration was later developed (Lassen, 1967) to measure the initial rate of hypoxanthine influx, and it was shown that at low concentrations this purine enters cells by a saturable process with a low K_m (0.4 mM), while at high concentrations uptake cannot be saturated. It was suggested that the nonsaturable process represents uptake by diffusion but, since both components showed the same high temperature dependence, it seems equally likely that a second carrier with a very high K_m may function at high concentrations of substrate. Lassen (1962, 1967) was unable to demonstrate counterflow of purines, so he suggested that the membrane carrier for purines may have properties different from those of the mobile carrier that had been described for sugars in human erythrocytes. However, alternative explanations are possible. The absence of counterflow of urate driven by hypoxanthine may be due to the fact that urate has a very low affinity for the carrier and hypoxanthine a very high affinity. A shorter incubation period may be required to detect a transient counterflow of radioactive hypoxanthine driven by a gradient of cold hypoxanthine.

A similar high permeability to purines and pyrimidines was demonstrated in cells of Ehrlich ascites tumors (Jacquez and Ginsberg, 1960; Jacquez, 1962) and in normal human leukocytes (Kessel and Hall, 1967), and the uptake of permeants was attributed to simple diffusion. However, the substrate concentrations used were far in excess of the range of K_m values measured subsequently in other cell lines, so that the nonspecific (or high K_m) components of transport may have obscured the specific system. It now seems that some carrier-mediated process similar to that described by Lassen in erythrocytes is generally present in mammalian cells.

For example, Hawkins and Berlin (1969) studied purine transport in rabbit polymorphonuclear leukocytes (PMN). Bases were found to enter PMN very rapidly, the period of linear uptake depending on the capacity of intracellular enzymes to metabolize the base and so maintain a concentration gradient of substrate. Therefore a technique was developed for taking initial rate samples as quickly as 10 seconds, using cell monolayers on glass cover slips. Figure 1 demonstrates the need for this rapid sampling. Uptake of 2 m*M* adenine which is partially metabolized to nucleotide is only linear for about 3 minutes, while for the nonmetabolized pyrimidine, arabinosyluracil, the period of linearity is reduced to about 30 seconds. Kinetic studies of adenine transport over a range of concentrations showed that two distinct mechanisms operated, one at low concentrations of substrate (low K_m, low V_{max}) and one at high concentrations (high K_m, high V_{max}) (Fig. 2). The low-K_m system was competitively inhibited by other purines, but not by nucleosides, pyrimidines, or nucleotides, while the high-K_m system was relatively insensitive to inhibition by other purines. Both systems were highly temperature-dependent, suggesting that a membrane carrier was involved in both cases. The uptake of xanthine was found to be considerably slower than uptake of adenine and was inhibited by adenine and by several compounds (pyrimidines, adenosine, adenine nucleotides, uric acid) that did not affect adenine influx. These data indicate that adenine and xanthine may have separate carrier systems. It was not possible to block more than 82% of xanthine uptake by any inhibitor, and so it was proposed that xanthine, like adenine, may also be able to cross the membrane of PMN cells via two distinct carriers with different kinetic properties.

Adenine uptake has also been studied in human platelets. Sixma *et al.* (1973) found that uptake measured at 5 minutes was saturable and temperature-dependent, indicating a carrier mechanism, but both K_m and V_{max} varied with the incubation medium (plasma or buffer), the length of storage of platelet-rich plasma, and the presence of ADP. It was suggested that these effects were related to changes in platelet shape under various conditions. Adenine uptake was competitively inhibited by other purines, but not by pyrimi-

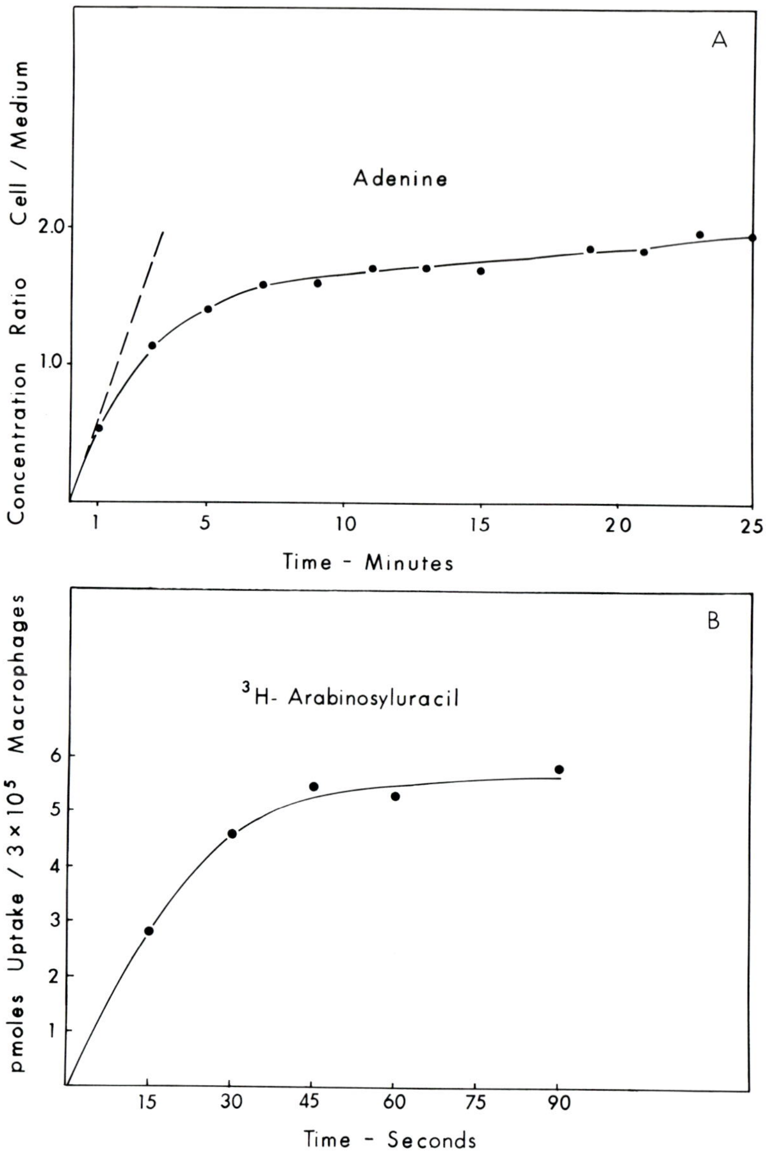

FIG 1. Time course of uptake in PMN using a rapid sampling technique. PMN obtained from sterile rabbit peritoneal exudates were suspended to a concentration of 4×10^6/ml in modified Hanks' medium. Monolayers were made by pipetting 0.5 ml of cell suspension onto round glass cover slips of 22 mm diameter and incubating for 30 minutes at 37°C on metal bars inside a humidified chamber. The cover slips were then drained, and about 350 μl of medium containing radioactive base [(A) 2 m*M* adenine-^{14}C (purine base) and (B) 0.01 m*M* arabinosyluracil-^{3}H (pyrimidine nucleoside)] was

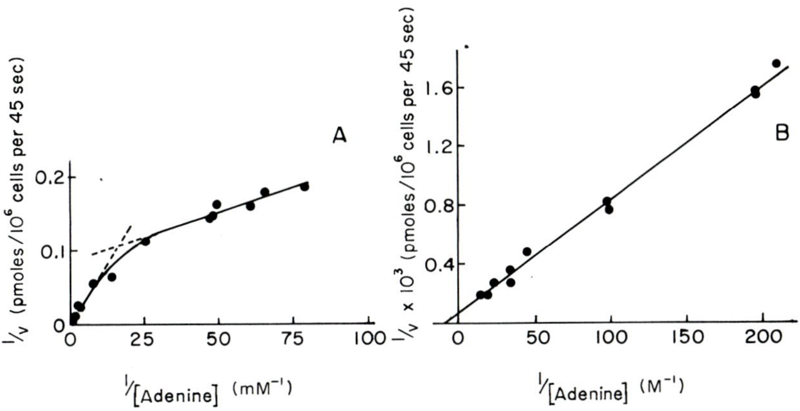

FIG. 2. Kinetics of adenine uptake. Cell monolayers were incubated with adenine-^{14}C in concentrations from 0.0125 to 73 mM for 45 seconds (in the period of linear uptake for all concentrations). The monolayers were rinsed and radioactivity counted as described in the legend for Fig. 1. All points plotted were averages of five determinations. Solutions over 10 mM were adjusted to isotonicity by omission of an appropriate amount of sodium chloride. From the double reciprocal plot (A), two additive entry mechanisms are apparent, one operating more efficiently at low concentrations ($K_m = 0.007$ mM; $V_{max} = 5.7$ pmoles/10^6 cells/45 seconds) and the other system predominating at high concentrations ($K_m = 100$ mM; $V_{max} = 13{,}700$ pmoles/10^6 cells/45 seconds). To display the saturation kinetics of the second system, the data are replotted at concentrations of adenine greater than 5 mM (B). (From Hawkins and Berlin, 1969, reproduced from *Biochim. Biophys. Acta* by permission of Elsevier, Amsterdam.)

dines, in agreement with Hawkins and Berlin's study on PMN. High concentrations of adenosine also inhibited adenine uptake, while inosine was without effect. It was suggested that this inhibition was due to competition between intracellular adenine and adenosine for ATP necessary for phosphorylation, since other compounds that reduce ATP levels (2-deoxy-D-glucose, papaverine, antimycin A, and prostaglandin E) also reduced the apparent rate of adenine uptake.

In view of the available data showing that the initial rate of purine transport is very short (on the order of 1–2 minutes; Hawkins and

---

placed over the monolayer. At various times the cover slips were drained, rinsed through a series of beakers containing cold medium, and collected in counting vials, and the cells were hydrolyzed with 0.5 N potassium hydroxide for 30 minutes. The solutions were neutralized with 10% perchloric acid, and the vials counted by liquid scintillation. [(A) From Hawkins and Berlin, 1969, reproduced from *Biochim. Biophys. Acta* by permission of Elsevier, Amsterdam, and (B) R. D. Berlin unpublished].

Berlin, 1969; Lassen, 1967), it is somewhat surprising that a recent study of hypoxanthine uptake in normal fibroblasts and fibroblasts that are partially or wholly deficient in hypoxanthine/guanine phosphoribosyl transferase, the enzyme that catalyzes the phosphoribosylation of hypoxanthine, employs considerably longer incubation periods. Benke *et al.* (1973) reported that the amount of hypoxanthine incorporated in 15 minutes by normal fibroblasts was higher at low cell densities than at high cell densities, while partially hypoxanthine/guanine phosphoribosyl transferase–deficient strains showed much lower incorporation at all stages of cell growth, with maximum incorporation into intracellular radioactivity at high cell density. This was interpreted as evidence for different rates of hypoxanthine transport as fibroblasts approach confluency, and for an intrinsically lower transport capacity in the enzyme-deficient cells. However, the rate of nucleic acid synthesis is also higher in growing cells than in confluent cells, and so these observations in normal fibroblasts are likely to reflect differences in intracellular phosphoribosylation and utilization of nucleotides during the 15-minute incubation period rather than differences in the rate of membrane transport of hypoxanthine.

These investigators also reported that totally enzyme-deficient (Lesch–Nyhan) cells showed a small uptake of hypoxanthine that did not increase beyond 5 minutes. Further, when cells were preincubated for 10 hours with aminopterin, an inhibitor of purine synthesis *de novo*, the uptake of hypoxanthine measured after 15 minutes was elevated in normal and partially enzyme-deficient cells but not in Lesch-Nyhan cells. From this they concluded that hypoxanthine/guanine phosphoribosyl transferase is essential for hypoxanthine transport and that aminopterin stimulates the enzyme-mediated uptake system. We discuss the relationship between transport and base phosphorylation in detail in Section III,B. At this point we mention only that these investigators were unable to demonstrate any membrane-associated hypoxanthine/guanine phosphoribosyl transferase and that aminopterin, by depleting cells of intracellular purine derivatives, would be expected to enhance the utilization of extracellular hypoxanthine in normal or partially deficient cells without necessarily influencing the transport process at the membrane level.

A recent report by Harris and Whitmore (1974) provides further evidence that hypoxanthine/guanine phosphoribosyl transferase and purine uptake are independent reactions. They isolated a phenotypically stable line of Chinese hamster ovary cells that had normal enzyme levels when grown at 34° and 38.5°C but showed a temperature-dependent reduction in uptake of guanine, hypoxanthine, azaguanine, and guanosine at the higher temperature. Uptake of adenine was the same in control and mutant cells at the permissive temperature (34°C) and the nonpermissive temperature (38.5°C), suggesting that the adenine transport carrier may be separate from that for guanine and hypoxanthine in this cultured cell line. Similar analyses are needed using other cultured cells to find whether or not there is generally more than one transport carrier for purine bases. Analysis of pyrimidine uptake is also required for all cell types.

It is interesting that a mediated transport system for purines and pyrimidines in some ways similar to those described for mammalian isolated cell systems has been established in the rat cestode *Hymenolepis diminuta*. MacInnis *et al.* (1965) showed a saturable uptake at low concentrations of purines and pyrimidines measured after 2 minutes, which was competitively inhibited by the presence of other bases but not by amino acids, nucleosides, or sugars. At high permeant concentrations a diffusional component of transport was more important. Hypoxanthine had the greatest affinity for the carrier site, followed by uracil and adenine; thymine and cytosine appeared to enter mainly by diffusion, even at low concentrations.

In contrast with these isolated cell systems, accumulative uptake of purine and pyrimidine bases appears to occur in intestine. That is, intestinal uptake involves active transport rather than facilitated diffusion.

This was first established for pyrimidines by Schanker and co-workers (Schanker and Tocco, 1960, 1962; Schanker *et al.*, 1963), using cannulated rat intestine *in vivo* and everted intestinal sacs *in vitro*. In the living animal absorption of radioactive thymine and uracil was found to be a saturable process which was inhibited by nonradioactive purine and pyrimidine bases, but not by sugars or amino acids. Studies with everted sacs showed that uracil and thymine can be transported from mucosa to serosa against a concentration gradient. At high permeant concentrations passive diffusion of these pyrimidines can also occur. The active transport requires oxygen and is competitively inhibited by a wide range of purine and pyrimidine bases, and accumulation is abolished by metabolic inhibitors. Czáky (1965) has presented evidence that sodium is essential for this accumulative uptake.

Schanker and co-workers were unable to study the uptake of hypoxanthine and xanthine, because of their rapid intracellular metabolism to uric acid catalyzed by xanthine oxidase. However, this difficulty was overcome by Berlin and Hawkins (1968a), who established that in isolated sacs of hamster small intestine xanthine oxidase occurs exclusively within the epithelial cells and can be blocked by the analog allopurinol, making possible transport studies of metabolically unaltered oxypurines. Their data show that—in contrast with pyrimidines—hypoxanthine, xanthine, and probably uric acid presented at the serosal side are secreted into the intestinal lumen against concentration gradients which are abolished by metabolic inhibitors (Fig. 3). The process could not be saturated, indicating a high K_m for transport. This active transport system is unidirectional, since mucosa-to-serosa fluxes are very low. Unlike xanthine the methylated derivative, caffeine, showed no active secretion from serosa to mucosa but rather a rapid flux in both directions, in keeping with its increased lipid solubility.

Having established that xanthine oxidase degrades oxypurines in-

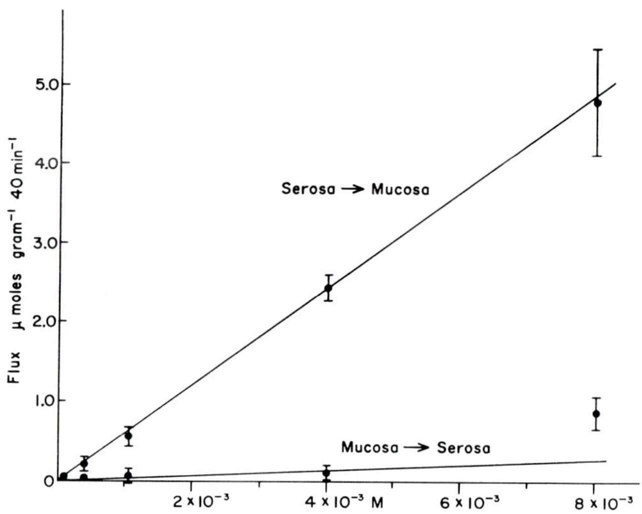

FIG. 3. Unidirectional transmural xanthine flux in isolated sacs of hamster small intestine. Gut sacs were incubated in gassed Krebs–Henseleit bicarbonate (without calcium) at 37°C in stoppered Erlenmeyer flasks. The medium contained various concentrations of xanthine-^{14}C (abscissa) plus allopurinol at a concentration (2×10^{-4} M) that inhibited xanthine oxidase by 95% without affecting xanthine flux. Inulin-^{3}H was included to allow for correction for contaminating medium. Flux (ordinate) was computed from the total purine within the sac after 40 minutes after correction for contamination by the medium. Bars through points depict 2 S.E. When xanthine was placed in equal concentrations on either side of uneverted sacs, it accumulated inside the sac; this indicates active secretion from serosa to mucosa. When everted sacs were used, there was little detectable transfer of radioactivity, indicating negligible absorptive flux from mucosa to serosa. (From Berlin and Hawkins, 1968a.)

tracellularly, Berlin and Hawkins (1968b) attempted to determine unidirectional fluxes across the separate (mucosal versus serosal) borders of the cell.

Since it is extremely difficult to measure initial rates of uptake in complex multicellular systems like epithelia, an alternative approach was developed to measure the rate of purine transport using perfused loops of intestine *in vitro*. Xanthine was presented in the medium bathing the loop, and transmural flux measured directly by collecting the medium perfusing through the loop and determining its concentration spectrophotometrically. However, it is apparent that the flux across a single membrane is the transmural flux (serosa to mucosa in uneverted loops; mucosa to serosa in everted loops) plus the backflux across the membrane where substrate is presented. The problem is to determine backflux. The method adopted was to establish a steady-state transmural flux of xanthine in the presence of a tracer quantity of radioactive hypoxanthine in the external medium. The hypoxanthine is transported and oxidized within the cell to radioactive xanthine which mixes with the unlabeled xanthine transported across the membrane. If it is assumed that xanthine derived from transport and from intracellular oxidation are thoroughly mixed, xanthine that is transported transmurally and xanthine that backfluxes must have the same specific activity. The former specific activity is measured from xanthine concentration and radioactivity in a transmural sample. Xanthine in the medium due to backflux is determined by measuring xanthine radioactivity after separation from hypoxanthine and calculating from this radioactivity and the specific activity obtained from the transmural sample.

Although somewhat cumbersome, this methodology provided an approach to determining the separate behavior of lumenal (brush border) and contralumenal (basal) membranes of the same cells. The results showed that the basal membrane is the major site of active purine secretion and is also the site where chemical species of purines are discriminated. The intracellularly generated xanthine and urate effluxes preferentially across the lumenal border, and both influx at the basal membrane and efflux at the mucosal membrane appear to be independent of extracellular sodium.

In contrast to the mammalian intestine, xanthine is actively absorbed (not secreted) by the chick embryo. Xanthine transport was determined in chick embryo intestine, in which the avian enzyme xanthine dehydrogenase is present in great excess, by quantitating the conversion of radioactive xanthine to urate (Taube and Berlin, 1970). The baseline urate production derived from noncarriermediated diffusion of xanthine was defined as the urate formed in the presence of a pyrimidine, 6-methyluracil. 6-Methyluracil competitively inhibited transport (so defined) but had no effect on xanthine dehydrogenase. This transport system was saturable, with a K_m for xanthine of 0.005 mM, and the carrier was located on the mucosal side of the epithelium. The technique lends itself to analysis in small amounts of tissue, and it was used to follow the embryological development of the system. Transport was clearly demonstrable 1 day prior to hatching.

Berlin (1969) also demonstrated active accumulation of purines in isolated rabbit choroid plexus that is saturable (K_m xanthine = 0.17 mM) and competitively inhibited by both purines and pyrimidines (Fig. 4).

Thus we can conclude that purine and probably pyrimidine transport is mediated by membrane carrier systems in all the cells and tissues that have been analyzed to date. At least in some cell types there seems to be an additional diffusional process or else a second carrier system with a very high K_m which operates at high concentrations of permeants. It is not established how many carriers for bases exist in mammalian cells. There is clearly some cross-reactivity between carrier systems for purines and pyrimidines, but some cells, for example PMN, appear to have more than one carrier for purines alone. Chinese hamster ovary cells may have adenine transport sites separate from those for guanine and hypoxanthine, and in intestine there is an opposite directionality and sodium requirement for the active movement of purines and pyrimidines which argues against a single transport system even though purines inhibit pyrimidine transport. It is particularly interesting that in all the isolated cell systems studied to date carrier-mediated transport does not lead to accumulation of metabolically unaltered substrate, while in the two epithelia that have been analyzed (intestine and choroid plexus) transport can occur against a concentration gradient. This is of course completely analogous to the transport of glucose in nonepithelial cells such as muscle and leukocytes as compared to epithelia.

We should note that urate transport has been extensively studied in kidney. This work has recently been reviewed (Mudge *et al.*, 1973) and is not covered in our treatment of purine transport.

B. The Relationship between Transport and Phosphorylation of Bases

It seems important to stress one point that has already been made, that base transport is a process separate from phosphorylation. We

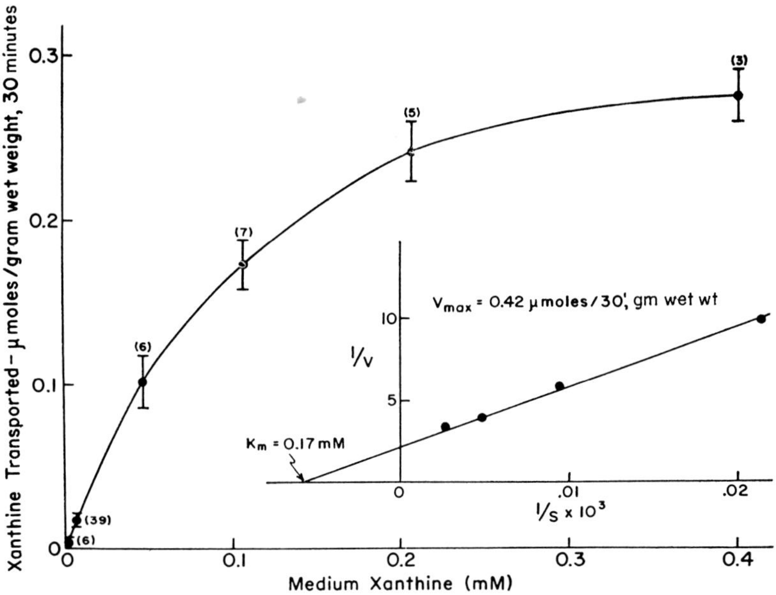

FIG. 4. Xanthine accumulation in choroid plexus. Choroid plexus from the lateral ventricle of rabbits was incubated at 37°C in gassed Krebs–Henseleit bicarbonate buffer containing xanthine-^{14}C plus inulin-^{3}H to allow correction for entrapped medium. After 30 minutes the tissue was removed, weighed, and hydrolyzed with potassium hydroxide, and radioactivity was measured by liquid scintillation. More than 95% of the xanthine recovered was metabolically unaltered by the tissue. The results show that xanthine accumulates intracellularly as a function of its concentration in the medium. The bars correspond to standard errors of the mean; the number of observations are given in parentheses. In media containing 63×10^{-6} *M* xanthine, at 30 minutes the ratio of tissue to medium xanthine concentration was 2.81 ± 0.31 (39 observations). The inner figure is a reciprocal plot of the data.

It should be emphasized that at 30 minutes the velocities are not initial rates. They represent instead an approximation to equilibrium between medium and tissue. The rate of approach to equilibrium may be inhibited by competition with analogs (as was shown), and saturation is also evident. However, the K_m is a complicated function of both influx and efflux from the tissue. Such data serve to demonstrate the existence of a transport system capable of active transport, but they cannot be used to derive quantitative information on the activity (or numbers) of carriers or the binding affinity of substrates. (From Berlin, 1969. Copyright 1969 by the American Association for the Advancement of Science.)

have noted that a certain amount of unnecessary confusion exists in this regard, partly as a result of failure to determine initial rates of transport. In addition to the results of Benke and co-workers, discussed above Hochstadt (1975) has also proposed that, by analogy to sugar transport by gram-negative bacteria involving concomitant phosphorylation (Roseman, 1969), purine uptake in animal cells is effected directly by the activity of phosphoribosyl transferases, the enzymes that convert bases to nucleotides. In essence it is suggested that the substrate interacts with the enzyme prior to its translocation across the membrane. The experimentally testable predictions of this hypothesis are: (1) the intracellular form of the purine is always the phosphorylated derivative (nucleotide), (2) the chemical specificities of transport and enzyme are identical, and (3) the enzyme is located in the membrane. The available evidence is not consistent with these predictions. Most phosphoribosyl transferases are clearly soluble enzymes. As noted, purine transport in erythrocytes can occur virtually without metabolism; and indeed, for erythrocytes of most species, the requisite enzymes are not present. Transport is also normal in Chinese hamster ovary cells lacking the appropriate phosphoribosyl transferase.

Some of the most convincing evidence against this theory of concomitant transport and phosphorylation comes from comparative studies of the substrate specificities of adenine transport and the adenine phosphoribosyl transferase of rabbit PMN. Berlin (1970) tested a series of adenine analogs derived by substitution of the 6-amino group (Table I, group A), by additions to the purine ring (Table I, group B), and by modification of the ring structure (Table I, group C), for their ability to inhibit adenine uptake in PMN and to inhibit adenine phosphoribosyl transferase in cell extracts. Transport was measured as described in Fig. 1, using 0.008 mM adenine-^{14}C plus the indicated amount of nonradioactive inhibitor as the substrate for a 45-second uptake. Binding of the test compound to the leukocyte enzyme was estimated from its ability to inhibit the initial rate of adenine conversion to AMP at limiting ($K_m = 0.002$ mM) concentrations of adenine and saturating concentrations of the second substrate, 5-phosphoribosyl-1-pyrophosphate.

Large differences between the specificites of the adenine carrier and the enzyme are apparent from the data in Table I. For example, both carrier and enzyme are relatively exacting for substituents at position 6 (amino group), although there is greater spatial tolerance by the carrier; thus 6-methyl and 6-dimethyladenine (analogs 6 and 7) have 20- to 40-fold greater affinities for transport than for the enzyme. However, 4-amino-5-imidazole carboxamide (analog 24), which lacks C-2, has no detectable affinity for the carrier but binds well with the enzyme. This indicates that C-2 or perhaps a bicyclic structure is essential for transport. The carrier is, however, less sensitive to C-2 substitution, so that 2-methyl and 2-methylaminoadenine (analogs 11 and 15) bind fairly strongly with the carrier but have no affinity for the enzyme.

TABLE I: INHIBITION OF ADENINE PHOSPHORIBOSYLTRANSFERASE AND ADENINE TRANSPORT IN PMN BY ADENINE ANALOGS[a,b]

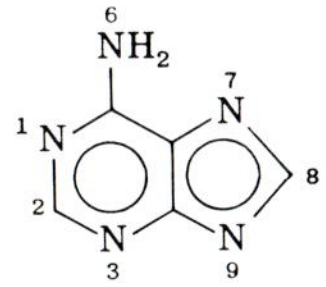

Adenine

Analog	C-6 Substituent	Enzyme: Concentration (mM)	Enzyme: Inhibition (%)	Transport: Concentration (mM)	Transport: Inhibition (%)
GroupA: Adenine analogs derived by substitution at the 6-position of the purine ring					
1	—H	1.0	39	1.0	30
2	—OH	1.0	0	1.0	18
3	—OCH_3	1.0	14	1.0	47
4	—Cl	1.0	7	1.0	66
5	—SH	1.0	12	0.1	22
6	—$NHCH_3$	1.0	33	0.06	50
7	—$N(CH_3)_2$	1.0	5	0.8	50
8	—C_6H_5	0.3	52	0.25	67
9	—$C_6H_5CH_2$	0.4	63	1.0	85
Group B: Adenine analogs derived by additions to the purine ring					
10	1-Methyladenine	2.0	0	1.0	30
11	2-Methyladenine	1.0	−5	0.25	35
12	3-Methyladenine	2.0	0	1.0	32
13	7-Methyladenine	1.0	48	1.0	12
14	9-Methyladenine	1.0	0	1.0	41
15	2-Methylaminoadenine	1.0	7	0.2	30
16	2-Aminoadenine	1.0	2	1.0	0
17	2-Hydroxyadenine	0.2	42	0.1	25
18	Guanine	0.72	32	0.1	25
19	Xanthine	1.0	3	1.0	11
20	8-Mercaptoadenine	1.3	35	1.0	71
21	8-Bromoadenine		0.9[c]		0.06[c]
Group C: Adenine analogs derived by modification of the purine ring structure[d]					
22	4,5,6-Triaminopyrimidine	1.0	11	1.0	12
23	2,5,6-Triaminopyrimidine	1.0	4	1.0	10
24	4-Amino-5-imidazolecarboxamide	1.0	56	1.0	5
25	Imidazole	20.0	8	—	—
26	Histamine	2.0	8	—	—
27	8-Azaadenine	1.0	0	1.0	28
28	4-Aminopyrazolo(3,4-*d*)pyrimidine		0.27[c]		0.23[c]
29	7-Deazaadenine		1.21[c]		0.14[c]

[a] From Berlin (1970) by permission. Copyright 1970 by the American Association for the Advancement of Science.

[b] Inhibitions of enzyme (adenine phosphoribosyltransferase) and transport were tested at the respective K_m values for adenine, 0.002 mM and 0.008 mM. See text.

[c] Value listed is K_i.

[d] See Table II, Group IV, for structural formulas of corresponding purine ribosides.

From consideration of all these data, Berlin has concluded that the electronic configuration about C-9 is critical for substrate binding to the carrier, while π bonding or interactions with positions 7 and 8 are more critical to the enzyme.

Since position 9 of adenine is the site of nucleophilic attack by the enzyme, it follows that the carrier, by binding at this site, protects the substrate from enzymic attack during transport. That is, far from transport and phosphorylation being simultaneous processes, they are probably mutually exclusive events. Perhaps the function of the very different specificities of transport and enzyme is to screen out potentially dangerous or nonutilizable extracellular compounds.

Thus, although it is true that in many cells it is difficult to detect chemically unaltered substrate intracellularly, this simply reflects the fact that transport is often rate-limiting for intracellular phosphorylation. There is little doubt that purine transport is a process separate from phosphorylation in animal cells.

IV. Nucleoside Transport

A. Mechanism

There is considerably more information available about nucleoside transport in mammalian cells than about transport of the corresponding purine and pyrimidine bases. Again, many reports are difficult to interpret, because transport and metabolism have not been differentiated. We start by establishing the general mechanisms of nucleoside transport in systems in which transport alone is being studied, and then consider reports dealing with both transport and metabolism.

Whittam (1960) showed that human erythrocytes are freely permeable to purine nucleosides, and Jacquez (1962) first found evidence to suggest mediated transport of several pyrimidine nucleosides by Ehrlich ascites carcinoma cells. However, it was not until 1968 that a careful analysis of pyrimidine nucleoside transport was published. Kessel and Shurin (1968) examined transport of cytosine arabinoside and deoxycytidine in a subline of L1210 murine leukemia unable to metabolize either nucleoside. Uptake measured at 1 minute was saturable, temperature-dependent, and at longer time intervals led to equilibration of the permeant across the membrane. Metabolic inhibitors did not affect uptake, but a variety of purine and pyrimidine nucleosides were competitive inhibitors. Free bases and sugars did not affect the transport process, indicating that it was specific for nucleosides. It was shown that the initial phase of nucleoside efflux was also saturable and temperature-dependent and, unlike influx, could be inhibited by uranyl ion. A second slow phase of efflux was unaffected by UO_2^{2+}. The mechanism of this unidirectional inhibition is inknown. It was concluded that nucleoside transport occurs by facilitated diffusion in these cells.

Uridine, a pyrimidine that is not metabolized by human erythrocytes, was used as the major permeant for analysis of red cell nucleoside transport by Oliver and Paterson (1971). Uptake was extremely rapid at 25°C, being complete in only about 40 seconds (Fig. 5; see also Fig. 10), and so a rapid sampling technique based on centrifugation through an inert oil (dibutyl phthalate) was used, which allowed measurement of the disappearance of nucleoside from the medium as early as 10 seconds after beginning the incubation. Transport was saturable, nonaccumulative, temperature-dependent, and competitively inhibited by a range of purine and pyrimidine nucleosides, but not by free bases, sugars, or amino acids. In addition, efflux of uridine from cells at equilibrium with uridine occurred when a second nucleoside was added to the incubation medium (Fig. 6). This demonstration of counterflow is strong evidence that all the nucleosides tested share a common membrane carrier. Radioactive thymidine could be substituted for uridine in all these experiments. In an extension of these studies, a wide range or purine and

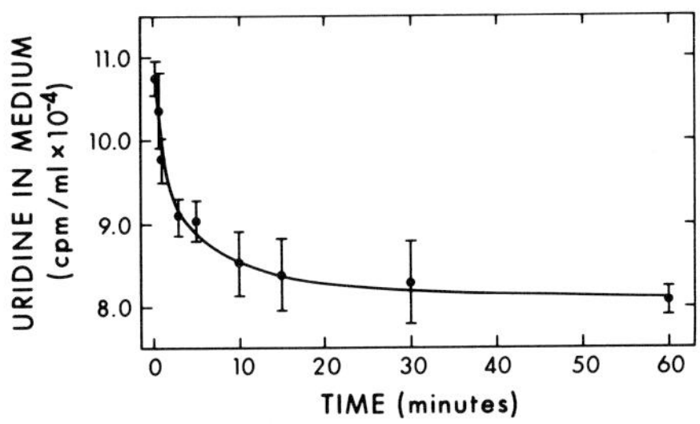

FIG. 5. Time course of uridine uptake by human erythrocytes, estimated from the rate of removal of uridine from the medium. Quadruplicate suspensions of washed erythrocytes (37.5% hematocrit) were incubated at 25°C in tris–ethane sulfonate (TES)-buffered saline (pH 7.4) containing Mg^{2+}, glucose, and uridine-2-^{14}C (2.96 mM, specific activity 3×10^4 cpm/μmole). At various times portions (about 0.5 ml) were removed into tubes containing 5 ml of di-1-butylphthalate and immediately centrifuged at 1700 g for 1.5 minutes. Cells rapidly sedimented through the inert oil, leaving an upper layer of cell-free aqueous medium. Portions (50 μl) of medium were counted by liquid scintillation. Averaged values are shown; bars indicate standard deviations. The uptake of uridine between 0 and 30 seconds (initial rate) was 0.68 μmole per minute per milliliter packed cells. The linear period of uptake does not extend beyond 1 minute. (From Oliver and Paterson, 1971.)

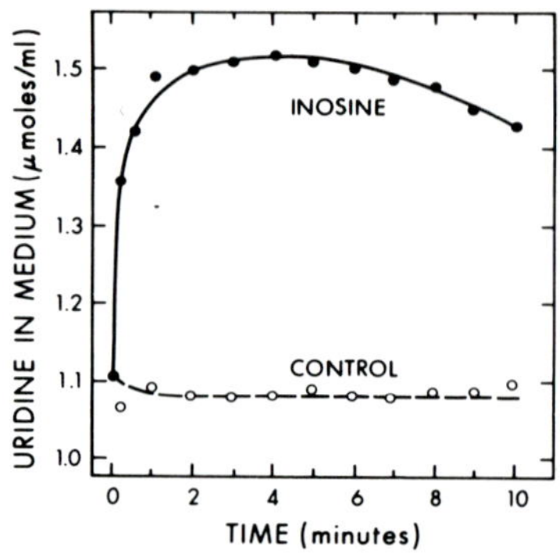

FIG. 6. Counterflow of uridine. A 45% suspension of washed erythrocytes was incubated with uridine-2-^{14}C at 37°C for 30 minutes to an equilibrium concentration of 1.01 m*M*. After cooling at 20°C duplicate 16-ml portions of the equilibrated cell suspension were incubated with the addition of 0.4 ml of medium (open circles) or 0.4 ml of 160 m*M* inosine (final concentration is 4.0 m*M* inosine) (solid circles). Uridine in 0.5-ml portions of the medium was measured at various times by the dibutylphthalate method described in the legend for Fig. 5. The inward gradient of inosine caused a rapid efflux of uridine out of the cells into the medium against a concentration gradient. (From Oliver and Paterson, 1971.)

pyrimidine nucleoside analogs with substituents on both the base moiety and the sugar moiety were tested for their ability to promote counterflow of uridine (i.e., to enter cells via the uridine carrier). It was shown that purine nucleosides were generally poorer substrates than pyrimidine nucleosides, nucleoside analogs with ionized substituents but not with uncharged substituents on the base portion showed reduced affinity, and substitution of the ribose at the 2′- and 3′-hydroxyl groups, or replacement of ribose with sugars other than arabinose, reduced affinity for the carrier (Cass and Paterson, 1972, 1973). Preliminary evidence for a common nucleoside carrier in erythrocytes was also provided by Lieu *et al.* (1971).

The mechanism and specificity of nucleoside transport has also been carefully analyzed in rabbit PMN. Taube and Berlin (1972) used a rapid sampling technique with cell monolayers, which allowed measurement of intracellular radioactivity as early as after 10 seconds of incubation.

As in erythrocytes, adenosine and thymidine were transported very rapidly by a saturable system which was competitively inhibited by a wide range or purine and pyrimidine nucleosides. The K_m value for adenosine was the same as its K_i value when adenosine was tested as an inhibitor of thymidine transport, and other nucleosides gave the same K_i values when tested against both adenosine and

TABLE II
INHIBITION OF ADENOSINE TRANSPORT BY ADENOSINE ANALOGS[a]

Group A: Purine ribosides

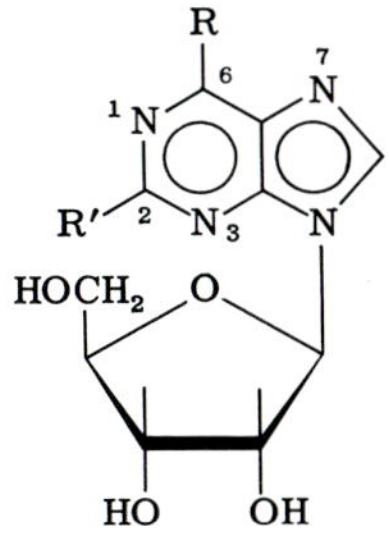

	Compound	R	R′	K_i (mM)
1	Adenosine	NH_2	H	0.010[b]
2	Isoguanosine	NH_2	OH	0.013
2a	Inosine	OH	H	0.022
3	Guanosine	OH	NH_2	0.033
4	6-Mercaptopurine riboside	SH	H	0.038
5	6-Methyladenosine	$NHCH_3$	H	0.107
6	6-Chloropurine riboside	Cl	H	0.12
7	6-Dimethyladenosine	$N(CH_3)_2$	H	0.18
8	Purine riboside	H	H	0.22
9	Xanthosine	OH	OH	0.91

Group B: Pyrimidine nucleosides

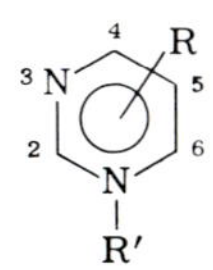

	Compound	R	R′	K_i (mM)
10	Uridine	2,4-Dihydroxy	1-Ribose	0.03
11	Cytidine	2-OH, 4-NH_2	1-Ribose	0.05
12	Thymidine	2,4-Dihydroxy-5-methyl	1-(2′-Deoxy)ribose	0.06
13	Pseudouridine	2,4-Dihydroxy-5-ribose	H	0.28

(Continued)

TABLE II (*Continued*)

Group C: 9-Pentosyladenine derivatives

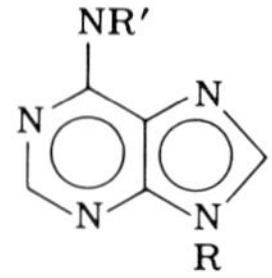

Compound	R	R′	K_i (m*M*)
1 Adenosine	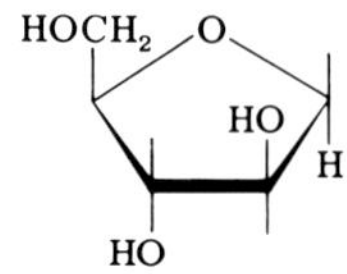	H_2	0.010
14 2′-Deoxyadenosine	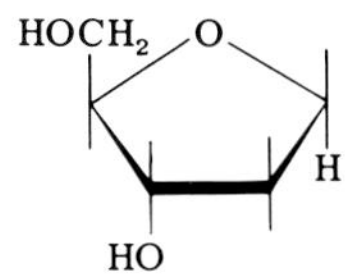	H_2	0.037
15 9-(β-D-Arabinofuranosyl)adenine	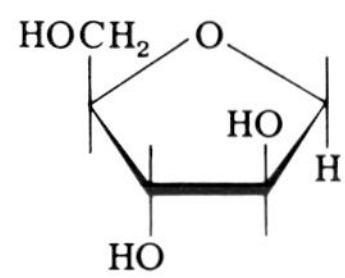	H_2	0.298
16 9-(β-D-3′-Deoxyribofuranosyl)-adenine, cordycepin	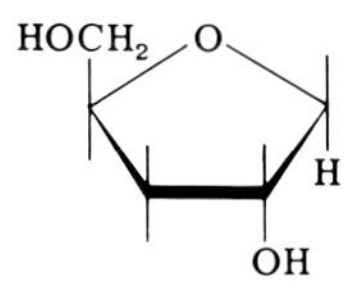	H_2	1.94
17 L-Adenosine	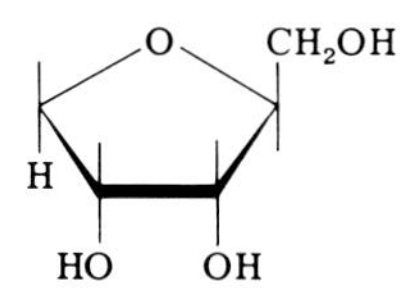	H_2	1.99
18 9-(β-D-Psicofuranosyl)adenine, psicofuranine	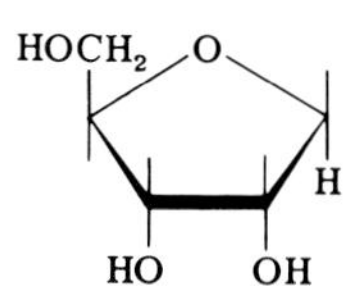	H_2	3.04
19 9-(β-D-Xylofuranosyl)adenine	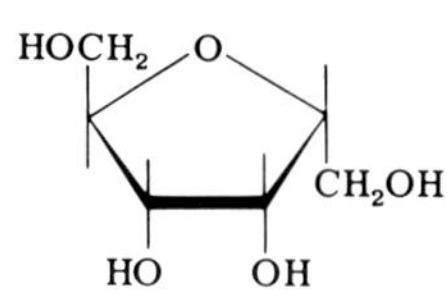	H_2	4.36

TABLE II (*Continued*)

Compound	R	R′	K_i (m*M*)
20 9-(β-D-Allofuranosyl)adenine	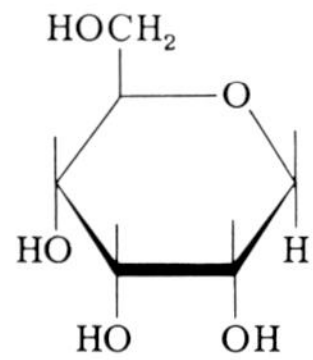	H_2	∞
21 Puromycin nucleoside	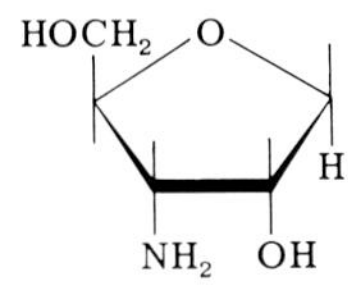	$(CH_3)_2$	1.88
22 Puromycin	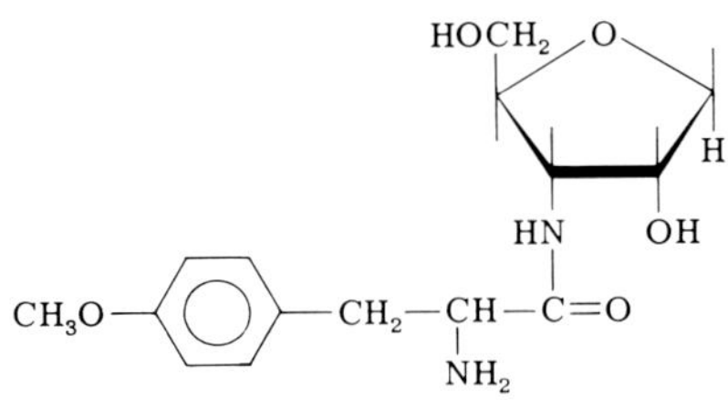	$(CH_3)_2$	2.47

Group D: Azalogs of Adenosine

Compound	Structure	K_i (m*M*)
23 Tubercidin (7-deazaadenosine)	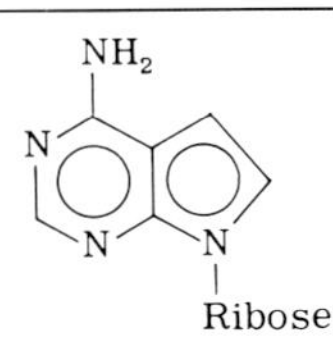	0.017
24 Formycin (8-aza-9-deazaadenosine)	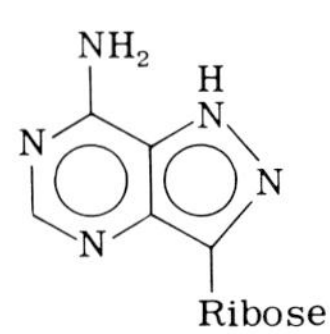	0.203
25 AICAR (4-amino-5-imidazole carboxamide riboside)	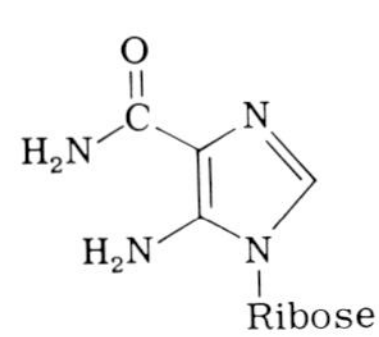	0.62

[a] From Taube and Berlin (1972). Reproduced from *Biochimica et Biophysica Acta* by permission of Elsevier, Amsterdam.

[b] K_m.

thymidine. This indicates a single membrane carrier for all the nucleosides tested. Free bases had essentially no affinity for the carrier.

Taube and Berlin investigated the substrate specificity of the carrier by determining the ability of a range of nucleoside analogs to inhibit transport of 0.007 m*M* adenosine-^{3}H measured over 45 seconds at 37°C as described in Fig. 1. The results, expressed as inhibition constants K_i, are grouped according to categories of molecular structure in Table II. Group A lists purine ribonucleosides in decreasing order of affinity; substitutions of bulk similar to that of the 6-amino group of adenine are well tolerated, but the poor affinity of purine riboside, with no substituted groups, and of xanthosine, with a 2-hydroxyl substituent, indicates that conjugated sites at positions other than C-6 may be significant for binding. The imidazole portion is nonessential, as shown in group B; uridine is comparable in affinity to the strongest purine inhibitor. However, the pyrimidine ring is important. Thus in group D it is seen that tubercidin, which is modified in the imidazole ring, binds well, while 4-amino-5-imidazolecarboxamide ribonucleoside, in which the pyrimidine ring is eliminated, binds poorly. Finally, with respect to the base–sugar linkage, the low affinity of pseudouridine and formycin (group D) suggests nitrogen-to-carbon bond specificity and with respect to the sugar itself (group C), the 3′-hydroxyl in the alpha-configuration appears to be the most important determinant. In support of this the introduction of a charged ion such as borate, which forms complexes with the 2′- and 3′-hydroxyl groups of ribose, strongly inhibits ribonucleoside transport (but not 2′-deoxyribonucleoside transport in which complex formation cannot occur).

The comparable affinities of purine and pyrimidine nucleosides for transport, despite the marked differences in spatial relationship of ribose to pyrimidine as compared with ribose to purine, were interpreted as evidence that binding occurs by an induced fit of a flexible carrier about a substrate. The spatial differences between purine and pyrimidine nucleosides are illustrated for cytidine and isoguanosine in Fig. 7. The base–ribose bonds are displaced approximately 2.5 Å and rotated 11° with respect to each other. Nevertheless, isoguanosine and cytidine have similar binding affinities. Since the foregoing analysis of specificity makes it exceedingly unlikely that critical structural groups can be displaced from the combining carrier site without marked effects on affinity, it seems that substrate-induced conformational changes of the carrier must occur to accommodate these structurally distinct molecules.

Further analysis similar to that already described for purine bases

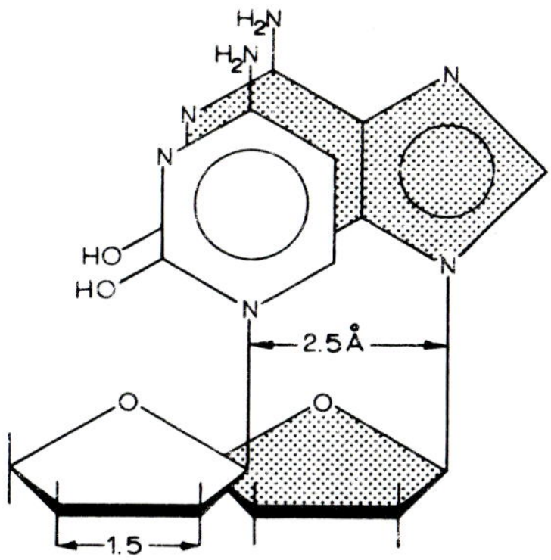

FIG. 7. The spatial relationship between base and sugar moieties of a pyrimidine nucleoside (cytidine) and a purine nucleoside (isoguanosine). Cytidine is shown with the atomic nuclei of its pyrimidine moiety superimposed on the corresponding pyrimidine moiety of isoguanosine. After superimposition, the carbon–nitrogen glycosidic bonds of the two nucleosides are separated by 2.5 Å. Individual hydroxyl groups of ribose, which make distinguishable contributions to the binding of the nucleosides to the carrier protein, are separated by only 1.5 Å. The representation does not indicate the difference in carbon–nitrogen bond angles with the bases, nor the preferred rotations of the ribose with respect to the plane of the bases. (From Taube and Berlin, 1972. Reproduced from *Biochim. Biophys. Acta* by permission of Elsevier, Amsterdam.)

established that binding specificities for the adenosine carrier and for the first major enzyme of adenosine metabolism in these cells, adenosine deaminase, are widely different. It was also shown that brief trypsinization of PMN does not affect adenosine transport, although it depresses the carrier-mediated uptake of lysine (Tsan *et al.*, 1973).

Berlin (1973) also analyzed the temperature dependence of nucleoside transport in leukocytes. There is a sharp transition temperature at approximately 25°C for both K_m and V_{max} of adenosine uptake in alveolar macrophages, which probably does not correspond to the temperature of phase transition for the bulk membrane lipids; and there is no transition temperature for nucleoside transport in PMN (Fig. 8). Colchicine, an inhibitor of adenosine transport, abolishes this transition, changing the slope of the plot over the entire temperature range to that normally measured only at low temperature (Fig. 9). In addition, the chemical specificity of transport is significantly altered at low temperature and after colchicine. For example, there is a large increase in K_i for puromycin nucleoside and other 3′-amino derivatives in cooled or colchicine-treated cells. From this, Berlin has proposed that the macrophage nucleoside carrier exists as an

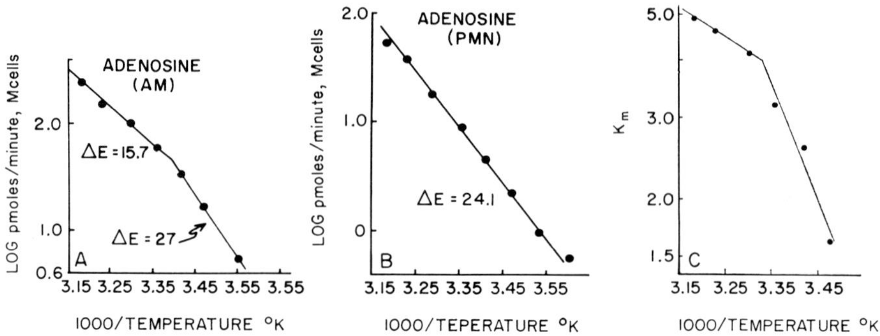

FIG. 8. Arrhenius plots of transport velocity and transport K_m for adenosine in leukocytes. Suspensions of rabbit PMN (4×10^6/ml) and alveolar macrophages (0.6×10^6/ml) were allowed to form monolayers at 37°C on 22-mm glass cover slips. The monolayers were drained and placed on aluminum bars in thermal equilibrium with a circulating bath at the desired temperature, and uptake of adenosine-^{3}H measured over 30 seconds as described in the legend for Fig. 1. (A) and (B) show log of transport velocity (picomoles per minute per million cells) against $1/T$ (the absolute temperature) at the K_m for transport at 37°C (0.04 mM for macrophages; 0.01 mM for PMN). ΔE is the activation energy defined from the slope of the plot taken equal to $-\Delta E/RT$, where R is the gas constant. The plot for macrophages shows a sharp break near 25°C; in contrast, the plot for PMN is linear. (C) is a plot of log K_m for transport versus $1/T$ measured in macrophages; it also shows a change in slope at about 25°C. (From Berlin 1973.)

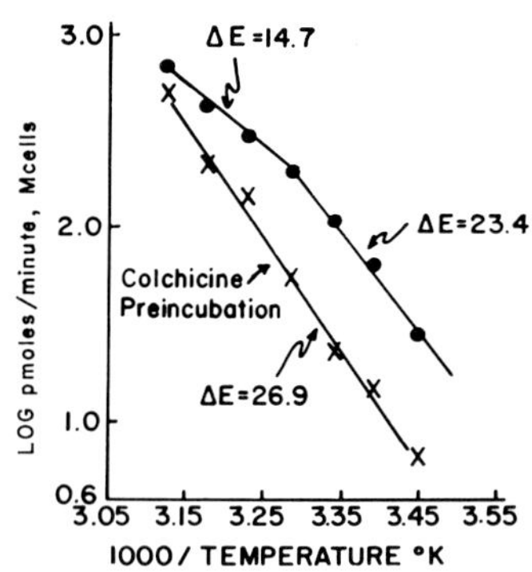

FIG. 9. Arrhenius plots of adenosine transport in alveolar macrophages with or without colchicine preincubation. Monolayers were formed in the presence and absence of 0.01 mM colchicine, and adenosine transport measured during 30-second incubations over a range of temperatures as described in the legend for Fig. 8. After colchicine pretreatment there is no thermal transition point, and the slope of the Arrhenius plot is essentially equal to that obtained without colchicine at temperatures below the transition point. (From Berlin, 1973.)

equilibrium between conformers favored by high and low temperatures. Their existence is shown operationally by characteristic substrate specificities and activation energies for translocation of substrates. Colchicine appears to stabilize the "low-temperature" conformer. The implications of these observations are discussed in Section V.

The process of nucleoside transport in tissue culture cells has also been extensively investigated and, as will become clear in later sections of this article, interpretations of many of these studies are somewhat controversial. We discuss some of the simplest cases here.

Plagemann and Erbe (1972a) showed that incubation of Novikoff hepatoma cells with cyanide rapidly depletes cells of intracellular adenine and uracil nucleotides and prevents phosphorylation of uridine taken up from the medium. That is, uridine acts as a nonmetabolizable transport substrate in these cells. The influx of uridine was linear over about 2 minutes at 18°C, was not accumulative, and was independent of a source of energy. Thus uridine appears to enter these cells by a facilitated-diffusion system similar to that described for other isolated cell systems.

Other reports by Plagemann have confirmed that the transport system in Novikoff hepatoma cells is very similar to that in red cells and leukocytes. For example, his studies of adenosine, uridine (Plagemann, 1970a), and thymidine (Plagemann and Erbe, 1972b) uptake and incorporation into nucleotide and nucleic acid at 5 and 10 minutes in hepatoma cells showed that uptake of all these substrates is saturable and strongly temperature-dependent at low substrate concentrations. When the extracellular nucleoside concentration is high, uptake is proportional to concentration and only weakly temperature-dependent, indicating that nucleosides can enter hepatoma cells by a carrier at low concentrations and by simple diffusion (or a second carrier with very high K_m) at high concentrations. These investigators express their results in terms of K_m and V_{max}; however, because of the long (5-minute) incubation period these numbers refer to uptake (the net resultant of transport and phosphorylation) and not to transport alone.

Plagemann (1971) has also reported differences between nucleosides with respect to transport kinetics (K_m and V_{max} of uptake and phosphorylation after a 5-minute incubation), extent of transport inhibition by nucleosides, and susceptibility of the transport process to heat shock (incubation at 47.5°C for 5 minutes) in hepatoma cells. He found that uptake of inosine and guanosine is inhibited to the same extent by all nucleosides tested, and by persantin, phenethyl alcohol, and heat shock; adenosine uptake is not inhibited by inosine and guanosine and is less susceptible to heat shock; and uridine and cytidine fluxes are inhibited in a fairly similar manner to each other, and are most susceptible to heat shock. From this he suggests there may be at least three different membrane carrier systems for nucleosides in these cells. In a similar analysis of deoxynucleoside transport, Plagemann and Erbe (1974) have suggested there are also several separate and specific carriers for these compounds. However, these data can also be accommodated by Berlin's proposal of a flexible common carrier with different affinities for its different substrates. By analogy with the effects of colchicine, heat shock may promote changes in carrier conformation which could significantly shift substrate specificity for transport. In addition, there is very likely to be competition between nucleosides for intracellular metabolism during the extended incubation times employed. Thus the lower incorporation of one radioactive nucleoside in the presence of a second nucleoside may be due to competition for a membrane transport carrier, which would reduce influx and also, if the two substrates share a common metabolic fate (hypoxanthine and guanine, for example), reduce intracellular trapping which may increase backflux.

In the single published analysis for a tissue, a facilitated diffusion mechanism also operates. Oliver (1971) studied the uptake of cytosine arabinoside in immature rat uterus, a tissue that does not metabolize this nucleoside, using a double-isotope technique to differentiate between nucleoside in the intracellular and extracellular space of the tissue. Uptake was saturable, nonaccumulative, and competitively inhibited by other nucleosides, but not by bases, amino acids, or sugars. In addition, a counterflow of arabinosylcytosine out of the tissue was induced by a gradient of uridine (Fig. 10). These data are compatible with transport by facilitated diffusion. An analogous transport system for sugars operates in this tissue (Roskoski and Steiner, 1967).

B. The Relationship between Transport and Phosphorylation of Nucleosides

In the discussion of purine and pyrimidine base transport, it was necessary to emphasize that permeation and phosphorylation are separate events in animal cells. It is necessary again to examine carefully the evidence concerning this point for nucleosides, since several investigators have favored the hypothesis that nucleoside transport is mediated by nucleoside kinases in animal cells in tissue culture.

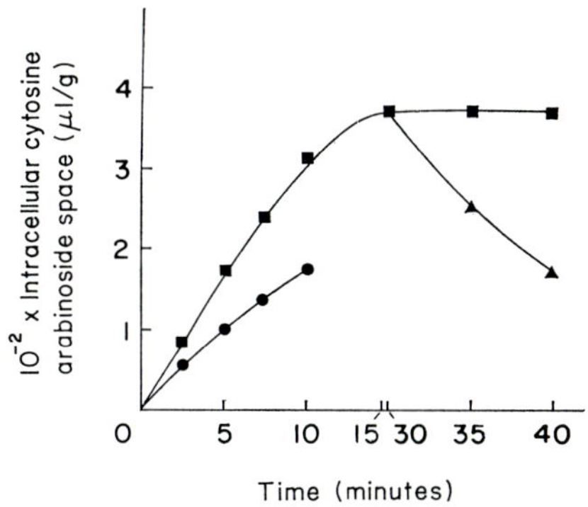

FIG. 10. Uptake of cytosine arabinoside in immature rat uterus. Groups of three uteri (about 50 mg of tissue) from immature female rats were incubated in stoppered tubes at 37°C in 1 ml of gassed Fischer's medium containing cytosine-^{3}H arabinoside (1 m*M*, 0.5 μCi/ml) and sucrose-^{14}C (1 m*M*, 0.4 μCi/ml). At various times the uteri were removed, rinsed briefly in cold saline, and blotted dry. Wet weights were recorded, and the tissue dissolved with 1 ml Nuclear Chicago solubilizer and counted. Portions of the media were also counted, and radioactivity from ^{14}C and ^{3}H determined. The *permeant space,* defined as the volume of tissue water required to maintain the permeant in the tissue at the concentration of the incubation medium, was calculated for both nucleoside and sucrose from the formula: permeant space (μl/gm) = (dpm of permeant/gm of tissue)/(dpm of permeant/μl of medium). Since sucrose is excluded from the intracellular space, intracellular cytosine arabinoside space (microliters per gram) = total cytosine arabinose space (microliters per gram) minus sucrose space (μl/gm). Total (extracellular plus intracellular) cytosine arabinoside space reached a maximum of 820 μl/gm after 10 minutes; the water content of the tissue measured by drying to constant weight was also 820 μl/gm, so the uterus does not concentrate cytosine arabinoside relative to the medium. Maximum extracellular space occupied by sucrose was 445 μl/gm. By difference the intracellular cytosine arabinose space is 375 μl/gm at equilibrium (squares). The rate of uptake into the intracellular space is depressed when a second nucleoside, uridine, is present at 10 m*M* in the medium (circles). Addition of 10 m*M* uridine to the medium after 30 minutes promotes a counterflow of cytosine arabinoside out of the tissue (triangles). This analysis demonstrates that nucleosides are transported by facilitated diffusion in rat uterus. It does not provide information about the kinetic properties of the carriers, which in any case are probably different in the different cell types that comprise the tissue. (From Oliver, 1971.)

The results described above showing rapid carrier-mediated uptake of nonmetabolizable nucleosides in erythrocytes and L1210 cells lacking the appropriate kinase, and inhibition of metabolizable nucleoside uptake by nonmetabolizable analogs (with low binding affinity to kinases) in leukocytes, constitute direct and unequivocal evidence that kinases are not involved in the transport process in these cells. In leukocytes it was also found that the first enzyme of adenosine metabolism is adenosine deaminase—not adenosine kinase.

In support of this, Breslow and Goldsby (1969) isolated mutants of Chinese hamster fibroblasts that fail to transport thymidine even though the kinase activity is still 50% of that of the parent cells. Similar thymidine transport mutants with active thymidine kinase have been isolated from haploid frog cells in culture (Freed and Mezger-Freed, 1973). It is interesting that both these cell lines transport uridine normally, suggesting there may be separate transport carriers for different nucleosides in these cells.

Studies with thymidine kinase–deficient cells have also demonstrated the separation between transport and phosphorylation. Steck *et al.* (1969) showed that thymidine is a competitive inhibitor of uptake of radioactive uridine, adenosine, and cytidine in cell lines lacking this enzyme, as well as in the normal parental strains.

Plagemann has provided further evidence for this from studies of uptake of uridine (Plagemann and Roth, 1969) and thymidine (Plagemann and Erbe, 1972b) in Novikoff hepatoma cells. In each case the apparent V_{max} for incorporation into whole cells (transport plus phosphorylation) is at least an order of magnitude lower than the corresponding values for *in vitro* phosphorylation in cell extracts. (Assuming of course that enzyme activity in extracts is comparable to activity in whole cells.) Further, competitive inhibitors of incorporation like adenosine, persantin, and phenethyl alcohol are not inhibitors of the corresponding kinases, kinase activity is not associated with the membrane, and heat shock depresses transport without reducing *in vitro* kinase activity.

Nevertheless, several investigators have held to this concept of kinase-mediated uptake in cultured cells. We indicate briefly how their data can be reinterpreted to be consistent with the complete separation of transport and phosphorylation.

Scholtissek (1968) first suggested a role of kinases in nucleoside uptake from his studies of transport inhibition by persantin [2,6-bis(diethanolamino)-4,8- dipiperidinopyrimido(5,4-*d*)- pyrimidine; dipyridamol] in chick fibroblasts. He employed metabolizable nucleosides as permeants and very long incubation times (15–90 minutes), so that all the intracellular radioactivity was recovered as nucleotides and nucleic acids. His results showed that incorporation of tracer amounts (of the order of 0.001 m*M*) of all nucleosides is inhibited in an apparently competitive manner by persantin, and that the block of incorporation of one nucleoside by persantin can be reversed by addition of a high concentration (on the order of 0.01–0.5 m*M*) of the same nucleoside or of a different nucleoside that is phosphorylated by the same kinase. From this it was proposed that nucleoside uptake is mediated by a nucleoside—kinase complex in the membrane, and that persantin competes with the nucleoside in the complex. However, persantin had no effect on kinase activities *in vitro*, and in any case Kessel and Hall (1970) showed that persantin inhibits influx and efflux of deoxycytidine in a subline of L1210 lacking deoxycytidine kinase. This seems to be strong evidence against the model for nucleoside transport based on a nucleoside-kinase complex in the membrane.

Hare (1970) and Schuster and Hare (1971) have also implicated kinases in transport from studies of thymidine uptake in normal, thymidine kinase–deficient (TK^-), and polyoma-transformed (elevated thymidine kinase) hamster cells. Uptake during 5 minutes at 25°C was greater than normal in tumor cells and less than normal in the mutant line, and all the thymidine was recovered as nucleotide in the TK^+ cells. When uptake was measured at 5°C, when kinase activity is reduced, all the cell lines transported thymidine at comparable low rates. From these data, Hare proposed that thymidine kinase regulates transport, perhaps by interacting reversibly or irreversibly with the membrane carrier. However, it seems extremely likely in these cells that the kinase is rate-limiting for intracellular utilization of thymidine and not for its transport across the membrane. No membrane-associated enzyme could be detected, a range of inhibitors or the uptake process in whole cells failed to inhibit the enzyme in cell extracts, and thymidine could still enter cells, although at a reduced rate, when incubated in the cold.

C. Inhibitors

So far we have mentioned several agents that inhibit nucleoside transport, but their specificity, reversibility, and mechanism of action have not been discussed. In this section we review agents that specifically or nonspecifically modify nucleoside uptake.

By far the largest group are nucleosides and nucleoside analogs. Since nucleosides generally seem to share a common membrane carrier, it is expected that they normally act as reversible, competitive inhibitors of transport. However, there is also a group of nucleoside analogs that are essentially irreversible, competitive inhibitors. These may be of particular importance in isolating and studying the nucleoside carrier, in studies of turnover of membrane carriers, and for experiments requiring selective blockage of nucleoside fluxes. The best studied are 6-thio ethers of purine or 2-aminopurine ribonucleoside. Paterson and Simpson (1965, 1966, 1967) found that a wide variety of these analogs, with substituents on the sulfur atom ranging from methyl to *p*-nitrobenzyl, all inhibited cleavage, exchange, and synthetic reactions involving nucleosides in intact human erythrocytes and Ehrlich ascites cells, but not in cell extracts. This indicated an effect on nucleoside transport rather than inhibition of intracellular nucleoside phosphorylase activity. In a later study, Paterson and Oliver (1971) confirmed this proposal by showing that the carrier-mediated uptake and efflux of uridine is reduced to zero on exposure of red cells to 10^{-6} *M* *p*-nitrobenzylthioguanosine. The inhibition was competitive and could not be reversed by washing the cells. The corresponding base, *p*-nitrobenzylthiaguanine did not affect uridine uptake (Fig. 11). The nucleoside analog and related derivatives are also inhibitors of nucleoside uptake in L5178Y murine lymphoma cells (Warnick *et al.*, 1972) and in rabbit lung macrophages and PMN (J. M. Oliver and R. D. Berlin, unpublished). The 6-alkylmercaptopurine ribosides are specific for nucleosides, since they do not affect uptake of purine and pyrimidine bases, sugar, or amino acids.

Radioactive nitrobenzylthioinosine has been used to obtain information about the nucleoside transport carrier in the erythrocyte membrane. The analog showed a high-affinity, saturable, binding component which was not removed by washing but could be displaced by a more potent transport inhibitor, 2-hydroxynitroben-

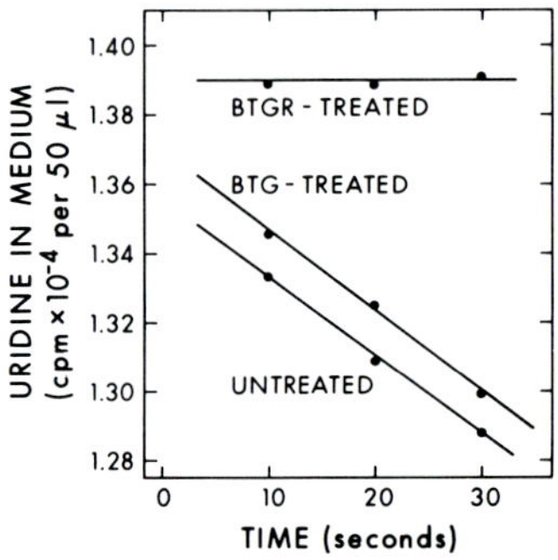

FIG. 11. Inhibition of uridine uptake (measured as decrease in medium concentration) in erythrocytes by *p*-nitrobenzylthioguanosine. Erythrocytes were incubated for 30 minutes at 37°C in TES-buffered saline (pH 7.4) containing no additive, 5×10^{-5} *M* *p*-nitrobenzylthioguanine (BTG-purine base), and 5×10^{-5} *M* *p*-nitrobenzylthioguanosine (BTGR-nucleoside). The cells were washed, suspended to 33% hematocrit, and assayed for their ability to take up 2.79 m*M* uridine-2-^{14}C at 25°C. The rapid sampling method was described in the legend for Fig. 5. Values are averages of seven replicate determinations. Uptake was linear over 30 seconds in control cells, and in cells incubated with the purine base analog. Cells preincubated with the nucleoside analog were completely impermeable to uridine. (From Paterson and Oliver 1971.)

zylthioguanosine, and a low-affinity component which was removed by washing. The extracted inhibitor was recovered as the unchanged molecule (Pickard *et al.*, 1973). Further analysis (Cass *et al.*, 1974) showed that the high-affinity binding was identical in intact erythrocytes and unsealed "ghosts," and that the nonsaturable binding component was greatly reduced by lysis and hypotonic washing of inhibitor-labeled red cells. Thus the saturable binding site is most likely a membrane component, and the low-affinity binding may be due at least in part to intracellular accumulation of the drug. It seems probable that the high-affinity binding is directly with a component of the nucleoside transport system, since the functional inhibition of uridine transport is strictly proportional to the amount of saturably bound inhibitor. Assuming this, Cass *et al.* have calculated there are between 1.0 and 1.5×10^4 nucleoside transport sites per cell, and a turnover number for uridine influx of 300–450 molecules per second at 25°C.

Another 6-substituted analog, N^6-isopentenyladenosine, has also been shown to specifically inhibit nucleoside uptake in a slowly reversible manner in mouse embryo cells (Hare and Hacker, 1972), and showdomycin is an essentially irreversible, competitive inhibitor of adenosine transport in rabbit alveolar macrophages (Strauss, 1974). High concentrations of colchicine and its photochemical derivative, lumicolchicine, also cause reversible, competitive inhibition of the uptake of nucleosides, but not of sugars and amino acids, in a range of cells in tissue culture (Mizel and Wilson, 1972a) and in rabbit lung macrophages (Berlin, 1973: see Section IV,A; Section V). This appears to be an effect of colchicine separate from its established effect on microtubular protein, since a photochemical derivative of colchicine which does not disrupt microtubules, lumicolchicine, also inhibits transport, while vinblastine, a structurally dissimilar alkaloid which disrupts microtubules, does not affect transport.

The antibiotic streptovaricin D is also a strong inhibitor of nucleoside uptake. Tan and McAuslan (1971) found that uptake of uridine in HeLa Cells measured as early as 20 seconds was competitively and reversibly inhibited by streptovaricin D but not by other streptovaricins or rifampicin. Uptake of adenosine and thymidine, but not of amino acids, was also inhibited. Similarly, 2-mercapto-1-(β-4-pyridethyl)benzimidazole (MPB) inhibited nucleoside incorporation into acid-soluble nucleotides measured after a 1-hour incubation in chick and rat embryo cells, 3T3, HeLa, and several tumor cell lines (Nakata and Bader, 1969). MPB had no effect on nucleoside kinase activity in cell extracts, nor on ^{32}P incorporation into RNA and DNA, indicating its effect is most likely on membrane transport rather than on intracellular nucleotide and nucleic acid metabolism. Some acridines (Scholtissek and Becht, 1966) may also inhibit nucleoside uptake in animal cells.

Collins and Roberts (1971) have reported that low concentrations of dimethyl sulfoxide (DMSO) inhibit nucleoside transport selectively in L cells; 5% DMSO reversibly inhibited the uptake of cytidine and uridine by 90%, without affecting adenosine uptake at this level. Inorganic and organic mercurials have also been identified as potent but probably nonspecific inhibitors of nucleoside transport (Tsan and Berlin, 1971; Schuster and Hare, 1971; Plagemann and Erbe, 1972b).

Another inhibitor of nucleoside transport mentioned briefly in Section IV,B is persantin or dipyridamol. This compound inhibits nucleoside uptake in heart (Kubler *et al.*, 1970), erythrocytes (Kubler and Bretschneider, 1964), hepatoma cells (Plagemann, 1971), and chick fibroblasts (Scholtissek, 1968). However, it also inhibits influx of inorganic phosphate (Gerlach *et al.*, 1964) and certain sugars (Deuticke *et al.*, 1964) into erythrocytes, indicating that its effect on nucleoside transport is not specific. Studies by Kessel and Dodd (1972) on murine leukemia cells have confirmed this lack of specificity. They showed that low concentrations of persantin inhibit influx of phosphate, fucose, deoxycytidine, and adenosine, and that efflux of these compounds is also inhibited at higher concentrations. In contrast, persantin seems to inhibit efflux of 3-*O*-methylglucose, uridine, uracil, and cycloleucine preferentially, and higher levels of inhibitor are required to block influx of these permeants. These experiments indicate that persantin can modify activity of a variety of membrane transport systems.

Phenethyl alcohol is also a nonspecific inhibitor of membrane transport, causing rapid and reversible blockage of uridine, thymidine, amino acid, and choline uptake in hepatoma cells (Plagemann, 1970b). Plagemann and Sheppard (1974) have reported similar competitive, reversible, but nonspecific inhibition of incorporation of nucleosides and other permeants (hypoxanthine, choline, and deoxyglucose) in hepatoma cells by the theophylline, papaverine, and prostaglandins. Benedetto and Cassone (1974) have presented further evidence for inhibition of uridine transport by theophylline in HeLa cells. These compounds may interact directly with transport carriers or other membrane components. However, they are all agents that elevate intracellular levels of cyclic AMP (cAMP), and so an alternative explanation for inhibition by these compounds may be that cAMP affects uptake and/or metabolism of transported substrates during the long incubation periods employed. Effects of cAMP on transport are discussed further in Section IV,D.

Mizel and Wilson (1972b) showed that low concentrations (8×10^{-8} *M*) of cytochalasin B, a substance best known for its putative effects on microfilaments, inhibits transport of hexoses in several tissue culture cell lines, but concentrations of 8×10^{-8}–3×10^{-6} *M* do not affect uptake of thymidine or uridine measured at 10 minutes in these cells. In contrast to this, Plagemann and Estensen (1972) claim that 4×10^{-6} *M* cytochalasin is a competitive inhibitor of uridine and thymidine uptake and incorporation into nucleotides and nucleic acids in hepatoma cells, as well as of hexose transport. This discrepancy may be due to different sensitivities to cytochalasin in different cell lines, since the highest concentration used by Mizel and Wilson was 3×10^{-6} *M*, or to the indirect effects of inhibition of sugar transport on nucleoside metabolism.

In general, one is impressed with the enormous variety of chemical agents, ranging from complex alkaloids to simple alcohols, that inhibits nucleoside transport in an apparently competitive manner. Some of these (the nucleoside analogs and colchicine) appear to act specifically on the nucleoside carrier, but many others profoundly influence the transport of other nutrients as well, suggesting that nonspecific interactions with membrane components may indirectly modify carrier activity. Some possible explanations for this extreme sensitivity of nucleoside transport to a variety of agents are discussed in Section V.

D. Physiological Modification

Several reports of changes in nucleoside transport associated with physiological state or alterations in the extracellular environment of cells have been published. Conditions that appear to modify transport include stage of the cell cycle, density of the cell population, loss of density-dependent growth inhibition, availability of serum factors, intracellular levels of cAMP, and lectin-induced transformation in lymphocytes. Close analysis suggests that some of these effects may be due to direct effects on membrane transport systems, while others may simply reflect changes in intracellular metabolism resulting in increases or decreases in nucleoside uptake.

In general, it is found that transport rates rise significantly from G_1 to S and drop after the G_2 phase of the cell cycle. Sander and Pardee (1972) showed that, in Chinese hamster ovary and L cells synchronized by incubation in isoleucine-free medium and collection of mitotic cells from monolayer cultures by shaking, uptake of radioactive uridine and thymidine at 15 minutes was low in early G_1, rose as the cell progressed through the cell cycle, and dropped again after mitosis. Everhart and Rubin (1974) also found a marked increase in thymidine uptake measured at 15 minutes in Chinese hamster ovary cells as they moved from G_1 into S phase. Similarly, Stambrook *et al.* (1973) measured a 10-fold increase in uridine uptake at 10 minutes in late S and G_2 compared with F_1 in Chinese hamster cells synchronized by colcemid. Uridine kinase activity fluctuated only 2.5-fold during the cell cycle. And in addition, Plagemann *et al.* (1974) showed that, in hydroxyurea-synchronized Novikoff cells, thymidine uptake doubles in S or late S phase and decreases at mitosis. At all stages intracellular radioactive thymidine was completely phosphorylated to dTTP, although pulse-chase experiments to examine the fate of the trinucleotide showed there was an appreciable rate of dephosphorylation back to thymidine, as well as removal of dTTP for DNA synthesis during S phase. It was not established if this degradation of dTTP resulted in a backflux of thymidine at some stages in the cycle that might compromise estimates of transport rate based on long (5-minute) incubation times. The finding that the apparent V_{max} was increased but the K_m unchanged suggested that new transport sites may operate in S phase. Of course, all these data must be interpreted with caution, since uptake was measured after relatively long incubation periods, when metabolism (which varies throughout the cell cycle) and not transport may be rate-limiting.

Transport in normal cells, but not transformed cells, is thought to be inversely related to cell population density. Plagemann *et al.* (1969) found there was about a 10-fold fluctuation in uridine uptake in Novikoff cells measured after a 30-minute incubation, being maximal in the exponential phase and minimal in the stationary phase. Similar fluctuations in uridine kinase activity were observed, but were thought not to influence the measured rate of uptake, since the rate of uridine phosphorylation measured in cell extracts at any stage of the cycle was always greater than the rate of uptake in intact cells. It is of course difficult to relate enzyme activities in extracts to the conditions within intact cells, particularly where activities are influenced by nucleotides (feedback inhibition). Similarly, Weber and Rubin (1971) reported a reduction in 30-minute uptake of uridine in chick embryo fibroblasts and mouse 3T3 cells at confluency. At very high substrate concentrations, incorporation was the same in both sparse and dense cultures. This was interpreted as evidence for a decreased transport rate into density-inhibited cells, which can be overcome by saturating amounts of substrate. Virally transformed 3T3 cells that no longer show density-dependent growth inhibition fail to show this decrease in transport rate in dense culture (Cunningham and Pardee, 1969). In addition, uridine transport in normal 3T3 is reduced by a dialyzable substance detected in growth medium from confluent 3T3. This substance does not inhibit transport when added to sparse cultures of polyoma virus–transformed 3T3 (Pariser and Cunningham, 1971).

These studies, like the reports of variations in transport throughout the cell cycle, suffer from the absence of data about the initial rate of uptake. In our experience the increased uptake of nucleoside in exponential cells is largely due to a rate of intracellular metabolism greater than that in confluent cells. For example, Fig. 12 shows that the initial rate (0–1 minute) of adenosine transport in exponential 3T3 cells is perhaps double the initial rate in density-inhibited 3T3, whereas the accumulation of adenosine after 10 minutes (transport plus metabolism) is eight times greater in the growing cells. These data indicate that intracellular metabolism of adenosine may be considerably more sensitive to cell density than the membrane transport system. We find the same is true when comparing adenosine uptake in 3T3 and SV3T3 cells; uptake is not greatly different in the first minute, but the greater rate of intracellular metabolism in the transformed cells gives a much greater accumulation of substrate in these cells after, for example, 10 minutes. Thus the experiments reviewed above need to be repeated during the initial phase of transport before firm conclusions can be drawn. This point needs to be emphasized, because several investigators have proposed that these apparent large fluctuations in membrane transport may be important in regulating cell division and in the process of malignant transformation (Pardee, 1971; Holley, 1972).

There appear to be factors in serum that can stimulate transport of several nucleosides. Addition of fresh serum to confluent monolayer cultures initiates a new round of cell division. Accompanying this, Cunningham and Pardee (1969) demonstrated a two to fourfold increase in uridine incorporation after a 15-minute incubation in confluent 3T3 cells preincubated for 10 minutes with fresh serum. Nonconfluent cells had a higher basal rate of uridine incorporation and were less sensitive to serum stimulation. No attempt was made in this study to distinguish between effects on uridine transport and intracellular phosphorylation. However, in a subsequent report, Cunningham and Remo (1973) showed that serum stimulated thymidine

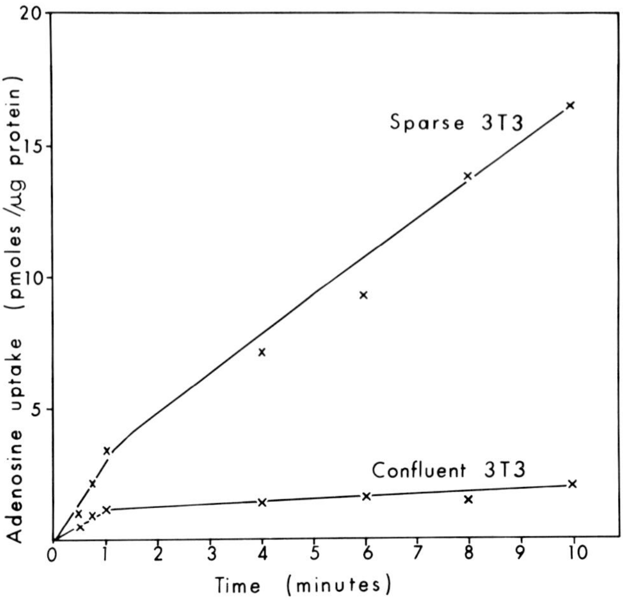

FIG. 12. Time course of adenosine transport in exponential and confluent 3T3 mouse fibroblasts. Sparse and dense cultures of 3T3 grown on 15-mm cover slips were incubated with 0.3 ml of 0.01 m*M* adenosine-^{3}H in phosphate-buffered saline (PBS) as described in the legend for Fig. 1. At various times the cover slips were drained, rinsed through four changes of cold PBS, collected in scintillation vials, and digested with 1 ml 0.4% sodium hydroxide–2% sodium carbonate overnight. Radioactivity was measured in 0.5-ml portions, and 0.5 ml was used for Lowry protein determination. Results are expressed as picomoles of adenosine per microgram of cell protein. The initial rate of transport (0–1 minute) is about doubled in exponential cells as compared with confluent cells, while the amount of adenosine accumulated after 10 minutes is eight times greater in growing cells (R. D. Berlin and J. P. Fera, unpublished.)

uptake even in cells lacking thymidine kinase. The increase, measured during a 10-minute incubation, was due to an increase in apparent V_{max} but no change in K_m, indicating increased numbers of functional carrier molecules in the membrane. Cycloheximide and cycloheximide plus actinomycin D antagonized this response to serum, suggesting that new protein synthesis was involved. The stimulation coincided in time with initiation of DNA synthesis and appeared to be specific for pyrimidine nucleosides and deoxynucleosides, since purine nucleosides and purine deoxynucleosides were not affected. This demonstration of changes in permeability in a system not complicated by metabolism is perhaps the best evidence available that physiological stimuli can affect transport activity directly in cultured cells. In support of these observations, DeAsua *et al.* (1974) found that the uptake of uridine measured after 2.5 minutes in confluent 3T3 cells is increased within 10 minutes after addition of 25% dialyzed serum and reaches a maximum at 25 minutes. An increase in phosphate transport and a decrease in intracellular cAMP occurred immediately on addition of serum. Hare (1972a,b) also showed an increase in uridine incorporation measured over 30 minutes when serumless mouse embryo cells are exposed to serum. This stimulation was readily reversible on removal of serum, and acquisition of uptake capacity was inhibited by cycloheximide and actinomycin D. Insulin and the serum glycoprotein fetuin similarly increased the apparent V_{max}, but did not affect the K_m of uridine incorporation in serumless cells. There was no significant difference in the activity of uridine kinase in cell extracts before or after the serum, and so it was proposed that serum affects the number or activity of the membrane transport carriers. Lemkin and Hare (1973) have reported that incubation of serumless cells with adenosine also stimulates uridine incorporation, but by a mechanism kinetically different from that for serum.

Effects of serum on transport are not limited to cells in tissue culture. Strauss and Berlin (1973) demonstrated that serum stimulates adenosine transport in rabbit alveolar macrophages. These investigators used a rapid sampling technique which allows analysis of changes in membrane transport as opposed to intracellular metabolic events. A 30-minute preincubation of cells with 0.5–5% serum stimulated the initial rate of adenosine transport, measured over 45 seconds, as much as 75%, and kinetic studies showed that serum increases V_{max} with little effect on apparent K_m of transport. The serum component was nondialyzable and was stable to heat at 65°C but not at 100°C. Migration inhibition factor (MIF), gamma globulin, macroglobulin, complement, and endotoxin were all without effect on adenosine transport. In a further analysis, Strauss (1974) showed that serum stimulation of adenosine transport is probably not due to interaction of serum with the active nucleoside binding site, since inclusion of thymidine or adenosine as protective substrates during preincubation does not alter the induced stimulation. Certain intracellular events are also unlikely to be involved, since serum stimulation is not affected by cycloheximide and chloramphenicol, and is not accompanied by changes in intracellular nucleotide levels. Experiments were also performed to determine whether serum activates existing transport sites or recruits new sites. Nucleoside transport is irreversibly inactivated when cells are preincubated with the nucleoside analog showdomycin, or with *p*-chloromercuribenzenesulfonate. Subsequent exposure to serum leads to an increase in transport, which is probably not due to removal of the inhibitors. This is evidence that new transport sites may be recruited in response to serum. Strauss has suggested a model in which serum might increase the number of functional transport carriers in the membrane without necessity for synthesis of new carriers. The model involves a coupling or clustering of transport sites, where only one site per group transports at any time and the others are in some fashion protected from interaction with showdomycin or other nucleoside substrates; according to this model, serum stimulates transport by uncoupling sites so that they can function independently.

In contrast to serum, cAMP generally depresses cell growth and antagonizes serum stimulation of division. Several investigators consider that this substance affects nucleoside transport at the membrane level. For example, Kram *et al.* (1973) used a 1-hour incubation period to study uptake of uridine and also amino acids and sugars in mouse fibroblasts deprived of serum or exposed to dibutyryl cAMP plus theophylline. Incorporation of all substrates was reduced by the cyclic nucleotide and elevated by serum. Inhibition was also induced by prostaglandin E, which elevates cAMP levels, and could be reversed with cyclic GMP (Kram and Tomkins, 1973). From this they suggest that uptake is regulated by cAMP, and that enhanced incorporation of nucleosides in malignant cells may be due to their abnormally low cAMP levels. They also propose that the cAMP effect on transport may be mediated via microtubules, since colcemid and vinblastine antagonize the inhibition. Initial rate studies of the membrane transport process are absolutely essential before these proposals can be seriously considered.

Rozengurt and DeAsua (1973) have provided indirect support for the proposal that cAMP levels may regulate transport. They found that uridine uptake measured after 5 minutes at 37°C in serum-starved chick embryo cells is increased by agents that decrease cyclic nucleotide levels (serum and insulin), and decreased by agents that elevate cAMP (prostaglandins and theophylline). Insulin also enhances 30-minute incorporation of uridine in isolated bone cells, and cortisol inhibits in this system, according to Peck and Messinger (1970); however, cAMP was not implicated by these investigators. In addition, Lingwood and Thomas (1974) reported decreased nucleoside uptake at 15 minutes in dibutyryl cAMP-treated P815Y cells, and Rubin and co-workers (Hauschka *et al.*, 1972; Everhart and Rubin, 1974) also showed reduced thymidine uptake both at 37° and 1°C in dibutyryl cAMP-treated Chinese hamster cells during long incubation periods. These investigators consider the inhibition to be due to concomitant inhibition of thymidine kinase, but the reduction in nucleoside uptake (3- to 21-fold) is much greater than the 2-fold inhibition of kinase measured in cell extracts.

In contrast with all these data, Roller *et al.* (1974) have reported that cAMP induces an increase in thymidine uptake measured at 15 minutes in normal and virally transformed monkey kidney cells. This contradictory observation underscores the need for studies that separate transport from metabolism before a conclusion can be reached as to whether or not cAMP levels directly affect transport carrier activity.

Somewhat more satisfactory data are available about transport changes that accompany transformation of lymphocytes with phytohemagglutinin (PHA). This process is accompanied by a marked increase in the incorporation of uridine into nucleotides and RNA. Hausen and Stein (1968) first proposed that the concomitant rise in uridine kinase activity may be primarily responsible for this stimulation. However, in a later publication, Peters and Hausen (1971) showed that increased accumulation of label within the cells occurred even at times when a significant increase in uridine kinase was not detectable in cell extracts, and the uptake rate of uridine in whole cells was two orders of magnitude lower than maximal rates of kinase activity when measured in cell extracts *in vitro*. This indicated that PHA may directly affect the rate-limiting process or uridine uptake and that the slower increase in kinase activity may act later to facilitate metabolism. The PHA-stimulated uptake of uridine into acid-soluble material was only slightly sensitive to actinomycin D, and inhibition of protein synthesis with cycloheximide did not interfere with the induction of uridine uptake. Thus activation does not seem to require RNA or protein synthesis, but may be a direct effect of PHA on the membrane. Kay and Handmaker (1970) confirmed that uridine incorporation is accelerated by PHA prior to detectable increases in the amount of uridine kinase. They suggested, however, that PHA may relax feedback inhibition of the kinase *in vivo*, perhaps by altering the intracellular concentration of pyrimidine nucleotides. Again, studies of the initial rate or uridine uptake are required to resolve this question.

Finally, Piatigorsky and Whitely (1965) have presented evidence that fertilization promotes changes in incorporation of uridine in sea urchin eggs. Unfertilized eggs are very poorly permeable to uridine, and fertilization is followed by rapid uptake and concentration of uridine in the cells in the form of nucleotides. It is not clear from these experiments whether transport carriers were absent before fertilization, or whether the methods used were not sufficiently sensitive to detect equilibration of uridine across the membrane in the absence of intracellular metabolism.

Thus we can conclude that nucleoside transport, like the transport of purine and pyrimidine bases, is mediated by membrane carriers that do not require direct input of energy and do not lead to a net accumulation of unchanged nucleoside. Nucleoside kinases do not form part of the transport system, although they clearly modify uptake by phosphorylating the substrates intracellularly, thereby maintaining a concentration gradient for the diffusion of nucleoside across the membrane. The evidence to date indicates that there is a single carrier that mediates uptake of all purine and pyrimidine nucleosides in red cells and leukocytes, but that several carriers with different affinities for different substrates may operate in some tissue culture cell lines. However, a detailed analysis of the effects of colchicine on the specificity of transport (Berlin, 1973), discussed in Section V, suggests that low-molecular-weight modifiers could readily shift the specificity of a single transport system so as to result in an apparent increase in activity for one nucleoside—for example, thymidine—and a decrease in another such as adenosine. Nucleoside transport is very likely subject to modification by many physiological stimuli, but in most cases firm conclusions cannot be drawn because of the lack of appropriate kinetic data.

V. Base and Nucleoside Carriers as Membrane Proteins

Thus far we have considered the transport of nucleosides and bases as physiological processes without focusing on the biochemical basis of transport. From the high degree of substrate specificity, susceptibility to inactivation by protein reagents and, in bacteria, the direct isolation of proteins that can be identified with transport systems, it is generally accepted that essential (if not all) components of base and nucleoside transport systems are proteins.

These proteins may vary widely in their properties (substrate affinities, substrate specificities, sensitivity to inhibitors) from species to species and cell to cell, indicating they may not be chemically identical even though they all transport by the same kinetic mechanism—facilitated diffusion. For example, we have observed that the 6-alkylmercaptopurine ribosides that are essentially irreversible inhibitors in erythrocytes (Paterson and Oliver, 1971) are fairly readily reversible competitive inhibitors of adenosine uptake in macrophages and PMN and are relatively ineffective as inhibitors of cytosine arabinoside uptake in immature rat uterus (J. M. Oliver, unpublished). In addition, the most effective inhibitor species differ between red cells and white cells. Rather significant differences in specificity exist even between neutrophils and macrophages of the same species; Berlin (1973) showed that adenosine transport in rabbit macrophages is highly sensitive to inhibition by puromycin and its nucleoside at 37°C ($K_i = 0.24$ mM and 0.37 mM, respectively) while in rabbit PMN these compounds are poor inhibitors ($K_i = 2.5$ mM and 1.9 mM, respectively). These differences in specificity clearly denote corresponding differences in the combining sites of the nucleoside carrier proteins in the two cell types, and possibly in their primary structure. As noted above (Section IV,A), Berlin also reported a transition temperature at which substrate specificity and activation energy for transport changes abruptly for macrophages but not for PMN. Pretreatment of macrophages with colchicine alkaloids results in a transport system that shows no transition temperature and has the specificity and activation energy of the untreated system in the low-temperature range. Thus the macrophage nucleoside carrier appears to exist in two or more conformers, one favored by high temperature and one by low temperature or colchicine, while the nucleoside carrier in PMN appears to have only one operational conformation. Extrapolating from these differences in carrier properties between two related cell types from the same animal, it is possible that the growth of cells dependent on exogenous bases or nucleosides (see Section VI) could be selectively inhibited by the development of analogs with specific affinity for the base or nucleoside transport system of the particular cell type.

In the light of modern theories of membrane structure and organization, it is proper to consider that transport activity may vary as a function not only of the chemical nature of the carrier, but also as a function of the arrangement or topography of the carrier in the lipid of the cell membrane and of the molecular association of carrier protein with membrane lipid.

Recent studies strongly support the concept developed by Singer and Nicolson (1972) that membrane lipids exist in a highly fluid state and form a matrix in which the membrane proteins are embedded, the two components interacting through hydrophobic bonds. Membrane proteins may be free to undergo both rotational and translational movement in the fluid lipid. Moreover, evidence developed in our laboratory and others suggests that subcellular structures, particularly microtubules, may impose additional restraints on the mobility of membrane components (Berlin *et al.*, 1974). Studies of the functional activities of nucleoside and base transport carriers during phagocytosis in macrophages and PMN have shown that, in keeping with other membrane components, transport carriers are in fact mobile proteins whose topographical distribution is determined by microtubular structures.

During phagocytosis, when a large fraction of the membrane is internalized, there is no change in nucleoside or base transport. Since it can be shown that new carriers are not inserted into the membrane under these conditions, it appears that transport carriers are normally excluded from internalized membrane (Tsan and Berlin, 1971). However, when phagocytes are pretreated with the alkaloid colchicine, which dissolves microtubules (but not with lumicolchicine, which does not), transport no longer remains constant but decreases—and decreases in proportion to the degree of phagocytosis (Ukena and Berlin, 1972). That is, carriers are not preserved on the cell surface after microtubular disruption. This result indicates that transport carriers may assume a variable distribution over the surface and apparent translational mobility within it. This mobility is determined, at least in part, by microtubular proteins.

It is very likely that the membrane lipids can also modify the mobility and functional activity of transport carriers. It is well established that the activities of membrane enzymes for example ATPase (Roelofsen and van Deenan, 1973) can be drastically changed by changes in the lipid environment. Thus it may be anticipated that some agents or conditions that alter the membrane lipid or hydrophobic bonding will affect the activity of transport proteins. In our opinion this is the most likely cause of inhibition of nucleoside transport by such diverse lipophilic agents as ethanol, phenethyl alcohol, and cytochalasin B (Section IV,C); these agents may have their primary effect on the fluidity or organization of membrane lipids, which in turn promotes changes in the conformation of the embedded membrane proteins and gives rise to changes in substrate affinity (competitive or mixed kinetics). These indirect effects may appear similar, by kinetic analysis, to the changes induced by agents such as substrate analogs and colchicine which most likely interact directly with carrier proteins.

The role of the lipid environment in regulating transport may be elucidated by studying transport in cells with modified membranes. It is now known that the fatty acid composition of the membrane lipids of cultured cells can be drastically modified by growth in media supplemented with specific fatty acids (Steele and Jenkin,

1972; Wisnieski *et al.*, 1973; Williams *et al.*, 1974). Further studies of transport at different temperatures may also help to establish the role of the lipid environment; perhaps the transition in activation energy for macrophage nucleoside transport at 25°C reflects a local change in the lipid environment, even though it does not correspond to the transition temperature of bulk membrane lipids. Ultimately, of course, one could hope to establish the role of the lipid environment by using isolated carrier proteins and artificial membranes of known composition in experiments analogous to those now in progress with membrane ATPase. Nucleoside transport carriers have not yet been isolated but, as noted above, Paterson and co-workers have made progress toward this goal using radioactively labeled nucleoside analogs with high binding affinities to tag the transport protein on the membrane of erythrocytes.

Clearly, then, it is not sufficient to consider transport systems only from the kinetic approach. The surface of a cell is not an inert substrate holding a number of transport proteins in place. Rather, it is a dynamic structure capable of responding to intracellular and extracellular forces with changes in lipid fluidity and organization, carrier conformation, hydrophobic bonding with lipids, rotational and translational mobility, and turnover rate. Much more information is needed in this regard to explain how physiological (serum, cAMP, etc.) and nonphysiological substances modify transport activity.

VI. The Physiological Role of Base and Nucleoside Transport Systems

It is clear that a source of circulating purine compounds is essential *in vivo*, since several tissues have now been identified that show high rates of purine turnover and yet are incapable of *de novo* synthesis. These include the erythrocyte (Mager *et al.*, 1967), leukocytes (Scott, 1962), bone marrow (Abrams and Goldfinger, 1951), and the gastrointestinal tract (MacKinnon and Deller, 1973). Since animals including man can be maintained indefinitely on a purine-free diet, it must be assumed that there is a large supply of nondietary purine available. Mammalian cells are impermeable to nucleotides, so that this purine must be supplied as a base or nucleoside. Evidence that the liver is the most likely source of this material has been presented (Lajtha and Vane, 1958; Pritchard *et. al.*, 1970, 1975). Thus the transfer of purines between tissues is of physiological importance, and membrane carrier systems for bases and nucleosides are very likely involved in this process. We have also seen that variations in nucleoside transport may be expressions of many physiological states, including malignant transformation, cell density, stage of the cell cycle, serum concentration, pH of the culture medium (Ceccarini and Eagle, 1971), PHA stimulation of lymphocytes, and so on. These compensatory responses to altered growth conditions in tissue culture systems also suggest that transport activity may be physiologically important. The preservation of transport sites during phagocytosis supports this proposal.

It should be remembered of course that many cells in tissue culture do not require exogenous purines or pyrimidines for growth, even though they invariably transport and utilize preformed bases and nucleosides supplied in the medium in preference to synthesizing them *de novo* (Murray, 1971). It is possible that a major function of carrier systems in cells growing in a purine- and pyrimidine-free medium is to excrete toxic metabolites. Several purine bases are known mutagens and, for example, Chan *et al.* (1973) showed that adenosine kinase–negative cells excrete adenosine metabolites presumably via the appropriate carrier. All cells in fact excrete at least hypoxanthine, xanthine, or urate, which are the end products of nucleoside catabolism; and phagocytic cells must excrete an additional load of purine and pyrimidine derived from digested microorganisms and other cellular debris. The serosa-to-mucosa directionality of active purine transport in rat intestine is further evidence for a secretory rather than an accumulative role of transport carriers in some cells.

Thus we cannot at this stage absolutely define the physiological role of membrane transport systems for purine and pyrimidine bases and nucleosides, but it seems highly likely that they are required for the transfer of essential nucleic acid precursors to some cells and probably by all cells for the excretion of nucleic acid catabolites.

VII. Concluding Remarks

There is increasing evidence that the growth of animal cells may be regulated, at least in part, by the availability of essential nutrients. For this reason there is a great deal of interest in membrane transport mechanisms and in changes in transport activity that parallel changes in cell growth properties.

In the case of purine and pyrimidine bases and nucleosides, transport mechanisms seem now to be well established. In general, these compounds enter animal cells by facilitated diffusion, and transport carriers are clearly separate from the enzymes responsible for subsequent intracellular metabolism. The carriers are proteins which have lateral mobility in the plane of the membrane. Very little is yet known about the molecular properties of carrier molecules and their relationship to other protein and lipid components of cell surfaces.

We have emphasized the chief difficulty in interpreting the literature on base and nucleoside transport—that very often the transport event has not been adequately separated from intracellular metabolism. Too many investigators, failing to measure initial rates of transport, have drawn conclusions about the activities of carriers, which cannot be properly evaluated. Thus there is a large literature on the subject of, for example, variations in transport activity as a function of cell density, and yet it is still not possible to state with certainty that transport rate is significantly reduced in confluent cells as compared with cells in the exponential phase of growth.

Nevertheless, it is very likely, from the huge range of different physiological and nonphysiological agents that appear to modify transport and/or metabolism or nucleosides, that carrier activity is a sensitive and tightly controlled function. Moreover, since there is almost always an excess of intracellular enzymes capable of metabolizing bases and nucleosides, carrier activity is a very logical control point for regulating growth properties, at least those of cells with no or limited capacity for *de novo* synthesis of these nucleic acid precursors.

The most important goal for future research is to define the relationship between transport activity and cell growth properties. This requires kinetic determination of initial rates of transport under various physiological conditions, and also information about the molecular properties of carriers (conformation, mobility, turnover rates, interactions with other membrane components) in these different physiological states. Suitable techniques, at least for the kinetic analyses, are now available.

It may be possible to apply this information clinically. Substances that inhibit transport in cells that depend on exogenous purines for nucleic acid synthesis may impose selective growth control. Malignancies of bone marrow (leukemia) and of the gastrointestinal tract could fall within this category. Such inhibitors should not damage the majority of cells and tissues that have the capacity to synthesize purines *de novo*. In addition, since carrier properties such as susceptibility to inhibition are highly variable among cells, inhibitors that are particularly effective against specific cell types may be developed.

References

Abrams, R., and Goldfinger, J. M. (1951). *Arch. Biochem. Biophys.* **30,** 261.

Benedetto, A., and Cassone, A. (1974). *Biochim. Biophys. Acta* **349,** 53.

Benke, P. J., Herrick, N., and Herbert, A. (1973). *Biochem. Med.* **8,** 309.

Berlin, R. D. (1969). *Science* **163,** 1194.

Berlin, R. D. (1970). *Science* **168,** 1539.

Berlin, R. D. (1973). *J. Biol. Chem.* **248,** 4724.

Berlin, R. D., and Hawkins, R. A. (1968a). *Amer. J. Physiol.* **215,** 932.

Berlin, R. D., and Hawkins, R. A. (1968b). *Amer. J. Physiol.* **215,** 942.

Berlin, R. D., Oliver, J. M., Ukena, T. E., and Yin, H. H. (1974). *Nature (London)* **247,** 45.

Breslow, R. E., and Goldsby, R. A. (1969). *Exp. Cell Res.* **55,** 339.

Cass, C. E., and Paterson, A. R. P. (1972). *J. Biol. Chem.* **247,** 3314.

Cass, C. E., and Paterson, A. R. P. (1973). *Biochim. Biophys. Acta* **291,** 734.

Cass, C. E., Gaudette, L. A., and Paterson, A. R. P. (1974). *Biochim. Biophys. Acta* **345,** 1.

Ceccarini, C., and Eagle, H. (1971). *Proc. Nat. Acad. Sci. U. S.* **68,** 229.

Chan, T. S., Ishii, K., Long, C., and Green, H. (1973). *J. Cell. Physiol.* **81,** 315.

Collins, F. D., and Roberts, W. K. (1971). *Biochim. Biophys. Acta* **233,** 459.

Cunningham, D. D., and Pardee, A. B. (1969). *Proc. Nat. Acad. Sci. U. S.* **64,** 1049.

Cunningham, D. D., and Remo, R. A. (1973). *J. Biol. Chem.* **248,** 6282.

Czáky, T. Z. (1965). *Annu. Rev. Physiol.* **27,** 419.

DeAsua, L. J., Rozengurt, E., and Dulbecco, R. (1974). *Proc. Nat. Acad. Sci. U. S.* **71,** 96.

Deuticke, B., Dumn, J., and Gerlach, E. (1964). *Pflueger's Arch. Gesamte Physiol. Menschen Tiere* **280,** 275.

Everhart, L. P., and Rubin, R. W. (1974). *J. Cell Biol.* **60,** 434.

Freed, J. J., and Mezger-Freed, L. (1973). *J. Cell. Physiol.* **83,** 199.

Gerlach, E., Deuticke, B., and Dumn, J. (1964). *Pflueger's Arch Gesamte Physiol. Menschen Tiere* **280,** 243.

Hare, J. D. (1970). *Cancer Res.* **30,** 684.

Hare, J. D. (1972a). *Biochim. Biophys. Acta* **255,** 905.

Hare, J. D. (1972b). *Biochim. Biophys. Acta* **282,** 401.

Hare, J. D., and Hacker, B. (1972). *Physiol. Chem. Phys.* **4,** 275.

Harris, J. F., and Whitmore, G. F. (1974). *J. Cell. Physiol.* **83,** 43.

Hausen, P., and Stein, H. (1968). *Eur. J. Biochem.* **4,** 401.

Haushka, P. V., Everhart, L. P., and Rubin, R. W. (1972). *Proc. Nat. Acad. Sci. U. S.* **69,** 3542.

Hawkins, R. A., and Berlin, R. D (1969). *Biochim. Biophys. Acta* **173,** 324.

Hochstadt, J. (1975). *Crit. Rev. Biochem.* (in press).

Holley, R. W. (1972). *Proc. Nat. Acad. Sci. U. S.* **69,** 2840.

Hori, M., and Henderson, J. F. H. (1966). *J. Biol. Chem.* **241,** 1406.

Jacquez, J. A. (1962). *Biochim. Biophys. Acta* **61,** 265.

Jacquez, J. A., and Ginsberg, F. (1960). *Proc. Soc. Exp. Biol. Med.* **105,** 478.

Kay, J. E., and Handmaker, S. D. (1970). *Exp. Cell Res.* **63,** 411.

Kessel, D., and Dodd, D. C. (1972). *Biochim. Biophys. Acta* **288,** 190.

Kessek, D., and Hall, T. C. (1967). *Biochem. Pharmacol.* **16,** 2359.

Kessel, D., and Hall, T. C. (1970). *Biochim. Biophys. Acta* **211,** 88.

Kessel, D., and Shurin, S. B. (1968). *Biochim. Biophys. Acta* **163,** 179.

Kram, R., and Tomkins, G. M. (1973). *Proc. Nat. Acad. Sci. U. S.* **70,** 1659.

Kram, R., Mamont, P., and Tomkins, G. M. (1973). *Proc. Nat. Acad. Sci. U. S.* **70**, 1432.
Kubler, W., and Bretschneider, H. J. (1964). *Pflueger's Arch. Gesamte Physiol. Menschen Tiere* **280**, 141.
Kubler, W., Spieckermann, P. G., and Bretschneider, H. J. (1970). *J. Mol. Cell. Cardiol.* **1**, 23.
Lajtha, L. G., and Vane, J. R. (1958). *Nature (London)* **182**, 191.
Lassen, U. V. (1961). *Biochim. Biophys. Acta* **53**, 557.
Lassen, U. V. (1962). *Biochim. Biophys. Acta* **57**, 123.
Lassen, U. V. (1967). *Biochim. Biophys. Acta* **135**, 146.
Lassen, U. V., and Overgaard-Hansen, K. (1962a). *Biochim. Biophys. Acta* **57**, 111.
Lassen, U. V., and Overgaard-Hansen, K. (1962b). *Biochim. Biophys. Acta* **57**, 118.
Lemkin, J. A., and Hare, J. D. (1973). *Biochim. Biophys. Acta* **318**, 113.
Lieu, T. S., Hudson, R. A., Brown, R. K., and White, B. C. (1971). *Biochim. Biophys. Acta* **241**, 884.
Lingwood, C. A., and Thomas, D. B. (1974). *J. Cell Biol.* **61**, 359.
MacInnis, A. J., Fisher, F. M., and Read, C. P. (1965). *J. Parasitol.* **51**, 260.
MacKinnon, A. M., and Deller, D. J. (1973). *Biochim. Biophys. Acta* **319**, 1.
Mager, J., Herschko, A., Zetlin-Beck, R., Shoshani, T., and Razin, A. (1967). *Biochim. Biophys. Acta* **149**, 50.
Mizel, S. B., and Wilson, L. (1972a). *Biochemistry* **11**, 2573.
Mizel, S. B., and Wilson, L. (1972b). *J. Biol. Chem.* **247**, 4102.
Mudge, C. H., Berndt, W. O., and Valtin, H. (1973). *In* "Handbook of Physiology," Sect. 8, Chapter 19, Amer. Physiol. Soc. (J. Field, ed.), Williams & Wilkins, Baltimore, Maryland.
Murray, A. W. (1971). *Annu. Rev. Biochem.* **40**, 811.
Nakata, Y., and Bader, J. P. (1969). *Biochim. Biophys. Acta* **190**, 250.
Oliver, J. M. (1971). *Biochem. J.* **121**, 83.
Oliver, J. M., and Paterson, A. R. P. (1971). *Can. J. Biochem.* **49**, 262.
Pardee, A. B. (1971). *In Vitro* **7**, 95.
Pariser, R. J., and Cunningham, D. D. (1971). *J. Cell Biol.* **49**, 525.
Paterson, A. R. P., and Oliver, J. M. (1971). *Can. J. Biochem.* **49**, 271.
Paterson, A. R. P., and Simpson, A. I. (1965). *Can. J. Biochem.* **43**, 1701.
Paterson, A. R. P., and Simpson, A. I. (1966). *Can. J. Biochem.* **44**, 1423.
Paterson, A. R. P., and Simpson, A. I. (1967). *Cancer Res.* **27**, 353.
Peck, W. A., and Messinger, K. (1970). *J. Biol. Chem.* **245**, 2722.
Peters, J. H., and Hausen, P. (1971). *Eur. J. Biochem.* **19**, 502.
Piatigorsky, J., and Whiteley, A. H. (1965). *Biochim. Biophys. Acta* **108**, 404.
Pickard, M. A., Brown, R. R., Paul, B., and Paterson, A. R. P. (1973). *Can. J. Biochem.* **51**, 666.
Plagemann, P. G. W. (1970a). *Arch. Biochem. Biophys.* **140**, 223.
Plagemann, P. G. W. (1970b). *J. Cell. Physiol.* **75**, 315.
Plagemann, P. G. W. (1971). *Biochim. Biophys. Acta* **233**, 688.
Plagemann, P. G. W., and Erbe, J. (1972a). *J. Cell. Physiol.* **81**, 101.
Plagemann, P. G. W., and Erbe, J. (1972b). *J. Cell Biol.* **55**, 161.
Plagemann, P. G. W., and Erbe, J. (1974). *J. Cell. Physiol.* **83**, 337.
Plagemann, P. G. W., and Estensen, R. D. (1972). *J. Cell Biol.* **55**, 179.
Plagemann, P. G. W., and Roth, M. F. (1969). *Biochemistry* **8**, 4782.
Plagemann, P. G. W., and Sheppard, J. R. (1974). *Biochem. Biophys. Res. Commun.* **56**, 869.
Plagemann, P. G. W., Ward, G. A., Mahy, B. W. J., and Korbecki, M. (1969). *J. Cell. Physiol.* **73**, 233.

Plagemann, P. G. W., Richey, D. P., Zylka, J. M., and Erbe, J. (1974). *Exp. Cell Res.* **83**, 303.

Pritchard, J. B., Chavez-Peon, F., and Berlin, R. D. (1970). *Amer. J. Physiol.* **219**, 1263.

Pritchard, J. B., O'Connor, N., Oliver, J. M., and Berlin, R. D. (1975). *Amer. J. Physiol.* (in press).

Roelofsen, R., and van Deenan, L. L. M. (1973). *Eur. J. Biochem.* **40**, 245.

Roller, B., Hirai, K., and Defendi, V. (1974). *J. Cell. Physiol.* **83**, 163.

Roseman, S. (1969). *J. Gen. Physiol.* **54**, 138.

Roskoski, R., and Steiner, D. F. (1967). *Biochim. Biophys. Acta* **135**, 717.

Rozengurt, E., and DeAsua, L. J. (1973). *Proc. Nat. Acad. Sci. U. S.* **70**, 3609.

Sander, G., and Pardee, A. B. (1972). *J. Cell. Physiol.* **80**, 267.

Schanker, L. S., and Tocco, D. J. (1960). *J. Pharmacol. Exp. Ther.* **128**, 115.

Schanker, L. S., and Tocco, D. J. (1962). *Biochim. Biophys. Acta* **56**, 469.

Schanker, L. S., Jeffrey, J. J., and Tocco, D. J. (1963). *Biochem. Pharmacol.* **12**, 1047.

Scholtissek, C. (1968). *Biochim. Biophys. Acta* **158**, 435.

Scholtissek, C., and Becht, H. (1966). *Biochim. Biophys. Acta* **123**, 585.

Schuster, G. S., and Hare, J. D. (1971). *In Vitro* **6**, 427.

Scott, J. L. (1962). *J. Clin. Invest.* **41**, 1304.

Singer, S. J., and Nicolson, G. L. (1972). *Science* **175**, 720.

Sixma, J. J., Holmsen, H., and Trieschnigg, A. C. M. (1973). *Biochim. Biophys. Acta* **298**, 460.

Stambrook, P. J., Sisken, J. E., and Ebert, J. D. (1973). *J. Cell. Physiol.* **82**, 267.

Steck, T. L., Nakata, Y., and Bader, J. P. (1969). *Biochim. Biophys. Acta* **190**, 237.

Steele, W., and Jenkin, H. M. (1972). *Lipids* **7**, 556.

Stein, W. D. (1967). "The Movement of Molecules across Cell Membranes." Academic Press, New York.

Strauss, P. R. (1974). *J. Cell Biol.* **60**, 571.

Strauss, P. R., and Berlin, R. D. (1973). *J. Exp. Med.* **137**, 359.

Tan, K. B., and McAuslan, B. R. (1971). *Biochem. Biophys. Res. Commun.* **42**, 230.

Taube, R. A., and Berlin, R. D. (1970). *Amer. J. Physiol.* **219**, 666.

Taube, R. A., and Berlin, R. D. (1972). *Biochim. Biophys. Acta* **255**, 6.

Tsan, M. F., and Berlin, R. D. (1971). *J. Exp. Med.* **134**, 1016.

Tsan, M. F., Taube, R. A., and Berlin, R. D. (1973). *J. Cell. Physiol.* **8**, 251.

Ukena, T. E., and Berlin, R. D. (1972). *J. Exp. Med.* **136**, 1.

Warnick, C. T., Muzik, H., and Paterson, A. R. P. (1972). *Cancer Res.* **32**, 2017.

Weber, M. J., and Rubin, H. (1971). *J. Cell. Physiol.* **77**, 157.

Whittam, R. (1960). *J. Physiol. (London)* **154**, 614.

Williams, R. E., Wisnieski, B. J., Rittenhouse, H. G., and Fox, C. F. (1974). *Biochemistry* **13**, 1969.

Wisnieski, B. J., Williams, R. E., and Fox, C. F. (1973). *Proc. Nat. Acad. Sci. U. S.* **10**, 3669.

Subject Index

Contents of Previous Volumes

Volume 7

Volume 8

Volume 14

Volume 15

Volume 16

Volume 17

Volume 18

Volume 19

Volume 20

Volume 21

Volume 22

Volume 23

Volume 24

Volume 25

Volume 26

Volume 27

Volume 28

Volume 29

Volume 30

Volume 31

Volume 32

Volume 33

Volume 34

Volume 35

Volume 36

Volume 37

Volume 38

Volume 39

Volume 40

Volume 41

A 5
B 6
C 7
D 8
E 9
F 0
G 1
H 2
I 3
J 4